Anisotropy Research: Applied Mathematics

Anisotropy Research: Applied Mathematics

Editor: Jessie McDonald

NY RESEARCH

P R E S S

New York

Published by NY Research Press
118-35 Queens Blvd., Suite 400,
Forest Hills, NY 11375, USA
www.nyresearchpress.com

Anisotropy Research: Applied Mathematics
Edited by Jessie McDonald

Cataloging-in-publication Data

Anisotropy research : applied mathematics / edited by Jessie McDonald.
    p. cm.
Includes bibliographical references and index.
ISBN 978-1-64725-446-9
1. Mathematics. 2. Anisotropy. 3. Applied mathematics. I. McDonald, Jessie.
QA36 .A55 2023
510--dc23

Contents

Preface

The purpose of the book is to provide a glimpse into the dynamics and to present opinions and studies of some of the scientists engaged in the development of new ideas in the field from very different standpoints. This book will prove useful to students and researchers owing to its high content quality.

Anisotropy refers to the property of a material that allows it to alter its properties or take on distinct properties in different directions. The property of anisotropy is observed in single crystals of solid materials such as solid compounds and elements. Anisotropy is utilized to describe circumstances in which properties change systematically depending on the direction. Anisotropic materials can have varying thermal conductivity and light polarization depending on the bonding. Anisotropy information can be visualized, processed and modeled using sophisticated applied mathematical constructs such as tensors and other higher-order descriptors. These mathematical constructs can also be applied to address problems in areas such as medical imaging, physical sciences and engineering. This book brings forth some of the current researches in anisotropy within the field of applied mathematics. Coherent flow of topics, student-friendly language, and extensive use of examples make it an invaluable source of knowledge.

At the end, I would like to appreciate all the efforts made by the authors in completing their chapters professionally. I express my deepest gratitude to all of them for contributing to this book by sharing their valuable works. A special thanks to my family and friends for their constant support in this journey.

Editor

Foundations

Continuous Histograms for Anisotropy of 2D Symmetric Piece-Wise Linear Tensor Fields

Talha Bin Masood and Ingrid Hotz

Abstract In this chapter we present an accurate derivation of the distribution of scalar invariants with quadratic behavior represented as continuous histograms. The anisotropy field, computed from a two-dimensional piece-wise linear tensor field, is used as an example and is discussed in all details. Histograms visualizing an approximation of the distribution of scalar values play an important role in visualization. They are used as an interface for the design of transfer-functions for volume rendering or feature selection in interactive interfaces. While there are standard algorithms to compute continuous histograms for piece-wise linear scalar fields, they are not directly applicable to tensor invariants with non-linear, often even non-convex behavior in cells when applying linear tensor interpolation. Our derivation is based on a sub-division of the mesh in triangles that exhibit a monotonic behavior. We compare the results to a naïve approach based on linear interpolation on the original mesh or the subdivision.

1 Introduction

Iso-contours or iso-surfaces of scalar fields, defined as the pre-image of an iso-value, play a central role in visualization. There is a large body of work centered around iso-contour computation and analysis with many applications. A prominent example is the *contour tree*, which keeps track of topological changes of iso-contours when changing the iso-value [5]. It provides a structural overview of the data, see Sect. 2.3 for more details. Complementary information is provided by the *continuous histogram*, which is an extension of the discrete histogram to encode the distribution of the scalar values of continuous functions, see Sect. 2.1. It can serve as a valuable quantitative signature [2] of a data set. Both structures are frequently used for inter-

T. B. Masood (✉) · I. Hotz
Department of Science and Technology (ITN), Linköping University, Norrköping, Sweden
e-mail: talha.bin.masood@liu.se

I. Hotz
e-mail: ingrid.hotz@liu.se

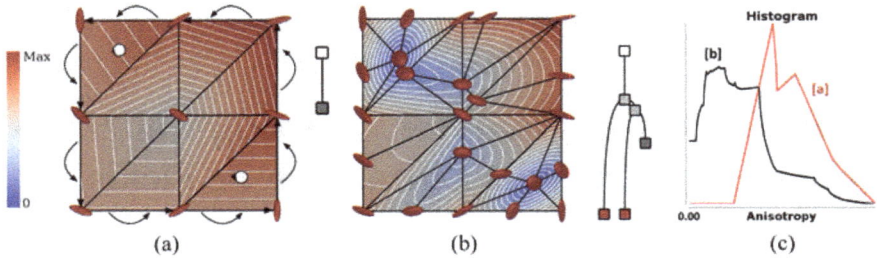

Fig. 1 An example demonstrating the behavior of the anisotropy field in a simple tensor field given on a triangulated domain. The ellipses at the vertices represent the tensors defining the field. The total rotation of the eigenvector when moving once around the domain is 2π. The white lines represent iso-contours. **a** The anisotropy is linearly interpolated within the original mesh. It only has one minimum and no zeros or isotropic points. **b** The consistent anisotropy assuming linear interpolation of tensor components has three minima and two isotropic points. The respective join trees for the anisotropy fields are shown next to the meshes. All zero-leaves, marked in red, in the tree represent isotropic points of the tensor field. **c** The histograms for anisotropy using the two approaches in comparison

action and filtering, for example, to determine interesting iso-values or for transfer function design [19]. Histograms have also been applied for tensor field visualization and exploration to display the distribution of scalar invariants derived from tensor fields [13]. This distribution can also be understood as statistics of iso-contours. Note that we use the term *iso-contour* to refer to the complete level set of a 2D scalar field for a given iso-value, not one of its connected component which is the case in some of the related contour tree literature.

There are many standard algorithms to compute and analyze iso-contours, which are mostly based on piece-wise linear or multi-linear behavior. This assumption is, however, often violated for tensor invariants when using component-wise tensor interpolation, for example, various anisotropy definitions or the determinant of the tensor. An illustrative example of the behavior of the contours of the anisotropy in piece-wise linear tensor field in comparison to a linear approximation of the anisotropy is shown in Fig. 1. It is important to note that, there is no continuous tensor field interpolating the tensors at the vertices exhibiting the piece-wise linearly interpolated anisotropy in Fig. 1a when assuming that the tensor field is fixed on the boundary of the domain. The number of rotations of the eigenvector field when moving once around the domain (Poincaré index) is a topological invariant that determines the minimum number of isotropic points in the field. In this example, there is a total rotation of 2π of the eigenvectors, which results in two zeros in the anisotropy field. These zeros in the anisotropy field are not present in the linear anisotropy field corrupting its histogram, Fig. 1c. Therefore, it is important to consider an anisotropy field which is consistent with the chosen tensor field interpolation when analyzing its behavior. Here we have chosen a piece-wise linear interpolation of tensor components as an example. In the following we refer to the anisotropy field resulting from this interpolation as the consistent anisotropy. In this chapter, we present the derivation of an exact closed-form expression for *continuous histograms* for quadratic tensor invariants that is consistent with a piece-wise linear interpolation of a tensor field given over a two-dimensional triangulated domain.

Here, we are especially interested in the analysis of the anisotropy, an important characteristic in many applications, defined as the difference between major and minor eigenvalue. While there are many other definitions of anisotropy, this version has a distinguished meaning when analyzing the stability of the critical points and isotropic points, when perturbing the tensor field using the Frobenius norm to measure the deviation of the tensor field [18]. Further, it is frequently used in mechanical engineering applications to analyze the failure of materials. The *join tree*, a subset of the contour tree, of this anisotropy can be used for a topology-preserving simplification of 2d tensor fields [11].

Our main contributions are:

- Framework for the analysis of scalar invariants that is consistent with a piece-wise linear interpolation of the tensor components. This includes topologically correct iso-contours, the contour tree, and continuous histograms.

- A generic closed-form formulation of histograms for continuous data based on a generalization of the cumulative histogram. This approach neither requires explicit computations of the length of iso-contours nor the computation of the gradient.

- Explicit solution for the histogram for piece-wise quadratic functions with positive Hessian over a two-dimensional domain. The anisotropy over a piece-wise linear component interpolation provides a relevant example.

- Comparison to naïve approaches using a linear interpolation of the anisotropy and a discussion of the results.

The relevant context and related work are summarized in Sect. 2, then an overview of the problem and its solution is given in Sect. 3. Fundamental definitions and notations are introduced in Sect. 4. A general expression for the anisotropy is developed in Sect. 5 and used in Sect. 6 to subdivide the domain in monotonous triangles. Section 7 derives the accurate histogram. After some notes on the implementation in Sect. 7.1 the results are discussed in Sect. 8.

2   Context and Related Work

In this section, we first summarize the development of the concept of histograms from a signature of discrete data to continuous fields. Then we summarize some aspects of tensor interpolation and invariants that motivated our work.

2.1 Continuous Histograms

Originally, *histograms* have been developed for discrete data. They go back to Pearson and refer to a graphical representation displaying the frequency of the data items as bars over the data range [10]. As such, a histogram provides a summary of the data and they are also used to compare and characterize data sets. Formally, the histogram for a discrete set of data values $F = \{f_i | i \in 1, \cdots, m\}$ is a one-dimensional bar chart displaying the frequency of each distinct element $f \in F$ as a bar with height

$$h(f) = \sum_i \delta(f - f_i)$$

where δ is the delta function which is equal to one if $f = f_i$ and zero otherwise. A related concept applicable to ordered data is the *cumulative histogram* which counts the cumulative number of data values smaller than a given value $f \in F$. This results in values

$$c(f) = \sum_{f_i \leq f} h(f_i).$$

Today, the term is often not only used for the graphical representation but also for the concept capturing the distribution of function values by binning the function range and counting the frequency of data samples within these bins. The resulting histograms, however, are strongly dependent on the binning size and the sampling strategy of the function in the domain. The continuous nature of the function between the sample points is not represented. For these reasons several concepts to extend histograms to continuous functions have been proposed resulting in a distribution of the function values. Bajaj et al. [2] have introduced the *contour spectrum* as a data-signature to find interesting iso-values for volume rendering. The contour spectrum assigns the geometric measures of the contour length or surface area to each scalar value. In addition, they consider areas and volumes of regions below or above a given iso-value. Bajaj et al. also propose a method to compute these values exactly under the assumption of a piece-wise linear interpolation and a piece-wise constant gradient. Later Carr et al. [4] investigated the relation between histograms and the contour spectrum based on a nearest-neighbor interpolation. Scheidegger et al. [16] completed this work and introduced a natural generalization of histograms to the continuous setting as the distribution given by the contour spectrum weighted by the *local isosurface density* expressed as the inverse gradient magnitude. Given a scalar field f and scalar value t the distribution is given as

$$\pi_f(t) = \int_{f^{-1}(t)} |\nabla f(x)|^{-1} dS$$

For the derivation of the distributions, they refer to Coarea formula that provides a relationship between the sum of area integrals and a global volume integral as used by Mullen et al. [14]. For piece-wise linear functions, this yields a more or less good approximation of the real distribution.

Similar concepts have also been considered for multivariate fields, for example, generalizations of scatterplotts [1] and parallel coordinates [9] to continuous data. In both cases, the mathematical model is based on mass conservation of a mapping from the function range to the scatterplot space and the parallel coordinates space respectively. The continuous histogram is a special case considering a one-dimensional range. While the mathematical model is generic, the proposed computational algorithms are limited to specific interpolation models in the spatial domain. The algorithm by Bachthaler et al. [1] assumes a linear interpolation in tetrahedral cells.

In our work, we focus on invariants with a quadratic behavior and propose an exact solution pursuing a slightly different approach than most of the previous methods. Instead of directly generalizing the concept of histograms, we start with the generalization of the cumulative histogram. This approach requires neither surface area/length of iso-contours nor gradient approximations and thus makes it possible to compute an accurate distribution for more complex interpolation schemes, demonstrated here for quadratic interpolations. The continuous histogram or scalar value distribution then follows as a derivative of the cumulative histogram.

2.2 Notes on Tensor Field Interpolation

In this section, we briefly summarize some notes related to tensor field interpolation without going much into detail. This has been a frequently discussed topic in many applications and it is agreed that this is a challenging topic.

While many theories and visualization models assume data given as fields on continuous domains, the data-reality are data sets sampled at discrete locations. Depending on the origin of the data this can be voxel-based data from imaging or meshes for simulation data. In any case, it is necessary to approximate and reconstruct the field from these samples. Thereby the most commonly used methods for tensor fields are based on a component-wise interpolation. Similar to other applications and other data types these are linear, bi-, or tri-linear interpolation depending on the grid type. Concerning such interpolations, a frequently discussed topic is the non-linear dependency of most tensor invariants on the tensor components [12]. There have been many methods proposed which explicitly try to preserve certain tensor characteristics [3, 17]. Recently there appeared a survey on interpolation methods for positive definite tensors [8].

We will, however, stay with the most simple interpolation methods, linear component-wise interpolation within tetrahedra which is also often the basis for finite element simulations. The non-linear dependence of the tensor invariants can lead to a non-convex behavior of the invariants inside the cells. An example we are especially interested in is the behavior of the anisotropy, defined as the difference between the maximum and minimum eigenvalue, which is typically decreased inside the cells. Sometimes this is referred to as "swelling-artifact", but we rather argue that this is a fundamental property of tensor behavior. It reflects the fact that the anisotropy is linked to the directional behavior of the tensor. This is expressed by the fact that the anisotropy is always zero at degenerate points. These are points where the eigenvalues are the same and there is no uniquely determined eigenvector direction. The number of degenerate points is a topological invariant of the field when fixing the boundary and thus determines the minimum number of zeros in the anisotropy field, see Fig. 1 for an example. Therefore the analysis of the anisotropy field must be handled consistently with the chosen interpolation method for the tensor field. This concerns the computation of iso-contours, the contour tree and also histograms.

For our consideration in this work, we assume that we have a continuous tensor field given on a two-dimensional triangulated domain. We further assume a piecewise linear, component-wise linear interpolation of the tensor field within these triangles. The anisotropy then has a quadratic behavior in the triangle. One can observe similar behavior for other invariants, for example, the determinant too. All the above-proposed methods for histogram computations fail when the interpolation of the data is not convex. Similarly, typical iso-contour or contour tree computation is also based on convex interpolation.

2.3   Contour Trees, a Topological Summary of Scalar Functions

The contour tree keeps track of the changes of sub- and super-level sets of a function, bounded by iso-contours, and thus provides a valuable summary of the function, compare Fig. 2. It can be assembled from the *join tree* tracing the changes in iso-contours when the function value is increased from $-\infty$ to ∞ and the *split tree* one where the function value is decreased from ∞ to $-\infty$. In more detail, the join tree tracks the creation and merging of sub-level sets, which are recorded in a tree. At the local minima of the function, new branches are created. As the function value increases, branches are extended and merged at saddle points where two sub-level

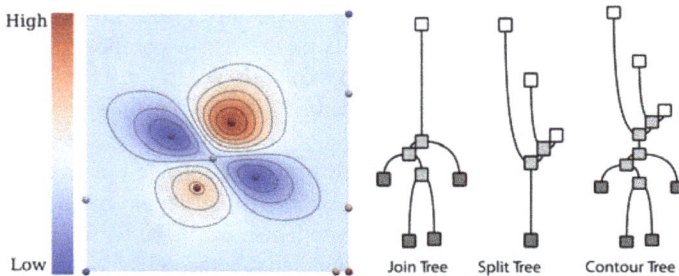

Fig. 2   For the scalar field on the left, the join tree, split tree and the contour tree are shown

sets merge. The global maximum of the function becomes the root of the tree. The split tree is constructed similarly while traversing the field from ∞ to $-\infty$. The most commonly used algorithm for contour tree computation assumes a piecewise linear interpolation of the scalar field where all critical points are located at the cell vertices [5]. In the case of the anisotropy, we are especially interested in the join tree, a sub-tree of the contour tree, since many of the minima correspond to the degenerate points in the tensor field. The join tree can be used to quantify the stability of these points [18]. For consistent results, assuming a specific interpolation, a join tree computation based on the quadratic behavior or the anisotropy is essential.

3 Problem Statement and Solution Overview

Given a tensor field sampled over a 2D triangular mesh, our goal is to compute the accurate contour-tree and histogram of the tensor anisotropy consistent with linear interpolation of tensor components. Anisotropy is a quadratic function under this interpolation.

Figure 3 gives an overview of our solution approach. For each triangle in the input mesh, we first determine the coefficients of the quadratic function for the anisotropy based on tensor values at the triangle vertices. We show that anisotropy is a special quadratic function which either has a single minimum equal to zero and elliptical iso-contours, or in a degenerate case, minima along a line and the iso-contours are

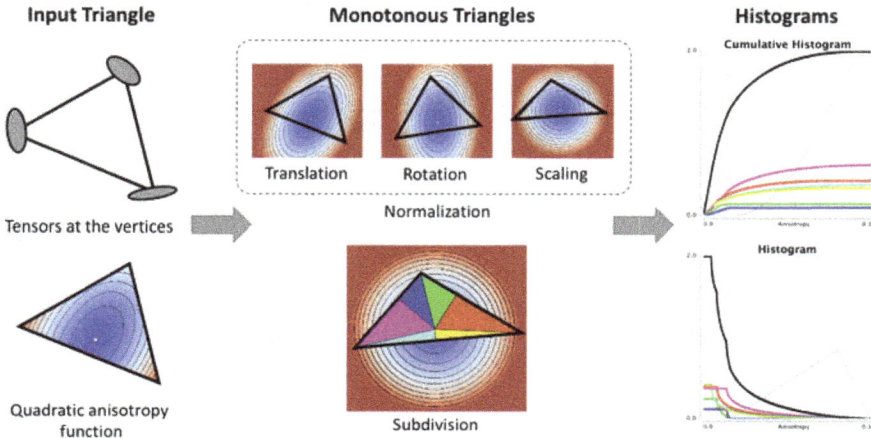

Fig. 3 Solution overview: First, the coefficients of the quadratic function are determined. Then a transformation is applied such that the minimum is at the origin and the iso-contours are circular, followed by a sub-division into monotonous triangles. For each monotonous triangle, the cumulative anisotropy histogram is computed using sub-level set areas, which are added to get the cumulative histogram for the original triangle. Finally, the derivative of the cumulative histogram yields the anisotropy histogram

pairs of parallel lines, refer to Sect. 5. The first step is the subdivision of the triangle into monotonous sub-triangles, which is the basis for the iso-contour extraction, the contour tree computation and the derivation of the histogram. We obtain the histogram as the derivative of the cumulative histogram, which is computed first from the area of the sub-level set for the isovalues. To compute the explicit area of the sub-level sets we apply a linear transformation to the coordinate system such that elliptical iso-contours become circular centered at the origin, refer to Sect. 5.1.

4   Background and Notations

In this section, we provide the notations and a brief mathematical background required for the proposed method discussed in sections later. First, in Sect. 4.1 we set up the notations for tensors and the invariant of interest to us that is tensor anisotropy. Then in Sect. 4.2, we describe the barycentric interpolation within a triangle, since this approach is used for linear interpolation of tensors within a triangle. Lastly, in Sect. 4.3, we discuss the general bi-variate quadratic function, its critical point, and the shape of its iso-contours.

4.1   Second Order Symmetric Tensors and Anisotropy

For a second order symmetric tensor, $T = \begin{pmatrix} e & f \\ f & g \end{pmatrix}$, the two eigenvalues are:

$$\lambda, \mu = \frac{1}{2}(e + g) \pm \sqrt{\frac{(e + g)^2}{4} - (eg - f^2)}$$

Depending on the application different measures for anisotropy are used. For positive definite tensors, relative measures like the fractional anisotropy are common. In mechanical engineering a typical measure is the difference between the maximum and the minimum eigenvalues $\alpha(T) = \lambda - \mu = \sqrt{(e + g)^2 - 4(eg - f^2)}$. It has been shown that this value is also related to the stability of degenerate points in tensor field topology [11] and is the measure we are mostly interested in. In the following, we will however consider the squared value of $\alpha(T)$, which has the same topological characteristics but simplifies the computations a lot.

$$v(T) = \alpha^2(T) = (\lambda - \mu)^2$$
$$= (e - g)^2 + 4f^2 \tag{1}$$

4.2 Barycentric Coordinates and Piece-Wise Linear Interpolation

For all our considerations in this paper we use a component-wise linear interpolation of the tensor field. We use barycentric coordinates for interpolation in the triangle. Since we require an explicit representation of the the tensor field for the derivation of the exact histogram we briefly review the main definitions and formulas in this section.

Refer to Fig. 4, for a triangle with vertices $p_1 = (x_1, y_1)$, $p_2 = (x_2, y_2)$ and $p_3 = (x_3, y_3)$, the barycentric coordinates $(\beta_1, \beta_2, \beta_3)$ of an arbitrary point $p = (x, y)$ within a triangle are given by

$$\beta_1 = \frac{(y_2 - y_3)(x - x_3) + (x_3 - x_2)(y - y_3)}{(y_2 - y_3)(x_1 - x_3) + (x_3 - x_2)(y_1 - y_3)},$$

$$\beta_2 = \frac{(y_3 - y_1)(x - x_3) + (x_1 - x_3)(y - y_3)}{(y_2 - y_3)(x_1 - x_3) + (x_3 - x_2)(y_1 - y_3)},$$

$$\beta_3 = \frac{(y_1 - y_2)(x - x_2) + (x_2 - x_1)(y - y_2)}{(y_2 - y_3)(x_1 - x_3) + (x_3 - x_2)(y_1 - y_3)}.$$

We assume that some scalar function s is sampled at the vertices of the triangle with scalar values s_1, s_2 and s_3 at vertices p_1, p_2 and p_3. Then, the scalar value s at an arbitrary point $p = (x, y)$ within the triangle can be determined as

$$s(x, y) = \beta_1 s_1 + \beta_2 s_2 + \beta_3 s_3.$$

This function $s(x, y)$ is linear in x and y, and can be alternatively written as

$$s(x, y) = s_x x + s_y y + s_c,$$

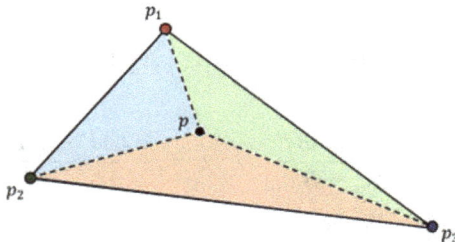

Fig. 4 Barycentric coordinates and interpolation. The point p within the triangle $\triangle p_1 p_2 p_3$ can be represented using barycentric coordinates $(\beta_1, \beta_2, \beta_3)$, where $\beta_1 = \text{Area}(\triangle p p_2 p_3)/\text{Area}(\triangle p_1 p_2 p_3)$, $\beta_2 = \text{Area}(\triangle p p_1 p_3)/\text{Area}(\triangle p_1 p_2 p_3)$ and $\beta_3 = \text{Area}(\triangle p p_1 p_2)/\text{Area}(\triangle p_1 p_2 p_3)$. Let s be some scalar quantity sampled on the vertices p_1, p_2 and p_3 as s_1, s_2 and s_3 respectively. Then the value of s at any point p inside the triangle is given by $s(p) = \beta_1 s_1 + \beta_2 s_2 + \beta_3 s_3$

with

$$s_x = \frac{(y_2 - y_3)s_1 + (y_3 - y_1)s_2 + (y_1 - y_2)s_3}{(y_2 - y_3)(x_1 - x_3) + (x_3 - x_2)(y_1 - y_3)}$$

$$s_y = \frac{(x_3 - x_2)s_1 + (x_1 - x_3)s_2 + (x_2 - x_1)s_3}{(y_2 - y_3)(x_1 - x_3) + (x_3 - x_2)(y_1 - y_3)}$$

$$s_c = \frac{(x_2 y_3 - x_3 y_2)s_1 + (x_3 y_1 - x_1 y_3)s_2 + (x_1 y_2 - x_2 y_1)s_3}{(y_2 - y_3)(x_1 - x_3) + (x_3 - x_2)(y_1 - y_3)}$$

4.3   Bivariate Quadratic Functions and Their Critical Points

As the anisotropy, as defined in Eq. (1), is a quadratic function we summarize some general facts about quadratic functions and define the notation that we will use later in the paper. In the following we assume that the quadratic function is not identical to zero.

A general quadratic function in two variables can be written as $s(x, y) = Ax^2 + Bxy + Cy^2 + Dx + Ey + F$ or in matrix form as

$$s(x, y) = (x, y) \cdot M \cdot \begin{pmatrix} x \\ y \end{pmatrix} + T \cdot \begin{pmatrix} x \\ y \end{pmatrix} + F \tag{2}$$

with $M = \begin{pmatrix} A & B/2 \\ B/2 & C \end{pmatrix}$ and $T = \begin{pmatrix} D \\ E \end{pmatrix}$. The critical point of the scalar function s is a point $p = (x, y)$ where the gradient $\nabla s(x, y) = 0$ is zero. The partial derivatives of $s(x, y)$ with respect to the two variables is:

$$\partial s / \partial x = 2Ax + By + D \tag{3}$$

$$\partial s / \partial y = 2Cy + Bx + E \tag{4}$$

The location of the critical point $p_c = (x_c, y_c)$ of s can be obtained after solving the linear equations $\partial s / \partial x = 0$ and $\partial s / \partial y = 0$ using Eqs. (3) and (4). The resulting coordinates are

$$x_c = \frac{-2CD + BE}{4AC - B^2}, \qquad y_c = \frac{-2AE + BD}{4AC - B^2}. \tag{5}$$

The function s and its critical point can be classified based on the sign of the determinant of the Hessian, H

$$H = 4AC - B^2. \tag{6}$$

If $H > 0$, then the critical point is either a maximum or minimum and the iso-contours are ellipses, compare Fig. 5a, b. The type of the critical point depends on

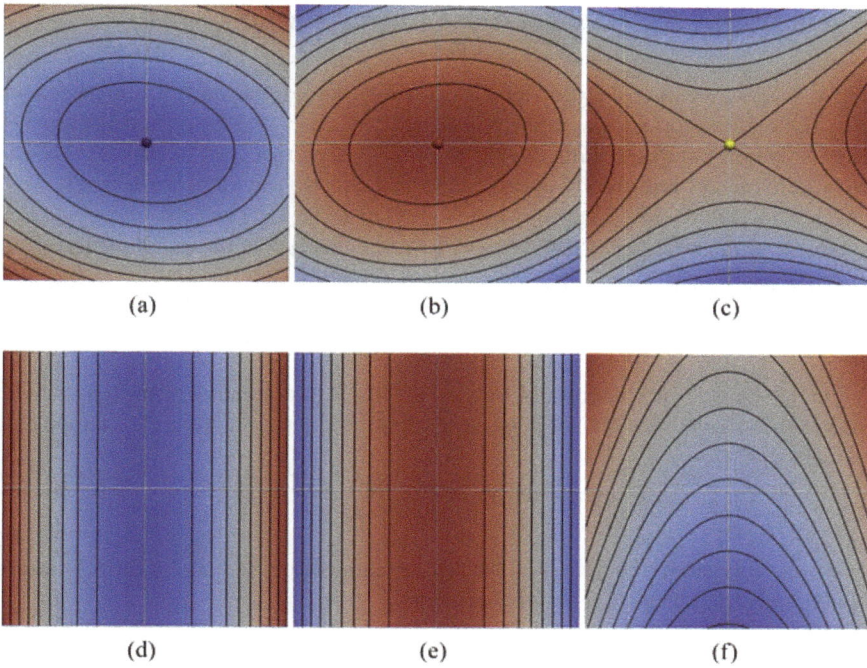

Fig. 5 The possible cases for a general quadratic function. **a** The critical point is a minimum and the iso-contours are elliptical. **b** The critical point is a maximum and the iso-contours are ellipses. **c** The critical point is a saddle and the iso-contours are hyperbolas. **d** The critical point does not exist and the function has a minimum value along a line, while the iso-contours are a pair of parallel lines. **e** The other case when the iso-contours are a pair of parallel lines, again the critical point does not exist but the function has a maximum value along a line. **f** The case when the critical point does not exist and the iso-contours are parabolic. We show that only the cases (**a**) and (**d**) are possible for the quadratic function for tensor anisotropy, the other four cases shown here are not possible

the sign of A and C. If $A, C > 0$, then the critical point is a minimum, otherwise it is a maximum. In the other case when $H < 0$, then the critical point is a saddle and the iso-contours are hyperbolas.

For the case when $H = 0$, there are no isolated critical points and the iso-contours are either parabolas, parallel or coincident lines. Note that it is possible that there are degenerate lines. The type of the iso-contour when $H = 0$ depends on the invariant I defined as

$$I = BDE - AE^2 - CD^2. \tag{7}$$

If $I \neq 0$, the iso-contours are parabolic, compare Fig. 5f, otherwise they are pair of parallel lines, compare Fig. 5d, e.

5 Anisotropy for 2D Piece-Wise Linear Tensor Fields

In this section, we have a closer look at the anisotropy function for a linear tensor field. The main observation is in the non-degenerate case, that the anisotropy always has exactly one minimum with a value of zero. This corresponds to the fact that the tensor field always has exactly one critical point. All iso-contours are ellipses.

Given a triangle with vertices $p_1 = (x_1, y_1)$, $p_2 = (x_2, y_2)$ and $p_3 = (x_3, y_3)$. Let the tensors at these vertices be T_1, T_2 and T_3, respectively. Let the components of the tensors be

$$T_1 = \begin{pmatrix} e_1 & f_1 \\ f_1 & g_1 \end{pmatrix}, \quad T_2 = \begin{pmatrix} e_2 & f_2 \\ f_2 & g_2 \end{pmatrix}, \quad T_3 = \begin{pmatrix} e_3 & f_3 \\ f_3 & g_3 \end{pmatrix}. \tag{8}$$

The barycentric coordinates can be used for linear interpolation of the tensor field within the triangle. Tensor T at an arbitrary point $p = (x, y)$ within the triangle can be found as:

$$T(p) = T(x, y) = \begin{pmatrix} e(x, y) & f(x, y) \\ f(x, y) & g(x, y) \end{pmatrix} \tag{9}$$

As described earlier in Sect. 4.2, the tensor components e, f and g linearly interpolated within the triangle as can be written as

$$e(x, y) = e_x x + e_y y + e_c, \tag{10}$$

$$f(x, y) = f_x x + f_y y + f_c, \tag{11}$$

$$g(x, y) = g_x x + g_y y + g_c. \tag{12}$$

The function, we are interested in is the anisotropy v, the explicit expression for which can be obtained by substituting the linear expressions for tensor components in Eq. (1) yielding

$$\begin{aligned} v(T) = & \left((e_x - g_x)^2 + 4f_x^2 \right) x^2 + 2 \left((e_x - g_x)(e_y - g_y) + 4f_x f_y \right) xy \\ & + \left((e_y - g_y)^2 + 4f_y^2 \right) y^2 + 2 \left((e_x - g_x)(e_c - g_c) + 4f_x f_c \right) x \\ & + 2 \left((e_y - g_y)(e_c - g_c) + 4f_y f_c \right) y + \left((e_c - g_c)^2 + 4f_c^2 \right). \end{aligned} \tag{13}$$

Comparing Eq. (13) with the general quadratic Eq. (2), we obtain the coefficients of the quadratic function in dependence of the tensor components:

$$A = \left((e_x - g_x)^2 + 4f_x^2 \right) \geq 0, \tag{14}$$

$$B = 2 \left((e_x - g_x)(e_y - g_y) + 4f_x f_y \right), \tag{15}$$

$$C = \left((e_y - g_y)^2 + 4f_y^2 \right) \geq 0, \tag{16}$$

$$D = 2 \left((e_x - g_x)(e_c - g_c) + 4f_x f_c \right), \tag{17}$$

$$E = 2 \left((e_y - g_y)(e_c - g_c) + 4f_y f_c \right), \tag{18}$$

$$F = \left((e_c - g_c)^2 + 4f_c^2 \right). \tag{19}$$

Substituting the above equations in Eq. (6), we derive the determinant of Hessian for the anisotropy to be (for derivation refer to Appendix A):

$$H = 16\big(f_x(e_y - g_y) - f_y(e_x - g_x)\big)^2 \geq 0. \tag{20}$$

From Eq. (20), we can conclude that the iso-contours of anisotropy function are never hyperbolic. In most cases, when H is strictly greater than 0, the contours are elliptical. Moreover, in that scenario, from Eqs. (14) and (16), we can deduce that the type of the critical point is always a minimum.

Appendix A contains the detailed analysis of v. We show that the bi-variate quadratic function v is a special function that has elliptical iso-contours or in some cases a pair of parallel lines, ruling out the possibility of hyperbolic or parabolic iso-contours.

5.1   Field Normalization Using Coordinate Transformations

In the following, we derive a coordinate transformation such that the bivariate quadratic scalar function determined for a triangle has a standard format. This allows for the application of a unified strategy for further analysis.

Equation (13) gives the expression of anisotropy in general bivariate quadratic form. We have also established that iso-contours of this function will are elliptical. Therefore we first transform the coordinates such that elliptical iso-contours are centered at the origin and aligned to the axes. This can be achieved by applying a translation and rotation, both rigid-body transformations preserving areas. Then, a transformation is applied such that iso-contours of the bivariate quadratic function become circles rather than ellipses. This can be achieved by applying a non-uniform scaling, a linear transformation which distorts the area by a constant factor given by the determinant of the transformation matrix. The approach described above is illustrated in Fig. 6.

Refer to Eqs. (13)–(19) for the detailed expression of anisotropy and refer to Eq. (2) for the general quadratic equation in matrix form. After translation such that the minimum $p_c = (x_c, y_c)$ falls into the origin, the expression for anisotropy can be written as

$$v(x_t, y_t) = (x_t, y_t) \cdot M \cdot \begin{pmatrix} x_t \\ y_t \end{pmatrix} + F_t \tag{21}$$

with the translated coordinates $x_t = x - x_c$, $y_t = y - y_c$ and $F_t = F + \frac{(BDE - AE^2 - CD^2)}{4AC - B^2}$. Using Eqs. (14)–(19), it can be shown that $F_t = 0$ for v. This implies that minimum value of v is zero at the critical point and this point is a degenerate point in the tensor field. Equation (21) becomes

$$v(x_t, y_t) = (x_t, y_t) \cdot M \cdot \begin{pmatrix} x_t \\ y_t \end{pmatrix} \tag{22}$$

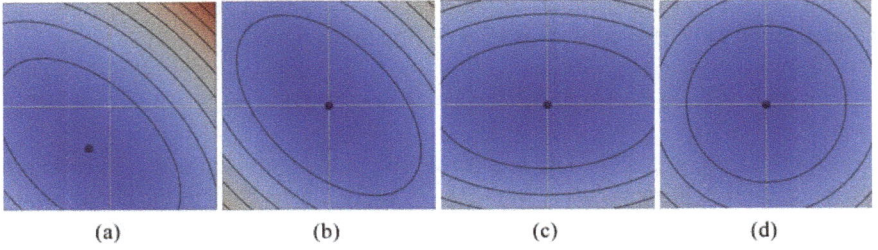

(a)   (b)   (c)   (d)

Fig. 6 Field normalization using coordinate transformations. **a** The original quadratic function, $f(x, y) = x^2 + xy + y^2 + 0.4x + 0.5y + 0.07$. **b** The function after applying translation such that the minimum is at the origin, $f(x_t, y_t) = x_t^2 + x_t y_t + y_t^2$. **c** After applying rotation such that elliptical iso-contours are axes aligned, $f(x_r, y_r) = 0.5x_r^2 + 1.5y_r^2$. **d** After scaling such that elliptical iso-contours are transformed into circular iso-contours, $f(x_s, y_s) = x_s^2 + y_s^2$

Now, we apply a rotation $O = (EV_1 | EV_2)$ to align the elliptic iso-contours with the axis using the eigenvectors EV_1 and EV_2 of the matrix M

$$\begin{pmatrix} x_r \\ y_r \end{pmatrix} = O \cdot \begin{pmatrix} x_t \\ y_t \end{pmatrix}, \tag{23}$$

which results in the diagonal representation of the anisotropy

$$v(x_r, y_r) = \lambda_1 x_r^2 + \lambda_2 y_r^2 \tag{24}$$

where λ_1 and λ_2 are the eigenvalues of M.

Lastly, we apply non-uniform scaling to obtain circular iso-contours:

$$v(x_s, y_s) = x_s^2 + y_s^2 \quad \text{where} \quad x_s = \sqrt{\lambda_1} x_r, \quad y_s = \sqrt{\lambda_2} y_r. \tag{25}$$

The area distortion because of this scaling is given by the factor $\sqrt{\lambda_1 \lambda_2}$.

6 Subdivision in Monotonous Sub-triangles

We consider a triangle with vertices $p_1 = (x_1, y_1)$, $p_2 = (x_2, y_2)$ and $p_3 = (x_3, y_3)$ and the tensors at these vertices are T_1, T_2 and T_3. The tensors are linearly interpolated within the triangle and the anisotropy is a quadratic function. For the extraction of iso-contours, the computation of the contour tree as well as for the histogram we require piece-wise monotonous behavior inside the triangles, which is given in this chapter. Similar, subdivisions have also been proposed by Dillard et al. [6] and by Nucha et al. [15]. There are five different cases depending on the location of the global minima and the local minima at the edges. Especially the cases when the minimum p_c lies within the triangle or outside the triangle need a different treatment for the computation of the anisotropy histogram.

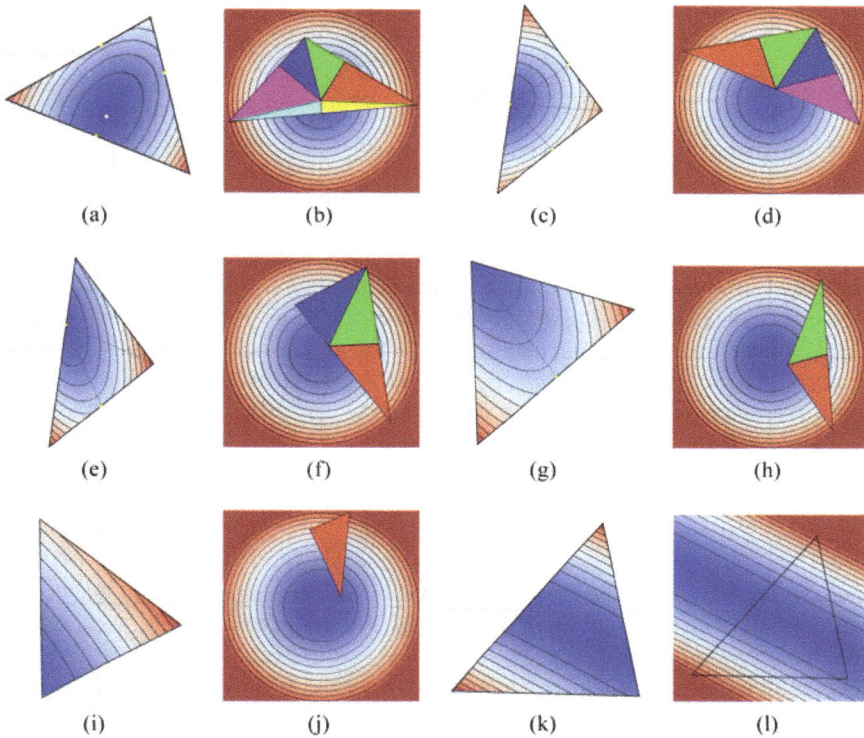

Fig. 7 All the possible cases and the corresponding sub-divisions into monotonous triangles. **a, b** The case when the minimum is inside the triangle. We can subdivide the triangle into six monotonous triangles. **c, d** The case when the minimum is outside the triangle, but the function restricted to the triangle edges has a minimum on all the edges. Four monotonous triangles can be generated. **e, f** The case when two of the edges have minima, three monotonous triangles are generated. **g, h** The case when only one triangle edge has a minimum, two monotonous triangles are created. **i, j** The case when none of the edges has a minimum, no triangle subdivision is required. **k, l** Lastly, the degenerate case when the function does not have a point minimum but has a degenerate minimum along a line. In this case, the iso-contours are a pair of parallel lines

Case A. The minimum p_c is within the triangle. In this case, the input triangle can be partitioned into six sub-triangles such that ν behaves monotonously within the triangle. See Fig. 7b.

Case B. The minimum p_c is outside the triangle. Here, we have four possibilities:

1. No triangle edge has an edge minimum. See Fig. 7i.
2. One triangle edge has an edge minimum. See Fig. 7g.
3. Two triangle edges have edge minima. See Fig. 7e.
4. All three triangle edges have edge minima. See Fig. 7c.

In all these cases, the triangle can be sub-divided into an appropriate number of monotonous triangles as described in Lines 14–30 of Algorithm 1. Although these sub-divided triangles are monotonous and look similar to Case A, an important difference is that in general none of the vertices lies at the origin.

7   Computation of the Histogram for v

In this section, we derive the exact continuous histogram of the anisotropy for linearly interpolated tensor fields. We use the term *histogram* here not strictly as it has been originally introduced but in the sense of the distribution of function values. It can also be considered as the *weighted contour spectrum* using the terminology of Bajaj et al. [2] as introduced by Scheidegger et al. [16]. In contrast to previous methods, we approach the problem by first computing the cumulative histogram CH and then derive the histogram from it by computing its derivative. This makes the exact computation of the histogram much more feasible.

$$CH(v_0) = Area(v \leq v_0) = \sum_{\triangle_i} Area_i(v \leq v_0)$$

where $Area_i(v \leq v_0)$ is the area of the sublevel set in triangle $\triangle_i$. Specifically, we consider here the anisotropy with its quadratic behavior, however the method also directly applies for linear fields.

   In our derivation of $Area_i(v \leq v_0)$ we consider the following setting. Given is a triangle $\triangle ABC$ with vertices $A = p_1 = (x_1, y_1)$, $B = p_2 = (x_2, y_2)$ and $C = p_3 = (x_3, y_3)$ and respective tensors T_1, T_2 and T_3, which are linearly interpolated within the triangle. Further we ordered the vertices such that the anisotropy values are $v_1 < v_2 < v_3$, using simulation of simplicity [7] to resolve the ties if required. To obtain the contribution of the triangle to the cumulative histogram at the value v we compute the area of the sublevel set at v within the triangle. In the following, we assume that anisotropy is monotonous within the triangles resulting from the subdivision introduced in Sect. 6. We further assume that the transformations as described in Sect. 5.1 have been applied. This means that the anisotropy has the form as given in Eq. (25) with a global minimum $p_c = (0, 0)$ at the origin and the level sets that are circles. The area within the transformed triangle is distorted by a constant factor, which can be appropriately multiplied to get the exact areas for the original triangle. We consider two cases depending on whether the vertex A of the triangle vertex lies at the origin or not.

Case 1: The global minimum p_c lies at one triangle vertex

Let's assume that the global minimum p_c lies in vertex $A = p_1$ of the triangle ABC. Then the shape of the sublevel set for v can have two different types depending on whether v is smaller or larger than $v_2 = r_{AB}^2$. If $v \leq v_2$ the sublevel set is a sector of a circle with radius $r = \sqrt{v}$ and opening angle θ_A, as shown in Fig. 8. The area of the sublevel set for $v \in [v_1, v_2]$ is then given by

$$Area(v) = \frac{\theta_A v}{2}. \tag{26}$$

where θ_A is angle between the edges AB and AC. The rate of change of the area is given by

$$\frac{\partial Area(v)}{\partial v} = \frac{\theta_A}{2}, \tag{27}$$

which is a constant not depending on the specific value of v.

If the isovalue v is greater than v_2 and less than v_3, the iso-contour is composed of a circular sector and an additional more complex shape $DBFE$ as illustrated in Fig. 8. This region is enclosed by two circular segments DB and EF and two line segments DE and BF and can be computed as

$$
\begin{aligned}
Area(BDEF) &= Area(BGHF) + Area(FHE) - Area(BGD) \\
&= \left(\frac{(r_{AB} \sin \theta_A + r \sin \theta)(r \cos \theta - r_{AB} \cos \theta_A)}{2} \right) \\
&\quad + \left(\frac{\theta r^2}{2} - \frac{r^2 \sin \theta \cos \theta}{2} \right) - \left(\frac{\theta_A r_{AB}^2}{2} - \frac{r_{AB}^2 \sin \theta_A \cos \theta_A}{2} \right) \\
&= \frac{\theta r^2 - \theta_A r_{AB}^2}{2} + r r_{AB} \left(\frac{\sin \theta_A \cos \theta - \cos \theta_A \sin \theta}{2} \right).
\end{aligned}
$$

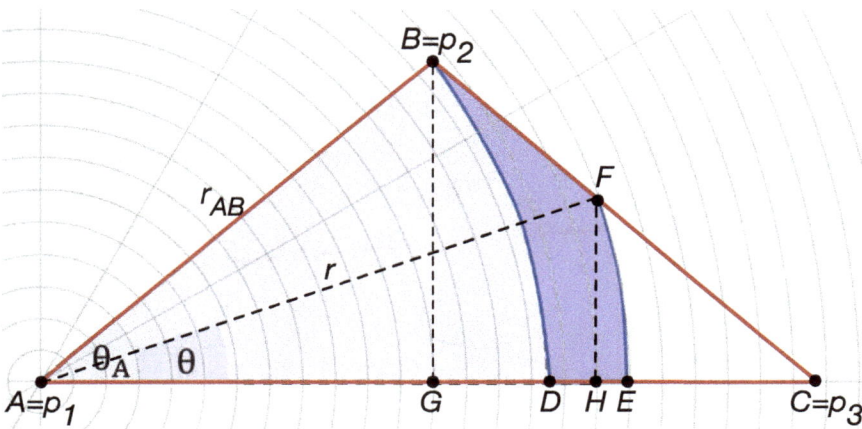

Fig. 8 A triangle with monotonous function behaviour where the global minimum p_c lies on one vertex $A = p_1$. Due to the normalization of the triangle, p_c is at the origin and the iso-contours are circular. If $v \leq v_2$ the sublevel set is a subset of the light blue area. If $v > v_2$ the sublevel area is composed of the complete light blue region and the area is highlighted in darker blue

With $r^2 = v$ and $r_{AB}^2 = v_2$ this results in

$$Area(BDEF) = \frac{\theta v - \theta_A v_2}{2} + \sqrt{v v_2} \frac{\sin(\theta_A - \theta)}{2}. \tag{28}$$

In the expression above, the given value of v determines the values of θ.

Case 2: The global minimum p_c does not lie at any triangle vertex

These triangles are the more generic case, we still assume that the anisotropy is monotonous inside the triangle and the vertices such that the anisotropy values are $v_1 < v_2 < v_3$. The triangles look similar to Case 1, but none of the vertices lie at the origin, compare Fig. 9. To compute the area of a sublevel set $ABFE$ within these triangles, we considering the sublevel sets in two triangles where we can use the algorithm of Case 1. In Fig. 9 they are the triangles $\triangle DBC$ and $\triangle DHC$. For the final area, we have in addition to consider the area of triangle $\triangle AHB$ which has to be added or subtracted depending on the exact position of A.

$$Area(ABFE) = Area(DBFG) - Area(DHEG) \pm Area(\triangle AHB) \tag{29}$$

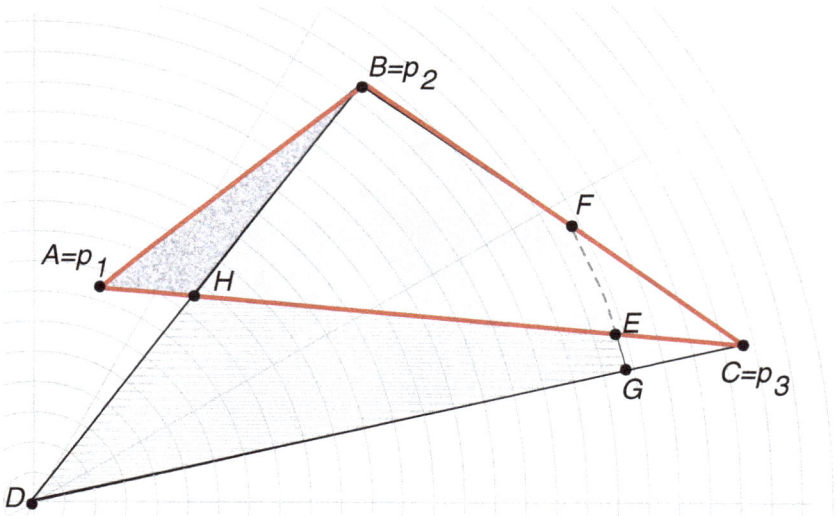

Fig. 9 A triangle $\triangle ABC$ with monotonous function behaviour where the global minimum p_c does not lie on any vertex

7.1 Implementation

We compute the cumulative histogram of anisotropy by computing the sublevel set area. The detailed algorithm is provided in Algorithm 1. For a given tensor mesh, we pre-process all the triangles to determine the coefficients of the quadratic function for anisotropy within the triangle, compare Sect. 5. Each triangle is appropriately subdivided into monotonous triangles, compare Sect. 6. Then for each anisotropy value in the histogram bin, we compute the sublevel set area by adding up the sublevel set area in all monotonous triangles. We also handle triangles with degenerate minimum appropriately.

Parallelization. The algorithm described in Algorithm 1 has lots of options for parallelization. The loops at Lines 4 and 38 are embarrassingly parallel because each triangle in the mesh can be pre-processed in parallel. Similarly, each bin in the histogram can be computed in parallel. Further, with atomic add operations, the computation of the sublevel set area for anisotropy value v corresponding to a particular histogram bin can be parallelized over the set of monotonous triangles. These are the loops at Lines 41 and 44 in Algorithm 1.

Numerical issues. Since there are a few floating-point operations and checks involved in the Algorithm 1, floating-point errors can be introduced which may propagate to the output resulting in a wrong histogram. We handle these errors by introducing reasonable checks, for example making sure that the cumulative histogram thus obtained is always a monotonically increasing function and sums up to the total area of the domain for the largest value of anisotropy. However, a deeper study of the errors and more robust computation is left for future work. We are confident that using a multi-precision library for computation will remove the floating-point errors.

Run times. We implemented the Algorithm 1 as a prototype in Java 1.8. This proof of concept implementation does not include extensive code and memory optimizations. The experiments reported in following Sect. 8 were performed on a workstation with 12 core Intel Xeon processor and 32 GB of RAM.

The running time depends on the size of the input mesh and the number of bins in the output histogram. For all the experiments we chose 500 as the number of bins in the histogram. The synthetic data used in Sect. 8.1 is a small data set with the mesh containing only 32 triangles. It took about 0.02 s to compute the continuous histogram for the meshes in this ensemble. The histogram computation for the two point load data reported in Sect. 8.2 took 0.2 s. This triangle mesh contained 1682 triangles. The run time improved to 0.1 s when using multi-threading. Lastly, for the DTI data set containing 722 triangles (Sect. 8.3), it took 0.09 and 0.17 s to compute the histogram for brain and noisy regions respectively. The times improved by a factor of 2 for both cases when using multi-threading.

Algorithm 1: Compute cumulative histogram

Data: A tensor mesh M, and histogram resolution B
Result: Cumulative histogram CH

1 Initialize histogram CH to zeros;
2 Initialize monotonous triangle list MT to be empty;
3 Initialize degenerate triangle list DT to be empty;
4 **foreach** *triangle* t *in* M **do** in parallel
5   Compute the quadratic function coefficients A, B, C, D, E, F;
6     $H = 4AC - B^2$;
7   **if** $H > 0$ **then**
8   Compute the location of minimum $p_c = (x_c, y_c)$;
9   Compute the number and locations of edge minima;
10   **if** p_c *is inside* t **then**
11   Generate six monotonous triangles from t and add to MT;
12   **else**
13   **switch** *Number of edge minima* **do**
14   **case** *0*
15   Add t to MT;
16   **case** *1*
17   Add an edge from edge minimum to opposite triangle vertex;
18   This divides t in two monotonous triangles;
19   Add the two monotonous triangles to MT;
20   **case** *2*
21   Add an edge between the two edge minima;
22   Add an edge from lowest valued edge minimum to opposite vertex;
23   These two edges divide t into three monotonous triangles;
24   Add the monotonous triangles to MT;
25   **case** *3*
26   Find the lowest valued edge minimum p_m;
27   Add edges between p_m and the two other edge minima;
28   Add an edge from lowest valued edge minimum to opposite vertex;
29   These three edges divides t in four monotonous triangles;
30   Add the monotonous triangles to MT;
31   **endsw**
32   **endsw**
33   **end**
34   **else**
35   Add t to DT;
36   **end**
37 **end**
38 **for** $i \leftarrow 1$ *to* B **do** in parallel
39     $v \leftarrow \texttt{Range(MT)}/i$;
40   CH $[i] \leftarrow 0$;
41   **foreach** *monotonous triangle* t *in* MT **do** in parallel
42   CH $[i] \leftarrow$ CH$[i]+$ $\texttt{GetSubLevelSetArea}$ (t, v);
43   **end**
44   **foreach** *degenerate triangle* t *in* DT **do** in parallel
45   CH $[i] \leftarrow$ CH$[i]+$ $\texttt{GetSubLevelSetArea}$ (t, v);
46   **end**
47 **end**
48 **return** CH;

8   Results

We apply the proposed approach of computing anisotropy histograms to three different case studies. We show results for synthetic, simulation and experimental data. For all the case studies we compute the continuous histograms for three different interpolations of the anisotropy.

- **Interpolation [a]**: Linear interpolation of the anisotropy within the original triangulation.
- **Interpolation [b]**: Linear interpolation of the anisotropy within the sub-divided monotonous triangles, compare Sect. 6.
- **Interpolation [c]**: Anisotropy based on the linear interpolation of the tensor components, resulting in a quadratic behavior or the anisotropy, compare Sect. 5. When using the term *accurate* we always refer to accuracy with respect to this interpolation.

The resulting anisotropy fields are directly shown using a color map where dark blue refers to an anisotropy value of zero and red assigned to the maximum value. We also show a set of iso-contours in these images as white lines. In addition, we computed the join tree for all cases, which are always the same for interpolation [b] and [c]. Finally, we computed the continuous histogram where the results for one data set are plotted in one image, interpolation [a] displayed as a red curve, interpolation [b] as a green curve and interpolation [c] as a black curve.

8.1   Synthetic Data

The first example is a synthetic data set where tensors with user specified tensor components are placed at grid locations of a 5×5 grid. This grid is triangulated to provide a mesh with 32 triangles. This simple example is well suited to demonstrate the differences between the histogram of the consistent anisotropy field [b], and the methods utilizing a linear approximation of the anisotropy on the original mesh [a], and on the subdivided mesh with monotonous triangles [b], as shown in Fig. 10. It can immediately be seen that for interpolation [a] the topology of the iso-contours is not in accordance with the tensor field. While the iso-contours based on the tensor interpolation differ between interpolation, they have the same topology [b] and [c]. This is also reflected in the corresponding join trees, which are the same for both fields, for more details see Sect. 8.

In the next step, we randomly perturb the directions of the tensors at the vertices without changing their eigenvalues to generate an ensemble of four tensor fields. Note that since we do not change the eigenvalues, the anisotropy at the vertices of the mesh remains unchanged after perturbation. Hence, for interpolation [a] all the fields, the iso-contours, the join trees, and their histograms will be the same as evident from Fig. 11a. However, if we use quadratic function resulting form the tensor interpolation

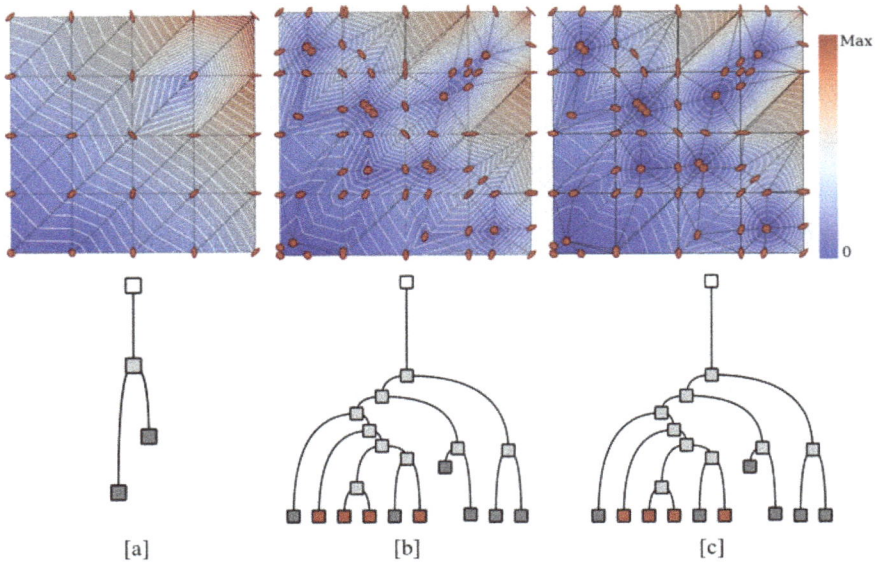

Fig. 10 Comparison of the behavior of the anisotropy field in a synthetic dataset. The anisotropy is linearly interpolated within the original mesh [a], and the subdivided monotonous triangles [b]. The anisotropy assuming linear interpolation of tensor components [c]. The upper row shows the resulting field as a color-map superimposed with contour lines. The ellipses at the vertices represent the tensors defining the field. The second row shows the respective join trees for the three anisotropy fields. The red squares correspond to the zeros inside of the original triangles. All zero-leaves in the tree represent degenerate points of the tensor field topology

[c] of anisotropy, we clearly observe the differences within the ensemble members as shown in Fig. 11c. Similarly, the histograms for the ensemble members are different as shown by black curves in the plots in Fig. 11d. While for interpolation [b], the subdivision into monotonous triangles helps in identifying the differences, Fig. 11b, the histograms are still not accurate and have a bias toward larger anisotropy values, Fig. 11d. The respective contour trees are shown in Fig. 12. The contour tree for the interpolation [a] is the same for all 4 fields and already given in Fig. 11a, so we only show the trees for the sub-division in monotonous triangles. The degenerate points of the tensor field where the anisotropy is zero appear also as minima in the join tree and are highlighted in red. The join trees provide an overview of the possible cancellations of degenerate points and thus their stability [18]. Comparing the trees it can be seen that the four data sets vary significantly for their topological structure and stability of its degenerate points. The join tree for the linear anisotropy, Fig. 11a, has no zero and is thus not consistent with the direction fields given by the tensors.

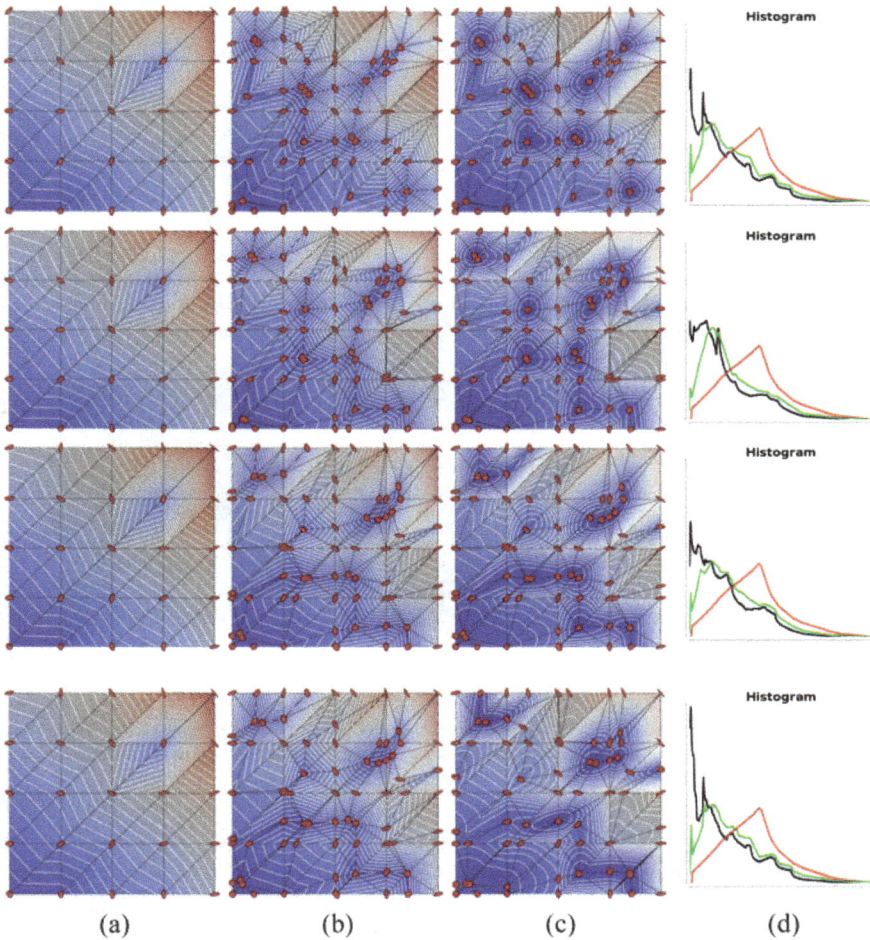

Fig. 11 Random directions tensor dataset with constant eigenvalues. **a** The original mesh with anisotropy computed at the mesh vertices and linearly interpolated within the triangles. **b** The subdivided mesh with monotonous triangles. Anisotropy is linearly interpolated within the monotonous triangles. **c** The anisotropy resulting from linear interpolation of tensor components. **d** summarizes all histograms with red curves for interpolation [a], green curves for interpolation [b] and black curves for interpolation [c]

8.2 Simulation Data

The difference between the three different interpolations for the anisotropy field gave significantly different results for the synthetic data. In the next step, we want to investigate the impact of these differences on real-world data. At first, we consider a simulation data set from mechanical engineering. It is a well-known data set, often

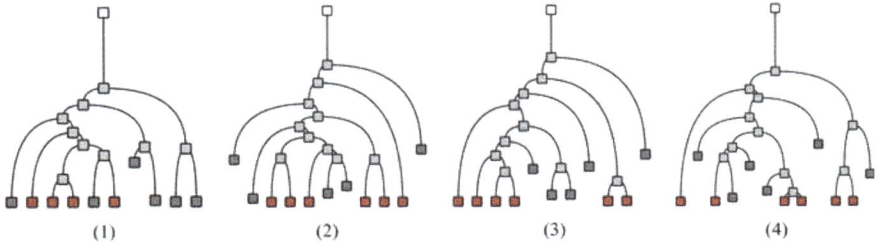

Fig. 12 The join trees for the four ensemble members shown in Fig. 11. The trees summarize the hierarchical structure of the minima in the anisotropy field and the degenerate points in the tensor field topology. Degenerate points inside the triangles are marked as red nodes in the tree

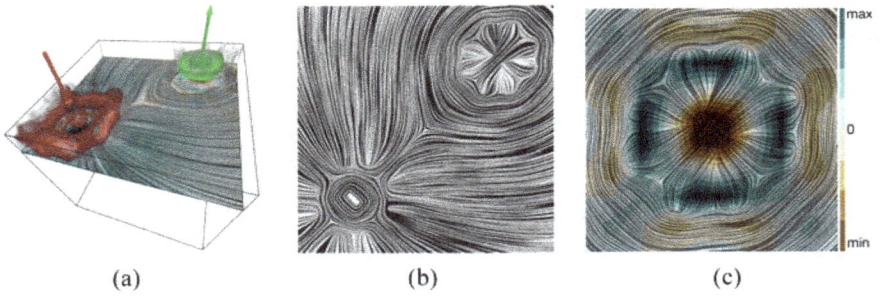

(a)　　　　　　　(b)　　　　　　　(c)

Fig. 13 Two-point load data: simulation of the stresses inside a solid block upon applying two forces. **a** shows a schematic of the set-up of the simulation, **b** shows the eigenvector directions in one slice, **c** is a close up with color showing the signed anisotropy using bi-linear interpolation on the original mesh

referred to as two-point load, that represents a stress field in a solid block resulting from the application of two external forces, Fig. 13. The data is given on a cubic mesh. The anisotropy measure used so far corresponds to the squared von Mises stress, which plays an important role in failure analysis of mechanical components. The direction field in one slice is shown in Fig. 13b. In our analysis, we consider a section of this slice, which is shown in Fig. 13c displaying one eigenvector field. Color represents the anisotropy using a bi-linear interpolation in the original mesh. In our analysis, we use a triangulated version of the data.

Figure 14 shows the anisotropy fields using the three different interpolations in comparison. In Fig. 14a one can observe the asymmetry introduced by the triangulation but is similar to Fig. 13c. The anisotropy fields in Fig. 14b, c are both based on the monotonous subdivision and are very similar to each other but differ strongly from Fig. 14a. The asymmetry due to the triangulation is largely reduced. The minima inside the original triangles in the field capture the locations of the degenerate points of the direction field. The corresponding histogram can be seen in Fig. 14d, e. As expected they differ strongly for very small values but are rather similar for large values of the anisotropy.

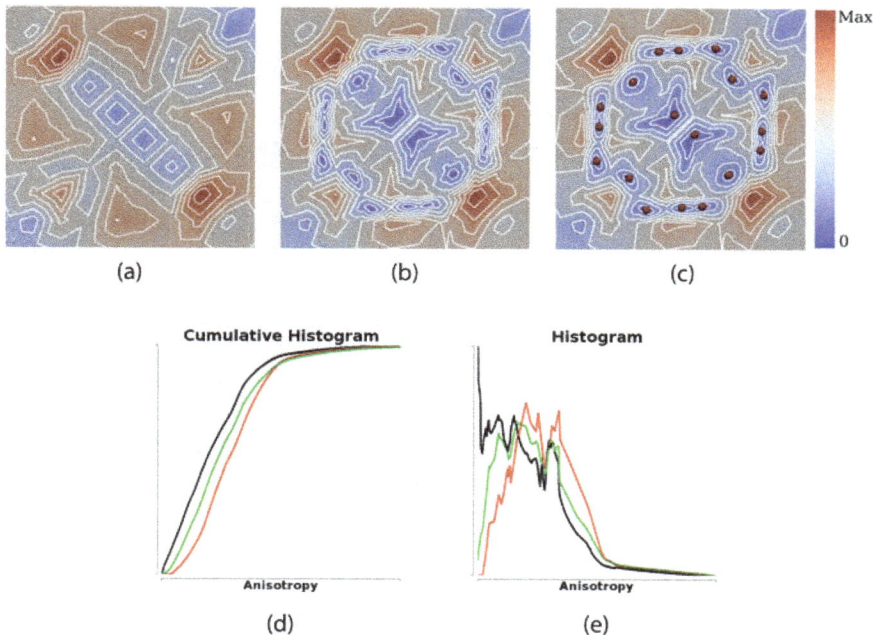

Fig. 14 Two point load simulation data with focus region around one of the loading points. **a** The original mesh with anisotropy computed at the mesh vertices and linearly interpolated within the triangles. **b** The subdivided mesh with monotonous triangles. Anisotropy is linearly interpolated within the monotonous triangles. **c** The anisotropy resulting from linear interpolation of tensor components. The isotropic points in the tensor field are highlighted by small red spheres. **d** the three cumulative histograms and (**e**) the histograms. The red curves correspond to (**a**), the green curves to (**b**) and the black curve to (**c**)

8.3 Measurement Data

As the last example, we examine two sections of a slice of 3D Diffusion Tensor Imaging data. Specifically, we compute and compare the anisotropy histograms of noisy regions outside the brain and a region inside the brain. The 2D slice from the data along with selected regions is shown in Fig. 15.

Figure 16 shows the histograms computed for the noisy region. With the interpolation approach [c] for computing histograms, we can observe a high contribution of anisotropy values near zero, which hints at the existence of a lot of degenerate points in the noisy region. This is not captured by the histogram computed with interpolation approach [a]. It should be noted that interpolation [b] (green plot) although yields better results than interpolation [a] (red plot), it is still quite different than the accurate histogram based on component-wise tensor interpolation (black plot). We

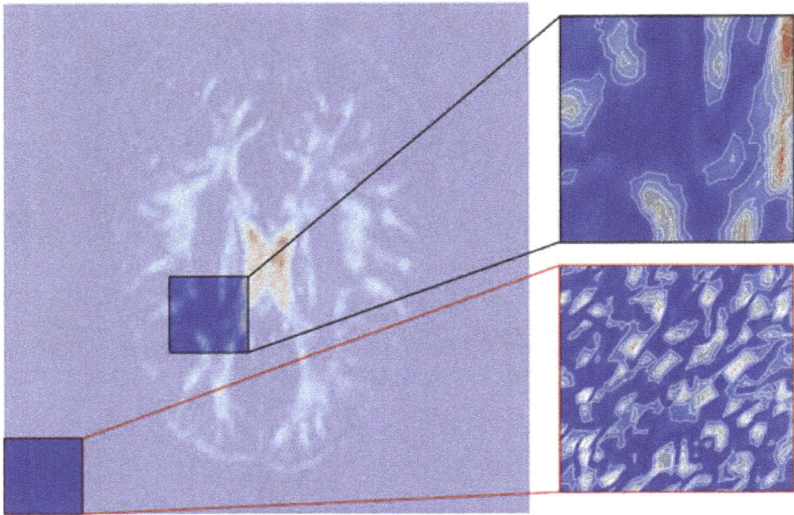

Fig. 15 Two segments selected from Diffusion Tensor Imaging data. The 20×20 2D grid shown in red outline is chosen from noisy region of the data while the selection shown in black outline is chosen from the region occupied by the brain. We plot the histograms for these selections in Figs. 16 and 17 respectively

did the same experiment for a region that is within the brain and hence contains less noise. The results are shown in Fig. 17. Here we observe, that histograms are very similar whether we use the accurate quadratic function for anisotropy of linearly interpolate it within the triangles.

9   Conclusions

In this paper, we explore the behavior of the anisotropy, as an example for a nonlinear derived tensor invariant, when applying a linear component-wise interpolation of the tensor field. We demonstrate that a linear interpolation of the invariant itself, the interpolation approach [a], leads to a topology of iso-contours that in many cases is not consistent with any tensor interpolation and leads to a bias in the histogram. With this analysis, we want to emphasize the importance of being consistent with the chosen interpolation in all analysis steps. For our analysis, we have chosen a linear interpolation of tensor components, which is the most commonly used method in simulations and provides a valid continuous field. We do not want to make a statement about the quality or suitability of different interpolation methods in specific applications, however we want to point out that the independent interpolation of the direction field and the anisotropy violates in many cases the preservation of topological invariants and does not result in a valid tensor field.

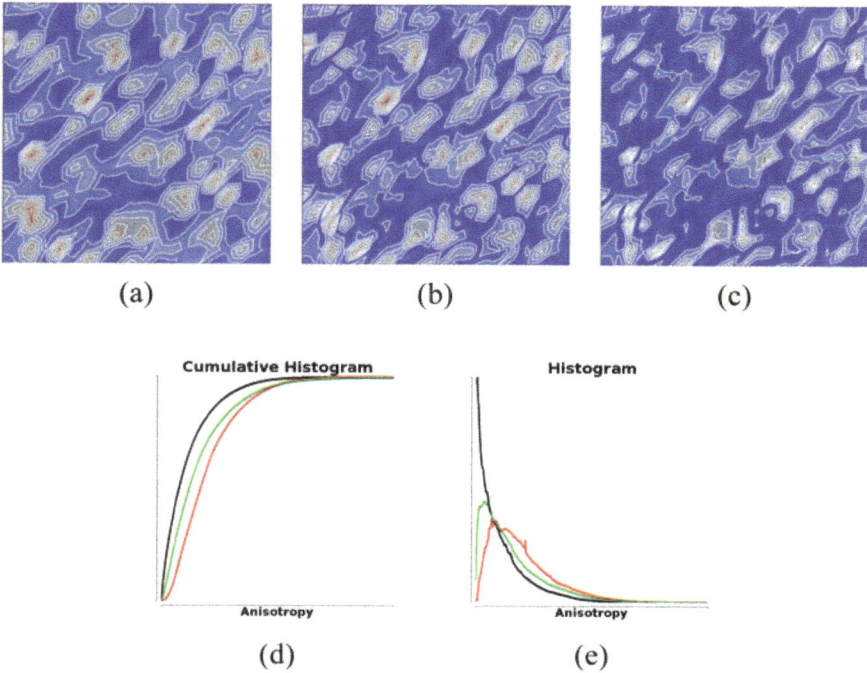

Fig. 16 Anisotropy fields in a small noisy subset of brain dataset. **a** Linear interpolation in the original mesh. **b** Linear interpolation within the subdivided mesh with monotonous triangles. **c** The anisotropy resulting from linear interpolation of tensor components. **d** The comparison of cumulative histograms under these settings. **e** The comparison of the corresponding histograms with red (**a**), green (**b**), and black (**c**)

More specifically we have presented a derivation for the computation of iso-contours and histograms in this setting. Component-wise linear interpolation of tensor components results in a quadratic function for anisotropy. The method is based on a subdivision of the mesh into triangles with monotonous behavior. This subdivision with a linear interpolation of the anisotropy, interpolation approach [b], already results in iso-contours that are topologically equivalent to the iso-contours based on the derived anisotropy field. However, the histograms are not accurate and show a bias towards larger anisotropy values. This is especially prominent in regions with many degenerate points. In areas of high anisotropy, interpolation [b] provides a good approximation. The method described in this chapter, the interpolation approach [c], can be used to compute an accurate continuous histogram for anisotropy using component-wise tensor interpolation.

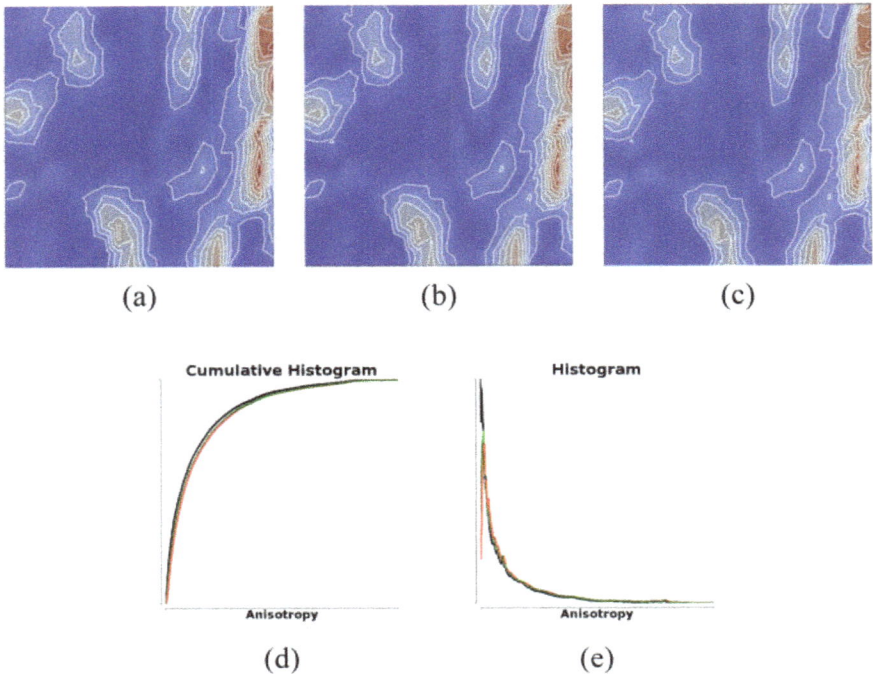

Fig. 17 A small subset of DTI brain dataset is selected within the brain region. **a** Linear interpolation in the original mesh, **b** Linear interpolation in monotonous triangles. **c** The anisotropy resulting from linear interpolation of tensor components. **d** and **e** display the respective cumulative histograms and histograms as (**a**) red, (**b**) green, (**c**) black plots

All derivations in this chapter are given for the anisotropy defined as the difference between the major and minor eigenvalue, for 2D tensor fields. Although not trivial, the extension of this work to 3D tensor fields is feasible. An extension to the determinant, which is also quadratic but not elliptic, would be possible in a similar way. A general extension to other non-linear tensor invariants, however, might not be possible in a closed form and will require a good approximation schema. Therefore, we plan to explore methods for efficient approximations to the distributions with clear error bounds. Also, computing continuous scatterplots to visualize the space of multiple invariants at once is an interesting topic and will be subject of future work.

Acknowledgments We thank Jochen Jankowai for creating Fig. 13 included in this chapter. This work was supported through the SeRC (Swedish e-Science Research Center) and the ELLIIT environment for strategic research in Sweden. We further want to thank the participants of the Seminar 18442 on Visualization and Processing of Anisotropy in Imaging, Geometry, and Astronomy, at Schloss Dagstuhl Leibniz-Zentrum for computer sciences, for fruitful discussions.

Appendix A

A. Detailed Analysis of Anisotropy and Its Iso-Contours

Proof of $H \geq 0$

Substituting the Eqs. (14)–(19) in Eq. (6), we can analyse the determinant of Hessian for anisotropy v:

$$
\begin{aligned}
H &= 4AC - B^2 \\
&= 4\left((e_x - g_x)^2 + 4f_x^2\right)\left((e_y - g_y)^2 + 4f_y^2\right) - 4\left((e_x - g_x)(e_y - g_y) + 4f_x f_y\right)^2 \\
&= 4\left((e_x - g_x)^2(e_y - g_y)^2 + 4f_x^2(e_y - g_y)^2 + 4f_y^2(e_x - g_x)^2 + 16f_x^2 f_y^2\right) \\
&\quad - 4\left((e_x - g_x)^2(e_y - g_y)^2 + 8f_x f_y(e_x - g_x)(e_y - g_y) + 16f_x^2 f_y^2\right) \\
&= 16\left(f_x^2(e_y - g_y)^2 + f_y^2(e_x - g_x)^2 - 2f_x f_y(e_x - g_x)(e_y - g_y)\right) \\
H &= 16\left(f_x(e_y - g_y) - f_y(e_x - g_x)\right)^2 \geq 0 \quad\quad\quad (30)
\end{aligned}
$$

Since $H \geq 0$, we have shown that the iso-contours of v are never hyperbolic.

Proof of $I = 0$ when $H = 0$

Let us analyze the case when $H = 0$ to complete the analysis of behaviour of anisotropy in all cases.

$$
\begin{aligned}
H &= 16\left(f_x(e_y - g_y) - f_y(e_x - g_x)\right)^2 = 0 \\
or, \quad &f_x(e_y - g_y) - f_y(e_x - g_x) = 0 \\
or, \quad &\frac{e_y - g_y}{e_x - g_x} = \frac{f_y}{f_x} \\
or, \quad &e_y - g_y = \frac{f_y}{f_x}(e_x - g_x) \quad\quad\quad (31)
\end{aligned}
$$

Substituting (31) in (16) and using (14):

$$
\begin{aligned}
C &= \left((e_y - g_y)^2 + 4f_y^2\right) \\
&= \frac{f_y^2(e_x - g_x)^2}{f_x^2} + 4f_y^2 \\
&= \frac{f_y^2}{f_x^2}\left((e_x - g_x)^2 + 4f_x^2\right) \\
or, \quad C &= \frac{f_y^2}{f_x^2} A \quad\quad\quad (32)
\end{aligned}
$$

Substituting (31) in (18) and using (17):

$$E = 2\big((e_y - g_y)(e_c - g_c) + 4f_y f_c\big)$$
$$= 2\left(\frac{f_y(e_x - g_x)(e_c - g_c)}{f_x} + 4f_y f_c\right)$$
$$= \frac{f_y}{f_x} \cdot 2\big((e_x - g_x)(e_c - g_c) + 4f_x f_c\big)$$

$$or, \quad E = \frac{f_y}{f_x} D \tag{33}$$

From Eqs. (32) and (33):

$$\frac{C}{A} = \frac{E^2}{D^2} = \frac{f_y^2}{f_x^2} \tag{34}$$

$$or, \quad AE^2 = CD^2 \tag{35}$$

Substituting (31) in (15) and using (16):

$$B = 2\big((e_x - g_x)(e_y - g_y) + 4f_x f_y\big)$$
$$= 2\left(\frac{f_x(e_y - g_y)^2}{f_y} + 4f_x f_y\right)$$
$$= 2\frac{f_x}{f_y}\left((e_y - g_y)^2 - 4f_y^2\right)$$

$$or, \quad B = 2\frac{f_x}{f_y} C \tag{36}$$

Multiplying Eqs. (33) and (36), we obtain:

$$BE = 2CD \tag{37}$$

Let us evaluate the Eq. 7 now:

$$I = BDE - AE^2 - CD^2$$
$$I = BDE - 2CD^2 \qquad\qquad using \quad (35)$$
$$I = D(BE - 2CD)$$
$$I = D(0) = 0 \qquad\qquad using \quad (37) \tag{38}$$

From Eq. (38), we conclude that when H is 0, I is also 0. This means that the iso-contours of v are never parabolic.

To conclude, based on Eqs. (30) and (38), we deduce that the contours of anisotropy v and hence α are either ellipses or a set of parallel lines.

References

1. Bachthaler, S., Weiskopf, D.: Continuous scatterplots. IEEE Trans. Visual. Comput. Graph. **14**(6), 1428–1435 (2008)
2. Bajaj, C.L., Pascucci, V., Schikore, D.: The contour spectrum. In: IEEE Visualization 1997. Proceedings. Phoenix, AZ, USA, pp. 167–174. IEEE Computer Society and ACM, Los Alamitos, CA, USA (1997)
3. Batchelor, P.G., Moakher, M., Atkinson, D., Calamante, F., Connelly, A.: A rigorous framework for diffusion tensor calculus. Magn. Reson. Med. **53**(1), 221–225 (2005)
4. Carr, H., Duffy, B., Denby, B.: On histograms and isosurface statistics. IEEE Trans. Visual. Comput. Graph. **12**(5), 1259–1266 (2006)
5. Carr, H., Snoeyink, J., Axen, U.: Computing contour trees in all dimensions. Comput. Geom. **24**(2), 75–94 (2003)
6. Dillard, S.E., Natarajan, V., Weber, G.H., Pascucci, V., Hamann, B.: Topology-guided tessellation of quadratic elements. Comput. Geom.: Theory Appl. **19**(2), 195–211 (2009)
7. Edelsbrunner, H., Mücke, E.P.: Simulation of simplicity: a technique to cope with degenerate cases in geometric algorithms. ACM Trans. Graph. **9**(1), 66–104 (1990)
8. Feragen, A., Fuster, A.: Geometries and interpolations for symmetric positive definite matrices. In: Schultz, T., Özarslan, E., Hotz, I. (eds.) Modeling, Analysis, and Visualization of Anisotropy, Mathematics and Visualization, pp. 85–114. Springer, Berlin (2017)
9. Heinrich, J., Weiskopf, D.: Continuous parallel coordinates. IEEE Trans. Visual. Comput. Graph. **15**(6), 1531–1538 (2009)
10. Ioannidis, Y.E.: The history of histograms (abridged). In: Freytag, J.C., Lockemann, P.C., Abiteboul, S., Carey, M.J., Selinger, P.G., Heuer, A. (eds.) Proceedings of 29th International Conference on Very Large Data Bases, VLDB 2003, pp. 19–30. Morgan Kaufmann, Berlin, Germany (2003)
11. Jankowai, J., Wang, B., Hotz, I.: Robust extraction and simplification of 2D symmetric tensor field topology. Comput. Graph. Forum **38**(3), 337–349 (2019)
12. Kindlmann, G., Estepar, R.S.J., Niethammer, M., Haker, S., Westin, C.F.: Geodesic-loxodromes for diffusion tensor interpolation and difference measurement. 10th International Conference on Medical Image Computing and Computer-Assisted Intervention (MICCAI'07). Lecture Notes in Computer Science 4791, pp. 1–9. Brisbane, Australia (2007)
13. Kratz, A., Meier, B., Hotz, I.: A visual approach to analysis of stress tensor fields. In: Hagen, H. (ed.) Scientific Visualization: Interactions, Features, Metaphors, Dagstuhl Follow-Ups, vol. 2, pp. 188–211. Schloss Dagstuhl-Leibniz-Zentrum fuer Informatik, Dagstuhl, Germany (2011)
14. Mullen, P., McKenzie, A., Tong, Y., Desbrun, M.: A variational approach to Eulerian geometry processing. ACM Trans. Graph. **26**(3), 66 (2007)
15. Nucha, G., Bonneau, G., Hahmann, S., Natarajan, V.: Computing contour trees for 2D piecewise polynomial functions. Comput. Graph. Forum **36**(3), 23–33 (2017)
16. Scheidegger, C.E., Schreiner, J.M., Duffy, B., Carr, H., Silva, C.T.: Revisiting histograms and isosurface statistics. IEEE Trans. Visual. Comput. Graph. (Proceedings Visualization 2008) **14**(6), 1659–1667 (2008)
17. Sreevalsan-Nair, J., Auer, C., Hamann, B., Hotz, I.: Eigenvector-based interpolation and segmentation of 2D tensor fields. In: Pascucci, V., Tricoche, X., Hagen, H., Tierny, J. (eds.) Topological Methods in Data Analysis and Visualization, Mathematics and Visualization, pp. 139–150. Springer, Berlin (2011)
18. Wang, B., Hotz, I.: Robustness for 2D symmetric tensor field topology. In: Hotz, I., Özarslan, E., Schultz, T. (eds.) Modeling, Analysis, and Visualization of Anisotropy, Mathematics and Visualization, pp. 3–27. Springer, Berlin (2017)
19. Weber, G.H., Dillard, S.E., Carr, H.A., Pascucci, V., Hamann, B.: Topology-controlled volume rendering. IEEE Trans. Visual. Comput. Graph. **13**(2), 330–341 (2007)

2

Variance Measures for Symmetric Positive (Semi-) Definite Tensors in Two Dimensions

Magnus Herberthson, Evren Özarslan, and Carl-Fredrik Westin

Abstract Calculating the variance of a family of tensors, each represented by a symmetric positive semi-definite second order tensor/matrix, involves the formation of a fourth order tensor R_{abcd}. To form this tensor, the tensor product of each second order tensor with itself is formed, and these products are then summed, giving the tensor R_{abcd} the same symmetry properties as the elasticity tensor in continuum mechanics. This tensor has been studied with respect to many properties: representations, invariants, decomposition, the equivalence problem et cetera. In this paper we focus on the two-dimensional case where we give a set of invariants which ensures equivalence of two such fourth order tensors R_{abcd} and $\widetilde{R}_{abcd}$. In terms of components, such an equivalence means that components R_{ijkl} of the first tensor will transform into the components $\widetilde{R}_{ijkl}$ of the second tensor for some change of the coordinate system.

1 Introduction

Positive semi-definite second order tensors arise in several applications. For instance, in image processing, a structure tensor is computed from greyscale images that captures the local orientation of the image intensity variations [10, 17] and is employed to address a broad range of challenges. Diffusion tensor magnetic resonance imaging (DT-MRI) [1, 5] characterizes anisotropic water diffusion by enabling the measurement of the apparent diffusion tensor, which makes it possible to delineate the fibrous structure of the tissue. Recent work has shown that diffusion MR measurements of

M. Herberthson (✉)
Department of Mathematics, Linköping University, Linköping, Sweden
e-mail: magnus.herberthson@liu.se

E. Özarslan
Department of Biomedical Engineering, Linköping University, Linköping, Sweden
e-mail: evren.ozarslan@liu.se

C.-F. Westin
Department of Radiology Brigham and Women's Hospital, Harvard Medical School, Boston, MA, USA
e-mail: westin@bwh.harvard.edu

restricted diffusion obscures the fine details of the pore shape under certain experimental conditions [11], and all remaining features can be encoded accurately by a confinement tensor [19].

All such second order tensors share the same mathematical properties, namely, they are real-valued, symmetric, and positive semi-definite. Moreover, in these disciplines, one encounters a collection of such tensors, e.g., at different locations of the image. Populations of such tensors have also been key to some studies aiming to model the underlying structure of the medium under investigation [8, 12, 18].

Irrespective of the particular application, let R_{ab} denote such tensors,[1] and we shall refer to the set of n tensors as $\{R_{ab}^{(i)}\}_i$. Our desire is to find relevant descriptors or models of such a family. One relevant statistical measure of this family is the (population) variance

$$\frac{1}{n}\sum_{i=1}^{n}(R_{ab}^{(i)} - \widehat{R}_{ab})(R_{cd}^{(i)} - \widehat{R}_{cd}) = \left(\frac{1}{n}\sum_{i=1}^{n}R_{ab}^{(i)}R_{cd}^{(i)}\right) - \widehat{R}_{ab}\widehat{R}_{cd} ,$$

where $\widehat{R}_{ab} = \frac{1}{n}\sum_{i=1}^{n}R_{ab}^{(i)}$ is the mean. (For another approach, see e.g., [8]). In this paper, we are interested in the first term, i.e., we study the fourth order tensor (skipping the normalization)

$$R_{abcd} = \sum_{i=1}^{n}R_{ab}^{(i)}R_{cd}^{(i)}, \quad R_{ab}^{(i)} \geq 0, \tag{1}$$

where $R_{ab}^{(i)} \geq 0$ stands for $R_{ab}^{(i)}$ being positive semi-definite. It is obvious that R_{abcd} has the symmetries $R_{abcd} = R_{bacd} = R_{abdc}$ and $R_{abcd} = R_{cdab}$, i.e., R_{abcd} has the same symmetries as the elasticity tensor [14] from continuum mechanics. The elasticity tensor is well studied [13], e.g. with respect to classification, decompositions, and invariants. In most cases this is done in three dimensions. The same (w.r.t. symmetries) tensor has also been studied in the context of diffusion MR [2].

In this paper we will focus on the corresponding tensor R_{abcd} in two dimensions. First, there are direct applications in image processing, and secondly, the problems posed will be more accessible in two dimensions than in three. In particular we study the equivalence problem, namely, we ask the question: given the components R_{ijkl} and $\widetilde{R}_{ijkl}$ of two such tensors do they represent the same tensor in different coordinate systems (see Sects. 2.1.2 and 4)?

1.1 Outline

Section 2 contains tensorial matters. We will assume some basic knowledge of tensors, although some definitions are given for completeness. The notation(s) used is

---

[1] For the notation of tensors used here, see Sect. 2.1.

commented on and in particular the three-dimensional Euclidean vector space $V_{(ab)}$ is introduced.

In Sect. 2.1.2, we make some general remarks concerning the tensor R_{abcd} and specify the problem we focus on. Section 2.1 is concluded with some remarks on the Voigt/Kelvin notation and the corresponding visualisation in $\mathbb{R}^3$.

Section 2.2 gives examples of invariants, especially invariants which are easily accessible from R_{abcd}. Also, more general invariant/canonical decompositions of R_{abcd} are given.

In Sect. 3, we discuss how the tensor R_{abcd} can (given a careful choice of basis) be expressed in terms of a 3×3 matrix, and how this matrix is affected by a rotation of the coordinate system in the underlying two-dimensional space on which R_{abcd} is defined.

In Sect. 4 we return to the equivalence problem and give the main result of this work. In Sect. 4.1.1 we provide a geometric condition for equivalence, while in Sect. 4.1.2, we present the equivalence in terms of a 3×3 matrix. Both these characterisations rely on the choice of particular basis elements for the vector spaces employed. In Sect. 4.1.3 the same equivalence conditions are given in a form which does not assume a particular basis.

2 Preliminaries

In this section we clarify the notation and some concepts which we need. Section 2.1 deals with the (alternatives of) tensor notation and some representations. The equivalence (and related) problems are also briefly addressed. Section 2.2 accounts for some natural invariants, traces and decompositions of R_{abcd}.

We will assume some familiarity with tensors, but to clarify the view on tensors we recall some facts. We start with a (finite dimensional) vector space V with dual V^*. A tensor of order (p,q) is then a multi-linear mapping $\underbrace{V \times V \cdots \times V}_{q} \times \underbrace{V^* \times \cdots \times V^*}_{p}$ $\to \mathbb{R}$. Moreover, a (non-degenerate) metric/scalar product $g : V \times V \to \mathbb{R}$ gives an isomorphism from V to V^* through $v \to g(v, \cdot)$, and it is this isomorphism which is used to 'raise and lower indices', see below. Indeed, for a fixed $v \in V$, $g(v, \cdot)$ is a linear mapping $V \to \mathbb{R}$, i.e., an element of V^*.

2.1 Tensor Notation and Representations

There is a plethora of notations for tensors. Here, we follow the well-adopted convention [16] that early lower case Latin letters ($T^a{}_{bc}$) refer to the tensor as a geometric object, its type being inferred from the indices and their positions (the abstract index notation). g_{ab} denotes the metric tensor. When the indices are lower case Latin letters from the middle of the alphabet, $T^i{}_{jk}$, they refer to components of $T^a{}_{bc}$ in a certain frame. The super-index i denotes a contravariant index while the sub-indices j, k are covariant. For instance, a typical vector (tensor of type $(1, 0)$) will be written v^a with components v^i, while the metric g_{ab} (tensor of type $(0, 2)$) has components g_{ij}. At a number of occasions, it will also be useful to express quantities in terms of components with respect to orthonormal frames, i.e., Cartesian coordinates. This is sometimes referred to as 'Cartesian tensors', and the distinction between contra- and covariant indices is obscured. In these situations, it is possible (but not necessary) to write all indices as sub-indices, and sometimes the symbol $\doteq$ is used to indicate that an equation is only valid in Cartesian coordinates. For example $T_i \doteq T_{ijk}\delta_{jk}$ instead of $T^i = T^i{}_{jk}g^{jk} = T^{ik}{}_k$. Often this is clear form the context, but we will sometimes use $\doteq$ to remind the reader that a Cartesian assumption is made. Here, the Einstein summation convention is implied, i.e., repeated indices are to be summed over, so that for instance $T^i = T^i{}_{jk}g^{jk} = T^{ik}{}_k = \sum_{j=1}^{n}\sum_{k=1}^{n} T^i{}_{jk}g^{jk} = \sum_{k=1}^{n} T^{ik}{}_k$ if each index ranges from 1 to n. We have also used the metric g_{ij} and its inverse g^{ij} to raise and lower indices. For instance, since $g_{ij}v^i$ is an element of V^*, we write $g_{ij}v^i = v_j$.

We also remind of the notation for symmetrisation. For a two-tensor $T_{(ab)} = \frac{1}{2}(T_{ab} + T_{ba})$, while more generally for a tensor $T_{a_1 a_2 \cdots a_n}$ of order $(0, n)$ we have

$$T_{(a_1 a_2 \cdots a_n)} = \frac{1}{n!} \sum_{\pi} T_{a_{\pi(1)} a_{\pi(2)} \cdots a_{\pi(n)}}$$

where the sum is taken over all permutations π of $1, 2, \ldots, n$. Naturally, this convention can also be applied to subsets of indices. For instance, $H_{a(bc)} = \frac{1}{2}(H_{abc} + H_{acb})$.

2.1.1 The Vector Space of Symmetric Two-Tensors

In any coordinate frame a symmetric tensor R_{ab} (i.e., $R_{ab} = R_{ba}$) is represented by a symmetric matrix R_{ij} (2×2 or 3×3 depending on the dimension of the underlying space). In the two-dimensional case, with the underlying vector space $V^a \sim \mathbb{R}^2$, this means that R_{ab} lives in a three-dimensional vector space, which we denote by $V_{(ab)}$. $V_{(ab)}$ is equipped with a natural scalar product: $< A_{ab}, B_{ab} > = A_{ab}B^{ab}$, making it into a three-dimensional Euclidean space. Here $A_{ab}B^{ab} = A_{ab}B_{cd}g^{ac}g^{bd}$, i.e, the contraction of $A_{ab}B_{cd}$ over the indices a, c and b, d, and the tensor product $A_{ab}B_{cd}$ itself is the tensor of order $(0, 4)$ given by $(A_{ab}B_{cd})v^a u^b w^c m^d = (A_{ab}v^a u^b)(B_{cd}w^c m^d)$ together with multi-linearity.

2.1.2 The Tensor R_{abcd} and the Equivalence Problem

As noted above, R_{abcd} given by (1) has the symmetries $R_{abcd} = R_{(ab)cd} = R_{ab(cd)}$ and $R_{abcd} = R_{cdab}$, and it is not hard to see that this gives R_{abcd} six degrees of freedom in two dimensions. (See also Sect. 2.1.3.) It is also interesting to note that

R_{abcd} provides a mapping $V_{(ab)} \to V_{(ab)}$ through

$$R_{ab} \mapsto R_{abcd} R^{cd},$$

and that this mapping is symmetric (due to the symmetry $R_{abcd} = R_{cdab}$). Given R_{abcd} there are a number of questions one can ask, e.g.,

- Feasibility—given a tensor R_{abcd} with the correct symmetries, can it be written in the form (1)?
- Canonical decomposition—given R_{abcd} of the form (1), can you write R_{abcd} as a canonical sum of the *form* (1), but with a fixed number of terms (cf. eigenvector decomposition of symmetric matrices)?
- Visualisation—since fourth order tensors are a bit involved, how can one visualise them in ordinary space?
- Characterisation/relevant sets of invariants—what invariants are relevant from an application point of view?
- The equivalence problem—in terms of components, how do we know if R_{ijkl} and $\widetilde{R}_{ijkl}$ represent the same tensor when they are in different coordinate systems?

We will now focus on the *equivalence problem* in two dimensions. This problem can be formulated as above: given, in terms of components, two tensors (with the symmetries we consider) R_{ijkl} and $\widetilde{R}_{ijkl}$, do they represent the same tensor in the sense that there is a coordinate transformation taking the components R_{ijkl} into the components $\widetilde{R}_{ijkl}$? In other words, does there exist an (invertible) matrix $P^m{}_i$ so that

$$R_{ijkl} = \widetilde{R}_{mnop} P^m{}_i P^n{}_j P^o{}_k P^p{}_l?$$

This problem can also be formulated when R_{ijkl} and $\widetilde{R}_{ijkl}$ are expressed in Cartesian frames. Then the coordinate transformation must be a rotation, i.e., given by a rotation matrix $Q^i{}_j \in SO(2)$. Hence, the problem of (unitary) equivalence is: Given R_{ijkl} and $\widetilde{R}_{ijkl}$, both expressed in Cartesian frames, is there a matrix (applying the 'Cartesian convention') $Q_{ij} \in SO(2)$ so that

$$R_{ijkl} = \widetilde{R}_{mnop} Q_{mi} Q_{nj} Q_{ok} Q_{pl}?$$

2.1.3   The Voigt/Kelvin Notation

Since (in two dimensions) the space $V_{(ab)}$ is three-dimensional, one can introduce coordinates, for example $R_{ij} = \left(\begin{smallmatrix} x & y \\ y & z \end{smallmatrix}\right) \sim \left(\begin{smallmatrix} x \\ y \\ z \end{smallmatrix}\right)$ and use vector algebra on $\mathbb{R}^3$. This is used in the Voigt notation [15] and the related Kelvin notation [6]. As always, one must be careful to specify with respect to which basis in $V_{(ab)}$ the coordinates $\left(\begin{smallmatrix} x \\ y \\ z \end{smallmatrix}\right)$ are taken. For instance, in the correspondence $R_{ij} = \left(\begin{smallmatrix} x & y \\ y & z \end{smallmatrix}\right) \sim \left(\begin{smallmatrix} x \\ y \\ z \end{smallmatrix}\right)$, the understood basis for $V_{(ab)}$ (in the understood/induced coordinate system) is $\{\left(\begin{smallmatrix} 1 & 0 \\ 0 & 0 \end{smallmatrix}\right), \left(\begin{smallmatrix} 0 & 1 \\ 1 & 0 \end{smallmatrix}\right), \left(\begin{smallmatrix} 0 & 0 \\ 0 & 1 \end{smallmatrix}\right)\}$.

Hence, in a Cartesian frame, where the index position is unimportant, we have for the matrices $\bar{\bar{T}} = T_{ij}$, $\bar{\bar{S}} = S_{ij}$

$$\bar{\bar{T}} = \sum_{i=1}^{n} \bar{\bar{R}}^{(i)}[\bar{\bar{R}}^{(i)}], \quad \bar{\bar{S}} = \sum_{i=1}^{n} \bar{\bar{R}}^{(i)} \bar{\bar{R}}^{(i)}.$$

To proceed there are two double traces (i.e., contracting R_{abcd} twice):

$$T = T_a{}^a = R_a{}^a{}_c{}^c = \sum_{i=1}^{n} R_a^{(i)a} R_c^{(i)c} = \sum_{i=1}^{n} [\bar{\bar{R}}^{(i)}]^2 \tag{5}$$

and

$$S = S_a{}^a = R_{ac}{}^{ac} = \sum_{i=1}^{n} R_{ac}^{(i)} R^{(i)ac} = \sum_{i=1}^{n} [(\bar{\bar{R}}^{(i)})^2]. \tag{6}$$

In two dimensions, the difference $T_{ab} - S_{ab}$ is proportional to the metric g_{ab}. Namely,

Lemma 1 *With T_{ab} and S_{ab} given by (3) and (4), it holds that (in two dimensions)*

$$T_{ab} - S_{ab} = \sum_{i=1}^{n} \det(\bar{\bar{R}}^{(i)}) g_{ab}.$$

Proof By linearity, it is enough to prove the statement when $n = 1$, i.e., when the sum has just one term. Raising the second index, and using components, the statement then is $T_i{}^j - S_i{}^j = \det(\bar{\bar{R}}^{(1)})\delta_i{}^j$. Putting $\bar{\bar{R}}^{(1)} = A$, we see that $T_i{}^j - S_i{}^j = A[A] - A^2$ while $\det(\bar{\bar{R}}^{(1)})\delta_i{}^j = \det(A)I$, and by the Cayley-Hamilton theorem in two dimensions, $A[A] - A^2$ is indeed $\det(A)I$. □

From lemma 1, it follows that $T - S = 2\sum_{i=1}^{n} \det(\bar{\bar{R}}^{(i)}) \geq 0$. In fact the following inequalities hold.

Lemma 2 *With T and S defined as above, it holds that $S \leq T \leq 2S$. If $T = S$, all tensors $R_{ab}^{(i)}$ have rank 1. If $T = 2S$, all tensors $R_{ab}^{(i)}$ are isotropic, i.e., proportional to the metric g_{ab}.*

Proof Again, by linearity it is enough to consider one tensor $\bar{\bar{R}}^{(1)} = A$. In an orthonormal frame which diagonalises A, we have $A = \begin{pmatrix} a & 0 \\ 0 & c \end{pmatrix}$ (with $a \geq 0$, $c \geq 0$, $a + c > 0$). Hence

$$S = a^2 + c^2 \leq a^2 + c^2 + 2ac = (a+c)^2 = T = 2(a^2 + c^2) - (a-c)^2 \leq 2S.$$

The first inequality becomes equality when $ac = 0$, i.e., when A has rank one. The second inequality becomes equality when $a = c$, i.e., when A is isotropic. □

Definition 1 We define the mean rank, r_m, by $r_m = T/S$, with T and S as above. Hence, in two dimensions, $1 \leq r_m \leq 2$.

2.2.2 A Canonical Decomposition

It is customary [3, 7] to decompose a tensor with the symmetries of R_{abcd} into a sum where one term is the completely symmetric part:

$$R_{abcd} = H_{abcd} + W_{abcd}, \text{ where } H_{abcd} = R_{(abcd)}, W_{abcd} = R_{abcd} - H_{abcd}.$$

It is also customary to split H_{abcd} into a trace-free part and 'trace part'. We start by defining $H_{ab} = H_{abc}{}^c$, $H = H_a{}^a$ and then the trace-free part of H_{ab}: $\mathring{H}_{ab} = H_{ab} - \frac{1}{2}Hg_{ab}$ so that $H_{ab} = \mathring{H}_{ab} + \frac{1}{2}Hg_{ab}$. (These decompositions can be made in any dimension, but the actual coefficients, e.g., $\frac{1}{2}$ above and $\frac{1}{8}$ and $\frac{3}{8}$ et cetera below depend on the underlying dimension.) It is straightforward to check that

$$\mathring{H}_{abcd} = H_{abcd} - g_{(ab}H_{cd)} + \frac{1}{8}Hg_{(ab}g_{cd)} = H_{abcd} - g_{(ab}\mathring{H}_{cd)} - \frac{3}{8}Hg_{(ab}g_{cd)}$$

is also trace-free. Hence we have the decomposition

$$H_{abcd} = \mathring{H}_{abcd} + g_{(ab}H_{cd)} - \frac{1}{8}Hg_{(ab}g_{cd)} = \mathring{H}_{abcd} + g_{(ab}\mathring{H}_{cd)} + \frac{3}{8}Hg_{(ab}g_{cd)}.$$

Moreover, due to the symmetry of R_{abcd}, we find that

$$H_{abcd} = \frac{1}{3}\left(R_{abcd} + R_{acbd} + R_{adbc}\right)$$

and therefore that

$$W_{abcd} = \frac{1}{3}\left(2R_{abcd} - R_{acbd} - R_{adbc}\right) \tag{7}$$

which implies that $H_{ab} = H_{abc}{}^c = \frac{1}{3}(T_{ab} + 2S_{ab})$ and $W_{ab} = W_{abc}{}^c = \frac{2}{3}(T_{ab} - S_{ab})$.

The degrees of freedom, which for R_{abcd} is six, is distributed, where $R_{abcd} \sim \{\mathring{H}_{abcd}, H_{ab}, W_{abcd}\}$, as

$$\underset{(6)}{R_{abcd}} \sim \underset{(2)}{\{\mathring{H}_{abcd}}, \underset{(3)}{H_{ab}}, \underset{(1)}{W_{abcd}\}} \sim \underset{(2)}{\{\mathring{H}_{abcd}}, \underset{(2)}{\mathring{H}_{ab}}, \underset{(1)}{H}, \underset{(1)}{W_{abcd}\}}.$$

For H_{ab} (or the pair $\mathring{H}_{ab}$, H) this is clear. The total symmetry of $\mathring{H}_{abcd}$ leaves only five components (in a basis), $\mathring{H}_{1111}, \mathring{H}_{1112}, \mathring{H}_{1122}, \mathring{H}_{1222}, \mathring{H}_{2222}$. However, the trace-free condition $\mathring{H}_{abcd}g^{cd} = 0$ imposes three conditions. (In an orthonormal frame, $\mathring{H}_{1122} = -\mathring{H}_{1111}$, $\mathring{H}_{2222} = -\mathring{H}_{1122}$ and $\mathring{H}_{1112} = -\mathring{H}_{1222}$.) That W_{abcd} has only one degree of freedom follows from the following lemma.

Lemma 3 *Suppose that W_{abcd} is given by (7), and put $W_{ab} = W_{abcd}g^{cd}$, $W = W_{ab}g^{ab}$. Then (in two dimensions)*

$$W_{abcd} = \tfrac{W}{4}\,(2g_{ab}g_{cd} - g_{ac}g_{bd} - g_{ad}g_{bc})$$

Proof By linearity, it is enough to consider the case when $R_{abcd} = A_{ab}A_{cd}$ for some (symmetric) A_{ab}. In terms of eigenvectors (to $A^a{}_b$) we can write $A_{ab} = \alpha x_a x_b + \beta y_a y_b$, where $x_a x^a = y_a y^a = 1$, $x_a y^a = 0$. In particular $g_{ab} = x_a x_b + y_a y_b$. From (7) we then get

$$
\begin{aligned}
W_{abcd} &= \tfrac{1}{3}\,(2R_{abcd} - R_{acbd} - R_{adbc}) \\
&= \tfrac{1}{3}\,(2A_{ab}A_{cd} - A_{ac}A_{bd} - A_{ad}A_{bc}) \\
&= \tfrac{1}{3}\,(2(\alpha x_a x_b + \beta y_a y_b)(\alpha x_c x_d + \beta y_c y_d) \\
&\quad - (\alpha x_a x_c + \beta y_a y_c)(\alpha x_b x_d + \beta y_b y_d) \\
&\quad - (\alpha x_a x_d + \beta y_a y_d)(\alpha x_b x_c + \beta y_b y_c)).
\end{aligned}
\tag{8}
$$

Expanding the parentheses, the components $x_a x_b x_c x_d$ and $y_a y_b y_c y_d$ vanish, leaving

$$
\begin{aligned}
&\frac{\alpha\beta}{3}(2x_a x_b y_c y_d + 2y_a y_b x_c x_d - x_a x_c y_b y_d \\
&\qquad - y_a y_c x_b x_d - x_a x_d y_b y_c - y_a y_d x_b x_c) \\
&= \frac{\alpha\beta}{3}\,(2g_{ab}g_{cd} - g_{ac}g_{bd} - g_{ad}g_{bc}),
\end{aligned}
\tag{9}
$$

where the last equality can be seen by inserting $g_{ab} = x_a x_b + y_a y_b$ (for all indices) and expanding. Taking one trace, i.e., contracting with g^{cd} gives $W_{ab} = \frac{2\alpha\beta}{3}g_{ab}$, and another trace gives $W = \frac{4\alpha\beta}{3}$, which proves the lemma. $\qquad\square$

3 R_{abcd} as a Quadratic Form on $\mathbb{R}^3$

Through the orthonormal basis for the space of symmetric two-tensors (in two dimensions) given by (2), the tensor R_{abcd} viewed as a quadratic form can be represented by a 3×3-matrix. Here, we will restrict ourselves to an orthonormal basis for $V_{(ab)}$, namely the basis $\{e^{(1)}_{ab}, e^{(2)}_{ab}, e^{(3)}_{ab}\}$ from Sect. 2.1.3, defined in terms of the orthonormal basis $\{\xi^a, \eta^a\}$ for V^a. Thus, given R_{abcd}, we associate the symmetric matrix M_{ij}, where (the choice of an orthonormal basis justifies the mismatch of the indices i, j)

$$M_{ij} \doteq R^{ab}{}_{cd}\,e^{(i)}_{ab}(e^{(j)})^{cd}, \quad 1 \le i, j \le 3.$$

It is instructive to see how the various derived tensors show up in M_{ij}. In terms of the basis (2) it is natural to look at the various parts of M_{ij} as follows

$$
M_{ij} \doteq \left(\begin{array}{cc|c} \times & \times & \times \\ \times & \times & \times \\ \hline \times & \times & \times \end{array}\right) \doteq \left(\begin{array}{c|c} A & \bar{v} \\ \hline \bar{v}^t & a \end{array}\right).
\tag{10}
$$

This splitting is natural for reasons which will become apparent in the next sections. Note, however, that with this representation it is tempting to consider coordinate changes in $\mathbb{R}^3$, which is not natural in this case. Rather, of interest is the change of basis in V^a and the related *induced* change of coordinates in the representation (10). See Sect. 3.2.

3.1 Representation of the Canonically Derived Parts of R_{abcd}

It is helpful to see how the components of the various tensors T_{ab}, S_{ab}, T, S, $\mathring{H}_{abcd}$, $\mathring{H}_{ab}$, H and W show up as components of M_{ij}. As for $\mathring{H}_{ab}$, e.g., $\mathring{T}_{ab}$ denotes the trace-free part of T_{ab}. Immediate is M_{33}:

$$M_{33} \doteq R^{ab}{}_{cd} e^{(3)}_{ab} (e^{(3)})^{cd} \doteq \frac{1}{2} R^{ab}{}_{cd} g_{ab} g^{cd} = \frac{1}{2} T_{cd} g^{cd} = \frac{1}{2} T. \tag{11}$$

Similarly, for $i = 1, 2$ we have

$$M_{i3} \doteq \frac{1}{\sqrt{2}} R^{ab}{}_{cd} e^{(i)}_{ab} g^{cd} \doteq \frac{1}{\sqrt{2}} T^{ab} e^{(i)}_{ab} \doteq \frac{1}{\sqrt{2}} \mathring{T}^{ab} e^{(i)}_{ab}, \tag{12}$$

where the last equality follows form the trace-freeness of $e^{(1)}_{ab}$ and $e^{(2)}_{ab}$. This means that the components of $\mathring{T}_{ab}$ (properly rescaled) goes into M_{ij} as the components of $\vec{v}$ (and $\vec{v}^t$) in (10). The same holds for $\mathring{S}_{ab}$ and $\mathring{H}_{ab}$, as $\mathring{S}_{ab} = \mathring{T}_{ab}$ by Lemma 1, which then implies that also $\mathring{H}_{ab} = \mathring{T}_{ab} = \mathring{S}_{ab}$. This latter relation follows from the trace-free part of the relation $H_{ab} = \frac{1}{3}(T_{ab} + 2S_{ab})$. Hence

$$M_{ij} \doteq \left(\begin{array}{c|c} A & \vec{\mathring{T}} \\ \hline \vec{\mathring{T}}^t & \frac{1}{2}T \end{array} \right) \doteq \left(\begin{array}{c|c} \frac{\sigma}{2}I + \mathring{A} & \vec{\mathring{T}} \\ \hline \vec{\mathring{T}}^t & \frac{1}{2}T \end{array} \right), \tag{13}$$

where $\vec{\mathring{T}} = \vec{\mathring{S}} = \vec{\mathring{H}}$ encodes the two degrees of freedom in $\mathring{T}_{ab} = \mathring{S}_{ab} = \mathring{H}_{ab}$. The matrix A is decomposed as $A = \frac{\sigma}{2}I + \mathring{A}$ where I is the (2×2) identity matrix and $\mathring{A}$ is trace-free part of A. In particular, $[A] = \sigma$.

To investigate $[M_{ij}] = M_{11} + M_{22} + M_{33}$, i.e., the trace of M_{ij} we note that for a general symmetric matrix $R_{ij} \doteq \left(\begin{smallmatrix} a & b \\ b & c \end{smallmatrix} \right)$ we have $R_{ij} e^{(1)}_{ij} \doteq \frac{a-c}{\sqrt{2}}$, $R_{ij} e^{(2)}_{ij} \doteq \frac{2b}{\sqrt{2}}$, $R_{ij} e^{(3)}_{ij} \doteq \frac{a+c}{\sqrt{2}}$. When M_{ij} is constructed from R_{abcd} which is an outer product $R_{ab} R_{cd}$ the trace is given by $M_{11} + M_{22} + M_{33} = (\frac{a-c}{\sqrt{2}})^2 + (\frac{2b}{\sqrt{2}})^2 + (\frac{a+c}{\sqrt{2}})^2 = a^2 + 2b^2 + c^2$ and from (6) this is S. Together with linearity, this shows that $[M] = M_{11} + M_{22} + M_{33} = S$ also when R_{abcd} is formed as in (1). Taking trace in (13), this gives

$$S = \sigma + \tfrac{1}{2}T, \quad \text{i.e.,} \quad \sigma = S - \tfrac{1}{2}T.$$

In addition, the relations below Eq. (7) show that

$$\begin{cases} H = \frac{1}{3}(T + 2S) \\ W = \frac{2}{3}(T - S) \end{cases} \quad \text{i.e.,} \quad \begin{cases} T = H + W \\ S = H - \frac{1}{2}W \end{cases} \quad \text{so that} \quad \sigma = \frac{1}{2}H - W.$$

The two degres of freedom in $\mathring{A}$ corresponds to the two degrees of freedom in $\mathring{H}_{abcd}$.

3.2 The Behaviour of M_{ij} Under a Rotation of the Coordinate System in V^a

The components of M_{ij} are expressed in terms of the orthonormal basis tensors given by (2), and these in turn are based on the ON basis $\{\hat{\xi}, \hat{\eta}\}$ for V. Putting the basis vectors in a row matrix $(\hat{\xi} \; \hat{\eta})$ and the coordinates in a column matrix $\binom{\xi}{\eta}$ so that a vector $\mathbf{u} = \xi\hat{\xi} + \eta\hat{\eta} = (\hat{\xi} \; \hat{\eta}) \binom{\xi}{\eta}$, and considering only orthonormal frames, the relevant change of basis is given by a rotation matrix $Q(v) = Q_v = \begin{pmatrix} \cos v & -\sin v \\ \sin v & \cos v \end{pmatrix}$, i.e., we consider the change of basis

$$(\hat{\xi} \; \hat{\eta}) \rightarrow (\hat{\tilde{\xi}} \; \hat{\tilde{\eta}}) = (\hat{\xi} \; \hat{\eta}) \begin{pmatrix} \cos v & -\sin v \\ \sin v & \cos v \end{pmatrix} = (\hat{\xi} \; \hat{\eta}) \, Q(v).$$

This means that for a vector $\mathbf{u} = (\hat{\tilde{\xi}} \; \hat{\tilde{\eta}}) \binom{\tilde{\xi}}{\tilde{\eta}} = (\hat{\xi} \; \hat{\eta}) \binom{\xi}{\eta}$, the coordinates transform as

$$\binom{\xi}{\eta} \rightarrow \binom{\tilde{\xi}}{\tilde{\eta}} = Q^{-1}(v) \binom{\xi}{\eta} = Q^t(v) \binom{\xi}{\eta} = Q(-v) \binom{\xi}{\eta}.$$

For the components of the basis vectors $e^{(1)}_{ab}, e^{(2)}_{ab}, e^{(3)}_{ab}$ we find (omitting the factor $1/\sqrt{2}$)

$$\begin{pmatrix} 1 & 0 \\ 0 & -1 \end{pmatrix} \rightarrow \begin{pmatrix} \cos v & \sin v \\ -\sin v & \cos v \end{pmatrix} \begin{pmatrix} 1 & 0 \\ 0 & -1 \end{pmatrix} \begin{pmatrix} \cos v & -\sin v \\ \sin v & \cos v \end{pmatrix} = \begin{pmatrix} \cos 2v & -\sin 2v \\ -\sin 2v & -\cos 2v \end{pmatrix}$$

$$\begin{pmatrix} 0 & 1 \\ 1 & 0 \end{pmatrix} \rightarrow \begin{pmatrix} \cos v & \sin v \\ -\sin v & \cos v \end{pmatrix} \begin{pmatrix} 0 & 1 \\ 1 & 0 \end{pmatrix} \begin{pmatrix} \cos v & -\sin v \\ \sin v & \cos v \end{pmatrix} = \begin{pmatrix} \sin 2v & \cos 2v \\ \cos 2v & -\sin 2v \end{pmatrix}$$

$$\begin{pmatrix} 1 & 0 \\ 0 & 1 \end{pmatrix} \rightarrow \begin{pmatrix} \cos v & \sin v \\ -\sin v & \cos v \end{pmatrix} \begin{pmatrix} 1 & 0 \\ 0 & 1 \end{pmatrix} \begin{pmatrix} \cos v & -\sin v \\ \sin v & \cos v \end{pmatrix} = \begin{pmatrix} 1 & 0 \\ 0 & 1 \end{pmatrix},$$

$$(14)$$

and this means that the components M_{ij} transform as

$$M_{ij} \doteq \left(\begin{array}{c|c} A & \bar{v} \\ \hline \bar{v}^t & a \end{array} \right) \rightarrow \tilde{M}_{ij} \doteq \left(\begin{array}{c|c} Q^t_{2v} A Q_{2v} & Q^t_{2v} \bar{v} \\ \hline \bar{v}^t Q_{2v} & a \end{array} \right). \qquad (15)$$

But this latter expression is just

$$\left(\begin{array}{c|c} Q_{2v}^t & \overline{0} \\ \hline \overline{0}^t & 1 \end{array}\right) \left(\begin{array}{c|c} A & \overline{v} \\ \hline \overline{v}^t & a \end{array}\right) \left(\begin{array}{c|c} Q_{2v} & \overline{0} \\ \hline \overline{0}^t & 1 \end{array}\right),$$

hence we have the following important remark/observation:

Remark 1 Viewing the matrix M_{ij} as an ellipsoid in $\mathbb{R}^3$, the effect of a rotation by an angle v in V^a corresponds to a rotation of the ellipsoid by an angle $2v$ around the z-axis in $\mathbb{R}^3$ (where the z-axis corresponds to the 'isotropic direction' given by g_{ab}).

4 The Equivalence Problem for R_{abcd}

The equivalence problem for R_{abcd} can be formulated in different ways (for an account in three dimensions, we refer to [3]). Given two tensors R_{abcd} and $\widetilde{R}_{abcd}$, both with the symmetries implied by (1), the question whether they are the same or not is straightforward as one can compare the components in any basis. However, R_{abcd} and $\widetilde{R}_{abcd}$ could live in different (but isomorphic) vector spaces, e.g. two tangent spaces at different points, and the concept of equality becomes less clear. On the other hand, in terms of components R_{ijkl} and $\widetilde{R}_{ijkl}$, one could ask whether there is a change of coordinates which takes one set of components into the other. If so, one can find a (invertible) matrix $P^i{}_j$ so that

$$R_{ijkl} = \widetilde{R}_{mnop} P^m{}_i P^n{}_j P^o{}_k P^p{}_l,$$

and the tensors are then said to be equivalent. As already mentioned, it is convenient to restrict the coordinate systems to orthonormal coordinates. This means that two different coordinate systems differ only by their orientation, i.e., the change of coordinates are given by a rotation matrix $Q \in \mathrm{SO}(2)$. Under the 'Cartesian convention' that all indices are written as subscripts, R_{abcd} and $\widetilde{R}_{abcd}$ are equivalent if there is a matrix $Q \in \mathrm{SO}(2)$ so that (their Cartesian components satisfy)

$$R_{ijkl} = \widetilde{R}_{mnop} Q_{mi} Q_{nj} Q_{ok} Q_{pl}.$$

4.1 Different Ways to Characterize the Equivalence of R_{abcd} and $\widetilde{R}_{abcd}$

In this section, we will discuss three ways to determine whether two tensors R_{abcd} and $\widetilde{R}_{abcd}$ are equivalent or not. In Sects. 4.1.1 and 4.1.2 we present two such methods briefly, while Sect. 4.1.3, which is more complete, contains the main result of this work.

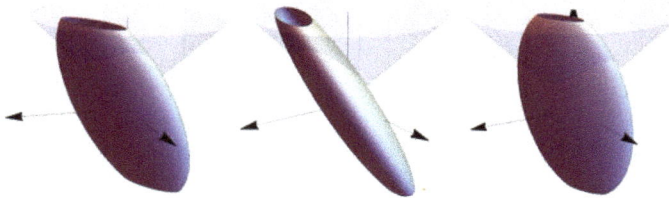

Fig. 2 Three identical (truncated) ellipsoids in $\mathbb{R}^3$ with different orientations. The two leftmost ellipsoids can be carried over to each other through a rotation around the (vertical in the figure) z-axis, which implies that they represent the same tensor R_{abcd} (up to the meaning discussed). The right ellipsoid, despite identical eigenvalues with the two others, represent a different tensor since the rotation which carries this ellipsoid to any of the others is not around the z-axis

As mentioned in Sect. 1.1, the results of Sects. 4.1.1 and 4.1.2, which may be used in their own rights, rely on particular choices of basis matrices for $V_{(ab)}$. The formulation in Sect. 4.1.3 on the other hand, is expressed in the components of R_{abcd} (in any coordinate system) directly.

4.1.1 Orientation of the Ellipsoid in $\mathbb{R}^3$

A necessary condition for R_{abcd} and $\widetilde{R}_{abcd}$ to be equivalent is that their corresponding 3×3-matrices M_{ij} and $\widetilde{M}_{ij}$ have the same eigenvalues. On the other hand, this is not sufficient since the representation in $\mathbb{R}^3$ should reflect the freedom in rotating the coordinate system in $V^a \sim \mathbb{R}^2$. With the coordinates adopted, this corresponds to a rotation of the associated ellipsoid around the z-axis in $\mathbb{R}^3$ (see Remark 1 in Sect. 3.2). This is illustrated in Fig. 2 where three ellipsoids, all representing positive definite symmetric mappings having identical eigenvalues, are shown. The two first ellipsoids can be rotated into each other by a rotation around the z-axis. This implies that the corresponding tensors R_{abcd} and $\widetilde{R}_{abcd}$ are equivalent. The third ellipsoid can also be rotated into the two others, but these rotations are around directions other than the z-axis, which means that this ellipsoid represents a different tensor.

In the generic case, with all eigenvalues different, it is easy to test whether two different ellipsoids can be transfered into each other through a rotation around the z-axis. This will be the case if the corresponding eigenvectors (of M_{ij} and $\widetilde{M}_{ij}$) have the same angle with the z-axis. Hence it is just a matter of checking the z-components of the three normalized eigenvectors and see if they are equal up to sign.

4.1.2 Components in a Canonical Coordinate System

In a sense, this is the most straightforward method. In a coordinate system which respects $e_{ab}^{(3)}$ as the z-axis in $V_{(ab)} \sim \mathbb{R}^3$, two tensors R_{abcd} and $\widetilde{R}_{abcd}$ are equivalent if there is a rotation matrix (in two dimensions) Q such that

$$\left(\begin{array}{c|c} A & \overset{\circ}{\overrightarrow{T}} \\ \hline \overset{\circ}{\overrightarrow{T}}{}^t & \frac{1}{2}T \end{array} \right) = \left(\begin{array}{c|c} Q^t \widetilde{A} Q & Q^t \overset{\approx}{\overrightarrow{T}} \\ \hline \overset{\approx}{\overrightarrow{T}}{}^t Q & \frac{1}{2}\widetilde{T} \end{array} \right). \tag{16}$$

Hence, equivalence can be easily tested by first checking that $T = \widetilde{T}$ and that $|| \overset{\circ}{\overrightarrow{T}} || = || \overset{\approx}{\overrightarrow{T}} ||$. If this is the case, (and if $|| \overset{\approx}{\overrightarrow{T}} || > 0$) one determines the rotation matrix Q which gives $\overset{\circ}{\overrightarrow{T}} = Q^t \overset{\approx}{\overrightarrow{T}}$, and equivalence is then determined by if $A = Q^t \widetilde{A} Q$ or not. If $|| \overset{\circ}{\overrightarrow{T}} || = || \overset{\approx}{\overrightarrow{T}} || = 0$, the equivalence of A and $\widetilde{A}$ can be determined directly, i.e., by checking whether $[A] = [\widetilde{A}]$ and $[A^2] = [\widetilde{A^2}]$ or not.

4.1.3   Equivalence Through (algebraic) Invariants of R_{abcd}

If a solution is found, this is perhaps the most satisfactory way to establish equivalence, in particular if the invariants are constructed by simple algebraic operations only. (For instance, to a symmetric 3×3-matrix A one can take the three eigenvalues as invariants or else for instance the traces of A, A^2 and A^3. The former set requires some calculations, but the latter is immediate.)

Examples of invariants are $T = R_{abcd}g^{ab}g^{cd}$, $S = R_{abcd}g^{ac}g^{bd}$ and the invariants $H = H_{ab}g^{ab}$, $W = W_{ab}g^{ab}$. To produce the invariants, we use the tensor R_{abcd} and the metric g_{ab}. However, if we regard $V^a \sim \mathbb{R}^2$ as oriented, so that the orthonormal basis $\{\hat{\boldsymbol{\xi}}, \hat{\boldsymbol{\eta}}\}$ for V^a also is oriented, then invariants can also be formed in another way. Namely, since the space of symmetric 2×2 matrices is 3-dimensional, and since the metric g_{ab} singles out a 1-dimensional subspace, it also determines a 2-dimensional subspace L; all elements orthogonal to g_{ab}. This subspace is the set of all symmetric 2×2 matrices which are also trace-free. L can be given an orientation through an area form, which in turn inherits the orientation from V^a.

In general, with right-handed Cartesian coordinates x^1, x^2, the area form ϵ is given by $\epsilon = dx^1 \wedge dx^2$ where $(\omega \wedge \mu)_{ab} = \omega_a \mu_b - \omega_b \mu_a$. With the orthonormal basis $\{\hat{\boldsymbol{\xi}}, \hat{\boldsymbol{\eta}}\}$ (for V^a) also right handed, we define, cf. (2),

$$e_{ab}^{(1)} = \frac{1}{\sqrt{2}}(\hat{\boldsymbol{\xi}}_a \hat{\boldsymbol{\xi}}_b - \hat{\boldsymbol{\eta}}_a \hat{\boldsymbol{\eta}}_b), \quad e_{ab}^{(2)} = \frac{1}{\sqrt{2}}(\hat{\boldsymbol{\xi}}_a \hat{\boldsymbol{\eta}}_b + \hat{\boldsymbol{\eta}}_a \hat{\boldsymbol{\xi}}_b). \tag{17}$$

The area form on L is then $\epsilon \sim e^{(1)} \wedge e^{(2)}$, or

$$\epsilon \sim E_{abcd} = e_{ab}^{(1)} e_{cd}^{(2)} - e_{ab}^{(2)} e_{cd}^{(1)}. \tag{18}$$

It is not hard to see that this definition is independent of the orientation of $\{\hat{\boldsymbol{\xi}}, \hat{\boldsymbol{\eta}}\}$. We observe that $2E_{abcd} = (\hat{\boldsymbol{\xi}}_a \hat{\boldsymbol{\xi}}_b - \hat{\boldsymbol{\eta}}_a \hat{\boldsymbol{\eta}}_b)(\hat{\boldsymbol{\xi}}_c \hat{\boldsymbol{\eta}}_d + \hat{\boldsymbol{\eta}}_c \hat{\boldsymbol{\xi}}_d) - (\hat{\boldsymbol{\xi}}_a \hat{\boldsymbol{\eta}}_b + \hat{\boldsymbol{\eta}}_a \hat{\boldsymbol{\xi}}_b)(\hat{\boldsymbol{\xi}}_c \hat{\boldsymbol{\xi}}_d - \hat{\boldsymbol{\eta}}_c \hat{\boldsymbol{\eta}}_d)$. By replacing $\hat{\boldsymbol{\xi}}$ by $\hat{\boldsymbol{\omega}} = \cos v\, \hat{\boldsymbol{\xi}} + \sin v\, \hat{\boldsymbol{\eta}}$ and $\hat{\boldsymbol{\eta}}$ by $\hat{\boldsymbol{\mu}} = -\sin v\, \hat{\boldsymbol{\xi}} + \cos v\, \hat{\boldsymbol{\eta}}$, i.e., a rotated orthonormal basis, it is straightforward to check that

$$(\hat{\omega}_a \hat{\omega}_b - \hat{\mu}_a \hat{\mu}_b)(\hat{\omega}_c \hat{\mu}_d + \hat{\mu}_c \hat{\omega}_d) - (\hat{\omega}_a \hat{\mu}_b + \hat{\mu}_a \hat{\omega}_b)(\hat{\omega}_c \hat{\omega}_d - \hat{\mu}_c \hat{\mu}_d)$$
$$= (\hat{\xi}_a \hat{\xi}_b - \hat{\eta}_a \hat{\eta}_b)(\hat{\xi}_c \hat{\eta}_d + \hat{\eta}_c \hat{\xi}_d) - (\hat{\xi}_a \hat{\eta}_b + \hat{\eta}_a \hat{\xi}_b)(\hat{\xi}_c \hat{\xi}_d - \hat{\eta}_c \hat{\eta}_d) \tag{19}$$

so that E_{abcd} is well defined. We recollect that area form E_{abcd} is defined, through the induced metric, on the plane L (which in turn is also defined through the metric g_{ab}) and the orientation on V^a. Hence E_{abcd} can be used when forming invariants.

We will now state the result of this work, namely the existence of six invariants which can be used to investigate equivalence of two tensors R_{abcd} and $\widetilde{R}_{abcd}$. We start by defining

$$
\begin{aligned}
S &= R_{abcd} g^{ac} g^{bd} \\
T &= R_{abcd} g^{ab} g^{cd} \\
J_0 &= R_{abcd} R^{abcd} \\
J_1 &= T^{ab} T_{ab} \\
J_2 &= R_{abcd} T^{ab} T^{cd} \\
J_3 &= T^{ab} R_{abcd} E^{cdef} T_{ef}.
\end{aligned}
\tag{20}
$$

where E_{abcd} is defined by (17) and (18). Similarly, we define $\widetilde{S}$, $\widetilde{T}$, $\widetilde{J}_0$, $\widetilde{J}_1$, $\widetilde{J}_2$ and $\widetilde{J}_3$ as the corresponding invariants formed from $\widetilde{R}_{abcd}$. We make the remark that for most of these invariants, their immediate interpretations still remain to be found. Rather, their value lie in the fact that they form a set which can be used to establish the equivalence in Theorem 1 below. On the other hand, some interpretations are possible. In particular, the quotient T/S (see Definition 1) lies in the interval $[1, 2]$ and has the meaning given by Lemma 2.

Theorem 1 *Suppose that $R_{abcd} = \sum_{i=1}^{n} R_{ab}^{(i)} R_{cd}^{(i)}$, with $R_{ab}^{(i)} \geq 0$ and that R_{ijkl} are the components of R_{abcd} in some basis. Suppose also that $\widetilde{R}_{abcd} = \sum_{i=1}^{\tilde{n}} \widetilde{R}_{ab}^{(i)} \widetilde{R}_{cd}^{(i)}$, with $\widetilde{R}_{ab}^{(i)} \geq 0$ and that $\widetilde{R}_{ijkl}$ are the components of $\widetilde{R}_{abcd}$ in some, possibly unrelated, basis. If (and only if) $S = \widetilde{S}$, $T = \widetilde{T}$, $J_0 = \widetilde{J}_0$, $J_1 = \widetilde{J}_1$, $J_2 = \widetilde{J}_2$, $J_3 = \widetilde{J}_3$, then there is a transformation matrix $P^i{}_j$ such that*

$$R_{ijkl} = \widetilde{R}_{mnop} P^m{}_i P^n{}_j P^o{}_k P^p{}_l.$$

Proof Since the invariants are defined without reference to any basis, it is sufficient to consider the components expressed in an orthonormal frame, and in that case we must prove the existence of a rotation matrix $Q \in SO(2)$ so that

$$R_{ijkl} = \widetilde{R}_{mnop} Q_{mi} Q_{nj} Q_{ok} Q_{pl}.$$

Since

$$R_{abcd} = M_{ij} e_{ab}^{(i)} e_{cd}^{(j)}, \tag{21}$$

we can consider the invariants formed from the components of

$$M_{ij} = \left(\begin{array}{c|c} A & \overline{u} \\ \hline \overline{u}^t & c \end{array}\right) \text{ and } \widetilde{M}_{ij} = \left(\begin{array}{c|c} \widetilde{A} & \widetilde{\overline{u}} \\ \hline \widetilde{\overline{u}}^t & \widetilde{c} \end{array}\right) \tag{22}$$

and we must demonstrate the existence of a rotation matrix $Q = Q_{2v}$ such that

$$\widetilde{A} = Q_{2v}^t A Q_{2v}, \quad \widetilde{\overline{u}} = Q_{2v}^t \overline{u}, \quad \widetilde{c} = c. \tag{23}$$

We make the ansatz

$$M_{ij} = \left(\begin{array}{cc|c} \frac{\sigma}{2}+a & b & x \\ b & \frac{\sigma}{2}-a & y \\ \hline x & y & c \end{array}\right), \quad \widetilde{M}_{ij} = \left(\begin{array}{cc|c} \frac{\widetilde{\sigma}}{2}+\widetilde{a} & \widetilde{b} & \widetilde{x} \\ \widetilde{b} & \frac{\widetilde{\sigma}}{2}-\widetilde{a} & \widetilde{y} \\ \hline \widetilde{x} & \widetilde{y} & \widetilde{c} \end{array}\right). \tag{24}$$

Through (21) it is straightforward to see that

$$S = \sigma + c, \quad T = 2c, \quad J_0 = 2(a^2 + b^2) + c^2 + \sigma^2/2 + 2(x^2 + y^2),$$
$$J_1 = 2(c^2 + x^2 + y^2)$$

so if $S = \widetilde{S}, T = \widetilde{T}, J_0 = \widetilde{J}_0, J_1 = \widetilde{J}_1$, it follows that $\sigma = \widetilde{\sigma}, c = \widetilde{c}, a^2 + b^2 = \widetilde{a}^2 + \widetilde{b}^2$ and $x^2 + y^2 = \widetilde{x}^2 + \widetilde{y}^2$. Since the isotropic part of A, i.e., $\frac{\sigma}{2}I$ is unaffected by a rotation of the coordinate system, we consider the traceless parts $\mathring{A} = \left(\begin{array}{cc} a & b \\ b & -a \end{array}\right)$, $\mathring{\widetilde{A}} = \left(\begin{array}{cc} \widetilde{a} & \widetilde{b} \\ \widetilde{b} & -\widetilde{a} \end{array}\right)$, and the task is to assert a rotation matrix Q such that

$$\left(\begin{array}{cc} a & b \\ b & -a \end{array}\right) = Q^t \left(\begin{array}{cc} \widetilde{a} & \widetilde{b} \\ \widetilde{b} & -\widetilde{a} \end{array}\right) Q, \quad \left(\begin{array}{c} x \\ y \end{array}\right) = Q^t \left(\begin{array}{c} \widetilde{x} \\ \widetilde{y} \end{array}\right),$$

if also $J_2 = \widetilde{J}_2, J_3 = \widetilde{J}_3$. Again it is straightforward to calculate the remaining invariants, and we find

$$J_2 = 4bxy + 2a(x^2 - y^2) + 2c^3 + (4c + \sigma)(x^2 + y^2)$$
$$J_3 = 4axy - 2b(x^2 - y^2).$$

and similarly for $\widetilde{J}_2, \widetilde{J}_3$. Hence, (since $\sigma = \widetilde{\sigma}, c = \widetilde{c}$)

$$a^2 + b^2 = \widetilde{a}^2 + \widetilde{b}^2$$
$$x^2 + y^2 = \widetilde{x}^2 + \widetilde{y}^2$$
$$2bxy + a(x^2 - y^2) = 2\widetilde{b}\widetilde{x}\widetilde{y} + \widetilde{a}(\widetilde{x}^2 - \widetilde{y}^2) \tag{25}$$
$$2axy - b(x^2 - y^2) = 2\widetilde{a}\widetilde{x}\widetilde{y} - \widetilde{b}(\widetilde{x}^2 - \widetilde{y}^2).$$

Suppose first that $x^2 + y^2 > 0$. The equality $x^2 + y^2 = \widetilde{x}^2 + \widetilde{y}^2$ then guarantees the existence of the rotation matrix Q which is determined via the relation $\left(\begin{array}{c} x \\ y \end{array}\right) = Q^t \left(\begin{array}{c} \widetilde{x} \\ \widetilde{y} \end{array}\right)$. This can also be expressed as $Q_1^t \left(\begin{array}{c} x \\ y \end{array}\right) = Q_2^t \left(\begin{array}{c} \widetilde{x} \\ \widetilde{y} \end{array}\right)$ for some rotation matrices Q_1, Q_2, where $Q = Q_2 Q_1^t$. We now choose the rotation matrix Q_1 so that in the untilded coordinates, $y = 0$. Similarly we choose Q_2 so that for the tilded coordinates, we get a frame where $\widetilde{y} = 0$. The equalities between the invariants in (25) then become

$$a^2 + b^2 = \widetilde{a}^2 + \widetilde{b}^2$$
$$x^2 = \widetilde{x}^2$$
$$ax^2 = \widetilde{a}\widetilde{x}^2$$
$$-bx^2 = -\widetilde{b}\widetilde{x}^2 \,,$$

so that $a = \widetilde{a}$, $b = \widetilde{b}$. This proves the theorem when $x^2 + y^2 > 0$. When $x^2 + y^2 = \widetilde{x}^2 + \widetilde{y}^2 = 0$, i.e., $x = y = \widetilde{x} = \widetilde{y} = 0$, the remaining equality $a^2 + b^2 = \widetilde{a}^2 + \widetilde{b}^2$ is sufficient since we can again choose frames in which $b = \widetilde{b} = 0$ and $a > 0, \widetilde{a} > 0$. It then follows that $a = \widetilde{a}$. □

5 Discussion

In this work, we started with a family of symmetric positive (semi-)definite tensors in two dimensions and considered its variance. This lead us to a fourth order tensor R_{abcd} with the same symmetries as the elasticity tensor in continuum mechanics. After listing a number of possible issues to address, we focused on the equivalence problem. Namely, given the components of two such tensors R_{abcd} and $\widetilde{R}_{abcd}$, how can one determine if they represent the same tensor (but in different coordinate systems) or not? In Sect. 4, we saw that this could be investigated in different ways. The result of Theorem 1 is most satisfactory in the sense that it is expressible in terms of the components of the fourth order tensors directly.

There are two natural extensions and/or ways to continue this work. The first is to apply the result to realistic families of e.g., diffusion tensors in two dimensions. The objective is then, apart from establishing possible equivalences, to investigate the geometric meaning of the invariants. The other natural continuation is to investigate the corresponding problem in three dimensions. The degrees of freedom of R_{abcd} will then increase from 6 to 21, leaving us with a substantially harder, but also perhaps more interesting, problem.

Acknowledgements The authors acknowledge the following sources for funding: Swedish Foundation for Strategic Research AM13-0090, the Swedish Research Council 2015-05356 and 2016-04482, Linköping University Center for Industrial Information Technology (CENIIT), VINNOVA/ITEA3 17021 IMPACT, Analytic Imaging Diagnostics Arena (AIDA), and National Institutes of Health P41EB015902.

References

1. Basser, P.J., Mattiello, J., LeBihan, D.: MR diffusion tensor spectroscopy and imaging. Biophys. J. **66**(1), 259–267 (1994)

2. Basser, P.J., Pajevic, S.: A normal distribution for tensor-valued random variables: applications to diffusion tensor MRI. IEEE Trans. Med. Imaging **22**(7), 785–94 (2003). https://doi.org/10. 1109/TMI.2003.815059

3. Boehler, J.P., Kirillov Jr., A.A., Onat, E.T.: On the polynomial invariants of the elasticity tensor. J. Elast. **34**(2), 97–110 (1994)

4. Burgeth, B., Didas, S., Florack, L., Weickert, J.: A generic approach to diffusion filtering of matrix-fields. Computing **81**, 179–197 (2007). https://doi.org/10.1007/s00607-007-0248-9

5. Callaghan, P.T.: Translational Dynamics and Magnetic Resonance: Principles of Pulsed Gradient Spin Echo NMR. Oxford University Press, New York (2011)

6. Helbig, K.: Review paper: What Kelvin might have written about elasticity. Geophys. Prospect. **61**, 1–20 (2013). https://doi.org/10.1111/j.1365-2478.2011.01049.x

7. Itin, Y., Hehl, F.W.: Irreducible decompositions of the elasticity tensor under the linear and orthogonal groups and their physical consequences. J. Phys.: Conf. Ser. **597**, 012046 (2015)

8. Jian, B., Vemuri, B.C., Özarslan, E., Carney, P.R., Mareci, T.H.: A novel tensor distribution model for the diffusion-weighted MR signal. NeuroImage **37**(1), 164–176 (2007). https://doi. org/10.1016/j.neuroimage.2007.03.074

9. Kanatani, K.: Group-Theoretical Methods in Image Understanding. Springer, Berlin (1990)

10. Knutsson, H.: Representing local structure using tensors. In: Proceedings of the 6th Scandinavian Conference on Image Analysis, pp. 244–251. Oulu University, Oulu (1989)

11. Özarslan, E., Yolcu, C., Herberthson, M., Westin, C.F., Knutsson, H.: Effective potential for magnetic resonance measurements of restricted diffusion. Front. Phys. **5**, 68 (2017)

12. Shakya, S., Batool, N., Özarslan, E., Knutsson, H.: Multi-fiber reconstruction using probabilistic mixture models for diffusion MRI examinations of the brain. In: Schultz, T., Özarslan, E., Hotz, I. (eds.) Modeling, Analysis, and Visualization of Anisotropy, pp. 283–308. Springer International Publishing, Cham (2017)

13. Slaughter, W.S.: The Linearized Theory of Elasticity. Birkhäuser, Basel (2002)

14. Thomson, W.: Xxi. elements of a mathematical theory of elasticity. Philso. Trans. R. Soc. Lond. **146**, 481–498 (1856)

15. Voigt, W.: Lehrbuch Der Kristallphysik. Vieweg + Teubner Verlag (1928)

16. Wald, R.M.: General Relativity. University of Chicago Press, Chicago (1984)

17. Weickert, J.: Anisotropic Diffusion in Image Processing. Teubner-Verlag, Stuttgart (1998)

18. Westin, C.F., Knutsson, H., Pasternak, O., Szczepankiewicz, F., Özarslan, E., van Westen, D., Mattisson, C., Bogren, M., O'Donnell, L.J., Kubicki, M., Topgaard, D., Nilsson, M.: Q-space trajectory imaging for multidimensional diffusion MRI of the human brain. NeuroImage **135**, 345–62 (2016). https://doi.org/10.1016/j.neuroimage.2016.02.039

19. Yolcu, C., Memiç, M., Şimşek, K., Westin, C.F., Özarslan, E.: NMR signal for particles diffusing under potentials: from path integrals and numerical methods to a model of diffusion anisotropy. Phys. Rev. E **93**, 052602 (2016)

20. Zucchelli, M., Deslauriers-Gauthier, S., Deriche, R.: A closed-form solution of rotation invariant spherical harmonic features in diffusion MRI, pp. 77–89. Springer, Cham (2019)

3

Degenerate Curve Bifurcations in 3D Linear Symmetric Tensor Fields

Yue Zhang, Hongyu Nie, and Eugene Zhang

Abstract 3D symmetric tensor fields have a wide range of applications in medicine, science, and engineering. The topology of tensor fields can provide key insight into their structures. In this paper we study the number of possible topological bifurcations in 3D linear tensor fields. Using the linearity/planarity classification and wedge/trisector classification, we explore four types of bifurcations that can change the number and connectivity in the degenerate curves as well as the number and location of transition points on these degenerate curves. This leads to four types of bifurcations among nine scenarios of 3D linear tensor fields.

1 Introduction

Tensor field visualization is an important topic in visualization, with many applications in medical imaging, solid and fluid mechanics, material science, earthquake engineering, and computer graphics.

Recent advances on tensor field visualization focus in topology-driven analysis and visualization of 3D symmetric tensor fields. Degenerate curves are one of the most fundamental topological features in a tensor field, and much research has focused on the understanding and efficient extraction of degenerate curves from piecewise linear tensor fields defined on a tetrahedral mesh [6–8].

Y. Zhang (✉)
School of Electrical Engineering and Computer Science 3117 Kelley Engineering Center, Oregon State University, Corvallis, OR 97331, USA
e-mail: zhangyue@eecs.oregonstate.edu

H. Nie
School of Electrical Engineering and Computer Science 1148 Kelley Engineering Center, Oregon State University, Corvallis, OR 97331, USA
e-mail: nieh@oregonstate.edu

E. Zhang
School of Electrical Engineering and Computer Science 2111 Kelley Engineering Center, Oregon State University, Corvallis, OR 97331, USA
e-mail: zhange@eecs.oregonstate.edu

In the book chapter, we focus on a problem that has received relatively little attention: bifurcations in tensor field topology. To make our investigation effective with potential application to real datasets, we focus on 3D linear tensor fields. We explore all the theoretically possible bifurcations. Moreover, we have conducted experiment to verify whether these bifurcations can occur.

The rest of the paper is structured as follows. Section 2 reviews past research in topology-driven analysis of symmetric tensor fields. In Sect. 3 we review relevant mathematical background and results on tensor fields. In Sect. 4 we report the findings of our exploration before concluding in Sect. 5.

2   Previous Work

Much research exists on 2D and 3D symmetric tensor fields, and we refer the readers to the survey by Kratz et al. [4] and Zhang et al. [11] for a more comprehensive review. In this book chapter we only refer to the research that is most relevant.

Delmarcelle and Hesselink [1] introduce the notion of *degenerate points* for 2D symmetric tensors, where eigenvector directions are not well-defined. Zhang et al. [12] explore the physical meanings of degenerate points in the stress tensor and strain tensor from continuum mechanics.

Hesselink et al. later extend this work to 3D symmetric tensor fields [3] and study the degeneracies in such fields. Zheng and Pang [16] point out that triple degeneracies are structurally unstable features. That is, an arbitrarily small perturbation to the field will remove such degenerate points. Zheng and Pang further show that double degeneracies, i.e., only two equal eigenvalues, form lines in the domain. In this work and subsequent research [18], they provide a number of degenerate curve extraction methods based on the analysis of the discriminant function of the tensor field. Furthermore, Zheng et al. [17] point out that near degenerate curves the tensor field exhibits 2D degenerate patterns and define separating surfaces which are extensions of separatrices from 2D symmetric tensor field topology. Tricoche et al. [9] convert the problem of extracting degenerate curves in a 3D tensor field to that of finding the ridge and valley lines of an invariant of the tensor field, thus leading to a more robust extraction algorithm. More recently, Palacios et al. [6] extract degenerate curves using an algorithm for algebraic surface extraction method called A-patches. Palacios et al. [5] introduce a number of topological editing operations with which a 3D tensor field can be edited for graphics applications.

Zhang et al. [13] describe a number of important properties of 3D linear tensor fields. They [15] show that in a 3D linear tensor field, there are at least two and at most four degenerate curves. Roy et al. [8] develop a parameterization with which all degenerate points in a 3D piecewise linear tensor field can be extracted efficiently and at any given accuracy. Zhang et al. [14] show that there are at most eight transition points in a 3D linear tensor field.

3   Background on Tensors and Tensor Fields

In this section we review the most relevant background on 2D and 3D symmetric tensors and tensor fields [14].

3.1   Tensors

A K-dimensional (symmetric) tensor $\mathbf{T}$ has K real-valued *eigenvalues*: $\lambda_1 \geq \lambda_2 \geq \ldots \geq \lambda_K$. The largest and smallest eigenvalues are referred to as the *major eigenvalue* and *minor eigenvalue*, respectively. When $K = 3$, the middle eigenvalue is referred to as the *medium eigenvalue*. An eigenvector belonging to the major eigenvalue is referred to as a *major eigenvector*. Medium and minor eigenvectors can be defined similarly. Eigenvectors belonging to different eigenvalues are mutually perpendicular.

The trace of a tensor $\mathbf{T} = (\mathbf{T}_{ij})$ is $trace(\mathbf{T}) = \sum_{i=1}^{K} \lambda_i$. $\mathbf{T}$ can be uniquely decomposed as $\mathbf{D} + \mathbf{A}$ where $\mathbf{D} = \frac{trace(\mathbf{T})}{K} \mathbb{I}$ ($\mathbb{I}$ is the K-dimensional identity matrix) and $\mathbf{A} = \mathbf{T} - \mathbf{D}$. The *deviator* $\mathbf{A}$ is a *traceless* tensor, i.e., $trace(\mathbf{A}) = 0$. Note that $\mathbf{T}$ and $\mathbf{A}$ have the same set of eigenvectors. Consequently, the anisotropy in a tensor field can be defined in terms of its deviator tensor field. Another nice property of the set of traceless tensors is that it is closed under matrix addition and scalar multiplication, making it a linear subspace of the set of tensors.

The magnitude of a tensor $\mathbf{T}$ is $||\mathbf{T}|| = \sqrt{\sum_{1 \leq i, j \leq K} T_{ij}^2} = \sqrt{\sum_i^K \lambda_i^2}$, while the determinant is $|\mathbf{T}| = \prod_{i=1}^{K} \lambda_i$.

A tensor is *degenerate* when there are repeating eigenvalues. In this case, there exists at least one eigenvalue whose corresponding eigenvectors form a higher-dimensional space than a line. When $K = 2$ a degenerate tensor must be a multiple of the identity matrix.

3.2   Tensor Field Topology

We now review *tensor fields*, which are tensor-valued functions over some domain $\Omega \subset \mathbb{R}^K$. A tensor field can be thought of as K eigenvector fields, corresponding to the K eigenvalues. A *hyperstreamline* with respect to an eigenvector field $e_i(p)$ is a $3D$ curve that is tangent to e_i everywhere along its path. Two hyperstreamlines belonging to two different eigenvalues can only intersect at the right angle, since eigenvectors belonging to different eigenvalues must be mutually perpendicular.

Hyperstreamlines are usually curves. However, they can occasionally consist of only one point, where there is more than one choice of lines that correspond to the eigenvector field. This is precisely where the tensor field is degenerate. A point

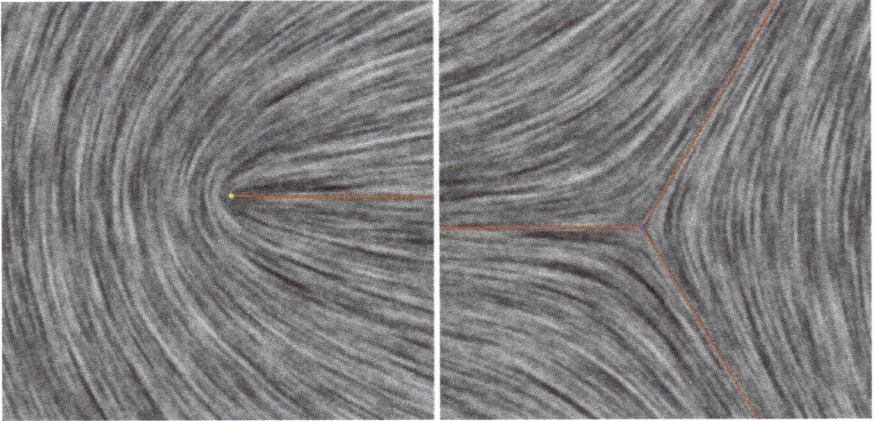

Fig. 1 A wedge (left) and a trisector (right)

$p_0 \in \Omega$ is a *degenerate point* if $\mathbf{T}(p_0)$ is degenerate. One important topological feature of a tensor field consists of its degenerate points.

In $2D$, the set of degenerate points of a tensor field consists of isolated points under numerically stable configurations, when the topology does not change given sufficiently small perturbation in the tensor field. An isolated degenerate point can be measured by its *tensor index* [10], defined in terms of the *winding number* of one of the eigenvector fields on a loop surrounding the degenerate point. The most fundamental types of degenerate points are *wedges* and *trisectors*, with a tensor index of $\frac{1}{2}$ and $-\frac{1}{2}$, respectively. Let $L\mathbf{T}_{p_0}(p)$ be the local linearization of $\mathbf{T}(p)$ at a degenerate point $p_0 = \begin{pmatrix} x_0 \\ y_0 \end{pmatrix}$, i.e.,

$$L\mathbf{T}_{p_0}(p) = \begin{pmatrix} a_{11}(x - x_0) + b_{11}(y - y_0) & a_{12}(x - x_0) + b_{12}(y - y_0) \\ a_{12}(x - x_0) + b_{12}(y - y_0) & a_{22}(x - x_0) + b_{22}(y - y_0) \end{pmatrix} \quad (1)$$

Then $\delta = \left| \begin{pmatrix} \frac{a_{11}-a_{22}}{2} & a_{12} \\ \frac{b_{11}-b_{22}}{2} & b_{12} \end{pmatrix} \right|$ is invariant under the change of basis [2]. Moreover, p_0 is a wedge when $\delta > 0$ and a trisector when $\delta < 0$. When $\delta = 0$, p_0 is a higher-order degenerate point. A *major separatrix* is a hyperstreamline emanating from a degenerate point following the major eigenvector field. A *minor separatrix* is defined similarly.

The total tensor index of a continuous tensor field over a two-dimensional manifold is equal to the *Euler characteristic* of the underlying manifold. Consequently, it is not possible to remove one degenerate point. Instead, a pair of degenerate points with opposing tensor indexes (a wedge and trisector pair) must be removed simultaneously [10]. Figure 1 shows a wedge pattern (left) and a trisector pattern (right), respectively.

In 3D, a degenerate point can have either two or three eigenvalues being the same, the latter of which is structurally unstable. Structurally stable degenerate points have two eigenvalues being the same. These eigenvalues are the *repeating eigenvalues*, while the third eigenvalue is the *dominant eigenvalue*. When the dominant eigenvalue is the major eigenvalue, the degenerate point is a *linear degenerate point* (L). On the other hand, when the dominant eigenvalue is the the minor eigenvalue, the degenerate point is a *planar degenerate point* (P). We refer to this linear/planar classification of a degenerate point as its L/P classification. Along a degenerate curve, its L/P classification does not change.

A degenerate point can also be classified in a different way, by projecting the tensor field onto the plane passing through the degenerate point and perpendicular to its dominant eigenvector. The degenerate point is also a degenerate point in the 2D projected tensor field. Consequently, the original 3D degenerate point is classified as either a wedge (W) or a trisector (T) based on the W/T type of the 2D degenerate point. Note that the projection of the 3D tensor field onto other planes not perpendicular to the dominant eigenvector does not necessarily have the same W/T classification. Along a degenerate curve, the W/T type can change, separated by degenerate points that are neither wedges nor trisectors. Such points are *transition points*.

Figure 2 illustrates these concepts with one example tensor field. The combination of the L/P and W/T classifications of a degenerate point leads to four combinations: LW (green), LT (blue), PW (yellow), and PT (red). Note that along a degenerate curve, the L/P classification is constant while the W/T classification can change. Therefore, a degenerate curve can consist of either green and blue segments, or yellow and red segments, but not other combinations of colors. Linear transition points appear between green and blue segments, while planar transition points separate yellow and red segments.

The projected tensor fields onto the plane of repeating eigenvalues are shown along the degenerate curve. Notice that at a W type point, the projected pattern shows a wedge, while at a T type point, the projected pattern shows a trisector. At a transition point, the projected pattern shows neither a wedge pattern nor a trisector pattern.

3.3   3D Linear Tensor Fields

A 3D linear tensor field is a 3D symmetric tensor field whose tensor entries are linear functions of the XYZ coordinates of the points in the domain. It has the following form:

$$T(x, y, z) = T_0 + x T_x + y T_y + z T_z \qquad (2)$$

where T_0, T_x, T_y, and T_z are 3D symmetric tensors. A 3D linear tensor field has all the aforementioned properties of a general tensor field. However, there are some important properties specific to 3D linear tensor fields.

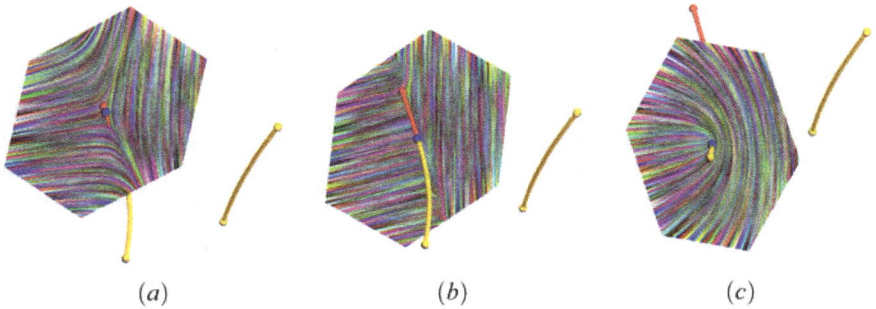

(a) (b) (c)

Fig. 2 Along a degenerate curve, the projection of the tensor field onto the repeating planes can exhibit 2D degenerate patterns such as a trisector (**a**) or a wedge (**c**). Between segments of wedges (yellow) and trisectors (red), transition points can appear (**b**)

First, there are either two or four degenerate curves in a 3D linear tensor field [15]. Half of these curves are L, and the other half are P.

Second, there are either zero, two, four, six, or eight transition points in a 3D linear tensor field [14].

As pointed out [8], the set of degenerate points in a 3D linear tensor field (including their asymptotic limits at infinity) is homeomorphic to a topological circle (a loop). Figure 3 illustrates this bijective map. The loop is situated on the unit sphere where the location of a degenerate point on the loop corresponds to its unit dominant eigenvector. Since eigenvectors have a sign ambiguity, there are two such loops. Every degenerate point corresponds to exactly one pair of antipodal points on the sphere. While L/P cannot change along a degenerate curve, such a switch can occur at infinity, which two degenerate curves approach in opposite directions. The number of such switch points, called ∞ points, is the same as the number degenerate curves in the field. Note that ∞ points are neither L nor P. On the other hand, they can be either W or T. This leads to nine scenarios [14]: (1) two WW curves, (2) two WT curves, (3) two TT curves, (4) four WW curves, (5) two WW curves and two WT curves, (6) one WW curve, two WT curves, and one TT curve, (7) four WT curves, (8) two WT curves and two TT curves, and (9) four TT curves.

Along a WW or TT curve, there must be an even number of transition points, while on a WT curve there must exist an odd number of transition points.

The color of the sphere represents the sign of the discriminant of the projected tensor field onto the plane whose normal is the displacement vector of the point on the sphere from the center of the sphere. The color is cyan if the discriminant is positive (wedge type) and magenta if the discriminant is negative (trisector type). Note the W/T classification of a degenerate point must match the sign of the discriminant on the sphere. That is, green and yellow segments must appear in the cyan region, while blue and red segments must appear in the magenta region. A transition point must appear on the boundary between cyan and magenta regions.

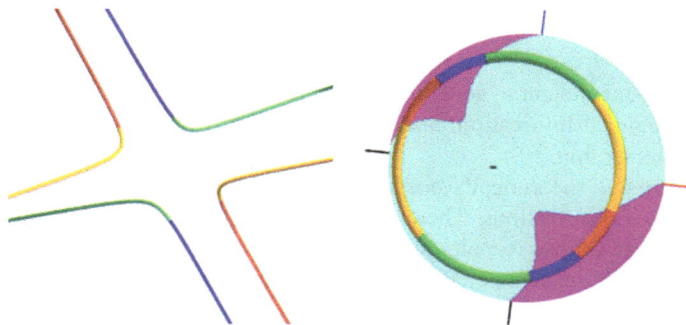

Fig. 3 This figure shows the parameterization of all degenerate points in a 3D linear tensor field by a topological circle. In the left are the degenerate curves of a 3D linear tensor field, and in the right is the topological circle superimposed on the unit sphere. Between the yellow and green segments are ∞ points. In addition, transition points occur either between yellow and red segments, or between green and blue segments. Moreover, we color a point in the sphere (right: representing a unit vector) cyan if the tensor projected onto the plane perpendicular to this vector has a wedge degenerate point. On the other hand, the point is colored magenta if the projected tensor field has trisector degenerate point. Notice that the transition points on the degenerate curves (between blue and green segments and between yellow and red segments) are precisely on the boundary between the cyan and magenta regions

4 Bifurcations

We now describe our findings of bifurcations in 3D linear tensor fields.

First, there can be either two or four degenerate curves. This means that some bifurcations can change the number of degenerate curves in the field. We refer to those bifurcations which reduce the number of degenerate curves from four to two as *degenerate curve removal* and those increase the number from two to four as *degenerate curve generation.*

Second, even when the number of degenerate curves does not change, the way the ∞ points are connected can change. We refer to this as *degenerate curve reconnection*, which is consistent with the topological editing operation of the same name from [5]. Note that two degenerate curves of opposite L/P types cannot be reconnected as this would generate a degenerate curve with changing L/P type. Consequently, when there are only two degenerate curves in the field, they cannot be connected. Degenerate curve reconnection can only occur for four degenerate curves.

Third, the number of transition points in the tensor field can change. Since this number must be even, we refer to such bifurcations as either *transition point pair cancellation* if the number is decreased by two or *transition point pair generation* if the number is increased by two.

Finally, even when the number of transition point does not change in a field, it is possible that some transition point has been moved from one degenerate curve to another. We refer to such a bifurcation as a *transition point relocation*.

Below we consider these bifurcations in the context of the nine scenarios described earlier.

4.1   Degenerate Curve Removal and Generation

Note that a degenerate curve generation bifurcation must be the inverse of a degenerate curve removal bifurcation, and vice versa. Consequently, we consider them together in this section.

Figure 4 lists all theoretically possible degenerate curve generation bifurcations and their inverse bifurcations. There is a total of nine scenarios, each of which is represented as an ellipse (topological disk) with the ∞ points marked along with their W/T types. A degenerate curve removal bifurcation must take a scenario with four degenerate curves (four ∞ points on the ellipse) to one with two degenerate curves (two ∞ points). Conversely, a degenerate curve generation bifurcation must take a scenario with two degenerate curves to one with four degenerate curves. In addition, when two degenerate curves are removed, two adjacent ∞ points are removed and two ∞ points remain. Each of the ∞ point is written as either W or T (called a symbol on the ellipse). This means that not any four-symbolled ellipse can connect with any two-symbolled ellipse. For example, while the box with four *W*'s can be mapped to either one with two *W*'s or one with one *W* and one *T*, it can *not* be connected with one with two *T*'s. Similarly, the box with one *W* and three *T*'s cannot be mapped to one with two *W*'s. Furthermore, when two symbols are removed, they need to be adjacent symbols. Consequently, it is not possible to connect the ellipse with two *W*'s and two *T*'s that are interleaved to one with two *W*'s or the one with two *T*'s.

These constraints give rise to a total of ten theoretically possible bifurcations, each of which is given as an edge in Fig. 4. The following figures provide examples of degenerate curve removal bifurcations.

In Fig. 5 (left), there are initially four TT curves, two of which are linear (blue) and two planar (red). After the bifurcation (right), the two linear degenerate curves become connected by a linear wedge segment (green), which replaces the lost planar

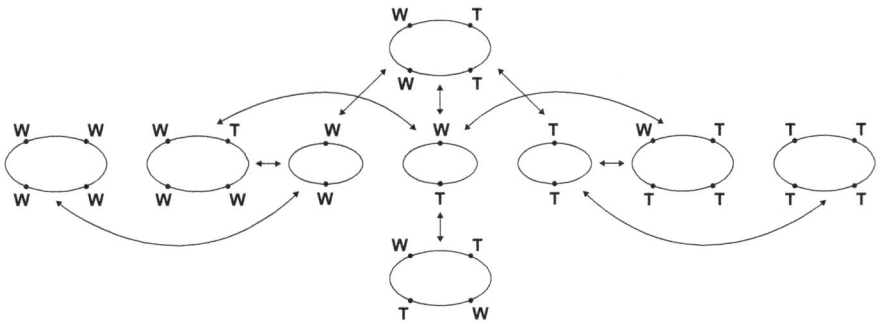

Fig. 4 This diagram shows all the theoretically possible degenerate curve removal and generation bifurcations in a 3D linear tensor field

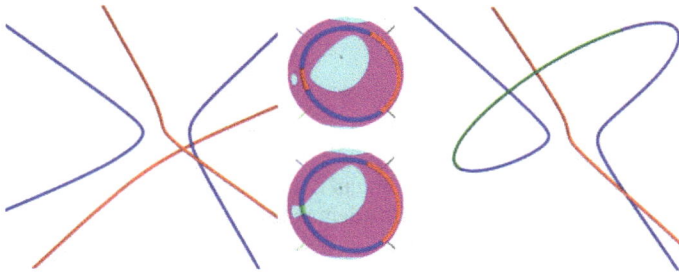

Fig. 5 This figure shows a degenerate curve removal bifurcation from the case of four TT curves (left) to two TT curves (right). The top-middle shows the non-repeating eigenvector manifold for the left, while the bottom-middle shows the non-repeating eigenvector manifold for the right. The inverse bifurcation, a degenerate curve generation exists, which swaps the before and after scenarios. Note that in these bifurcations the number of segments do not change after the bifurcation

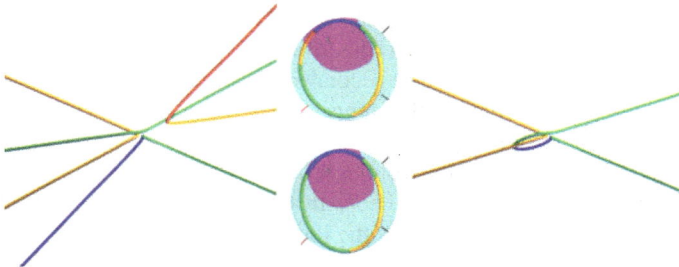

Fig. 6 Another degenerate curve removal bifurcation takes the case of two WW curves and two WT curves (left) to two WW curves (right). Note that the number of segments is decreased by two after the bifurcation, i.e. two transition points are lost as a result of the bifurcation

TT curve (red). Furthermore, two transition points appear on the newly connected linear degenerate curve.

The middle column compares the two fields using their parameterization: (upper-middle) corresponding to the field in the left, and (lower-middle) corresponding to the tensor field in the right. Through the bifurcation, two infinity points are replaced by two transition points, resulting in the switch of a red segment (middle-top) to a green segment (middle-bottom), which leads to the loss of two degenerate curves and the creation of two transition points.

Figure 6 provide an additional example of such types of bifurcations. Note that in this example, the loss of two degenerate curves also results in the loss of two transition points, i.e. the loss of two colored segments. This is a different way of achieving degenerate curve removal than the example bifurcation in Fig. 5, in which the number of segments does not change.

4.2   Degenerate Curve Reconnection

Degenerate curve reconnection bifurcations do not change the number of degenerate curves. Instead, they change how the ∞ points are connected. As mentioned earlier, two degenerate curves can be connected only if they have the same L/P type. When there are only two degenerate curves, one is L and the other P, which cannot be connected after reconnection. Consequently, degenerate curve reconnection bifurcations only occur for the case of four degenerate curves.

There is a total six scenarios with four degenerate curves, as shown in Fig. 7. Since degenerate curve reconnection bifurcations do not change the number and type of ∞ points, there is at least one theoretically possible bifurcation that connects a scenario to itself. Furthermore, the two scenarios with two W's and two T's are connected through reconnection bifurcations.

However, it is worth noting that when reconnection occurs, the ∞ points on the two degenerate curves to be reconnected cannot be arbitrarily connected after bifurcation. For an example, consider the case of two W's and two T's that appear on the ellipse in an alternating fashion (the lower-middle ellipse) in Fig. 7. Suppose the top and bottom segments are to be reconnected while the left and right segment remain. Then one cannot connect the upper-left W ∞ point with the lower-left T ∞ point, as it would cause the ellipse to be broken into two topological disks, a structurally unstable case. This means that it is *not* possible to have a reconnection bifurcation that takes this scenario to itself. Instead, the only way to reconnect is to connect the upper-left W ∞ point with the lower-right W ∞ point and connect the upper-right T ∞ point with the lower-left T ∞ point, leading to the scenario of two adjacent W and two adjacent T ∞ points on the ellipse (the upper-middle ellipse).

In fact, the above argument is true in general. That is, when reconnecting, the upper-left ∞ point must be connected to the lower-right ∞ point, and the upper-right ∞ point must be connected to the lower-left ∞ point. This analysis shows that there are five degenerate curve reconnection bifurcations that takes one scenario to itself, and there are two more that occur between two different scenarios (the upper-middle ellipse and the lower-middle ellipse).

Figure 8 shows the reconnection in the scenario of four WW curves, in which two linear degenerate curves (green) are reconnected, resulting in two linear degenerate curves. The two planar degenerate curves (yellows) are not reconnected. This bifurcation does not increase or decrease the number of transition points.

Figure 9 presents another degenerate curve reconnection bifurcation in which two planar WT degenerate curves are reconnected into one planar WW degenerate curve and one planar TT degenerate curve. As a result of this bifurcation, two transition points are removed, which leads to a decrease in the total number of transition points in the field. Notice the order of colored segments (middle column of Fig. 9) is changed due to the reconnection.

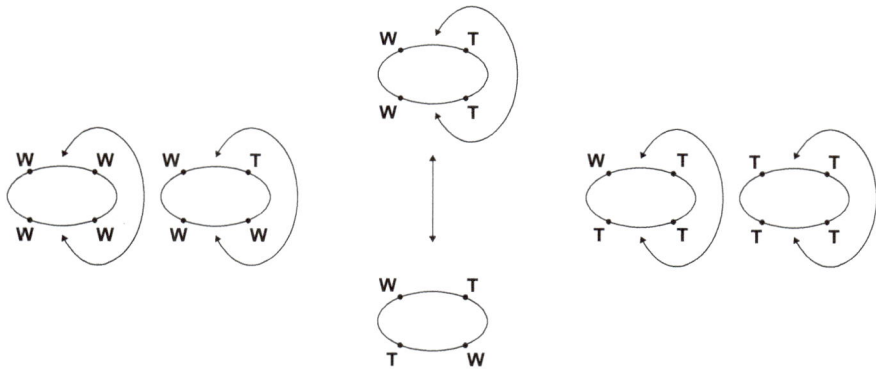

Fig. 7 This diagram shows all the possible degenerate curve reconnection bifurcations in a 3D linear tensor field

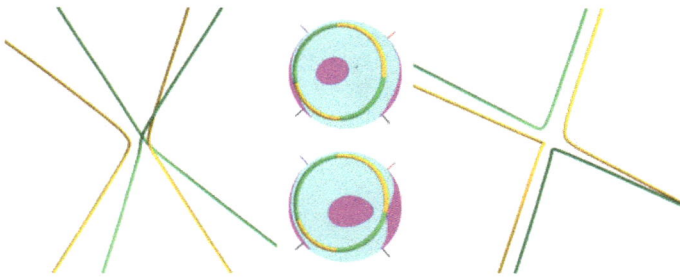

Fig. 8 This figure shows a degenerate curve reconnection bifurcation from the case of four WW curves (left) to four WW curves (right). The number of transition points does not change after the bifurcation

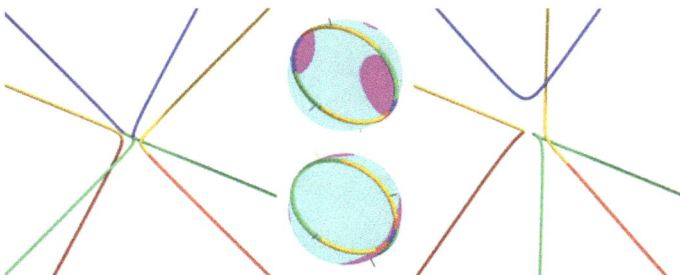

Fig. 9 Another degenerate curve reconnection bifurcation takes the case of one WW curve, 2WT curves, and one TT curve (left) to four WT curves (right). In this case, the number of degenerate curve bifurcation is decreased by two

4.3 Transition Point Pair Cancellation and Generation

The total number of transition points in a 3D linear tensor field is an even number between zero and eight. That means that any fundamental bifurcation that changes the number of total transition points must either increase it by two or decrease it by two. The former is the *degenerate curve pair generation* bifurcation, and the latter *degenerate curve pair cancellation* bifurcation. They are inverse bifurcations and thus discussed together in this section.

Two examples of transition point pair cancellation bifurcations are shown in Figs. 11 and 12, for the cases of two degenerate curves and four degenerate curves, respectively. The transition point pair cancellation and generation bifurcations do not change the scenario. Consequently, for each of the nine scenarios, there is a bifurcation that takes this scenario to itself, as shown in Fig. 10. There is a total of nine transition point pair cancellation bifurcations and nine transition point pair generation bifurcations.

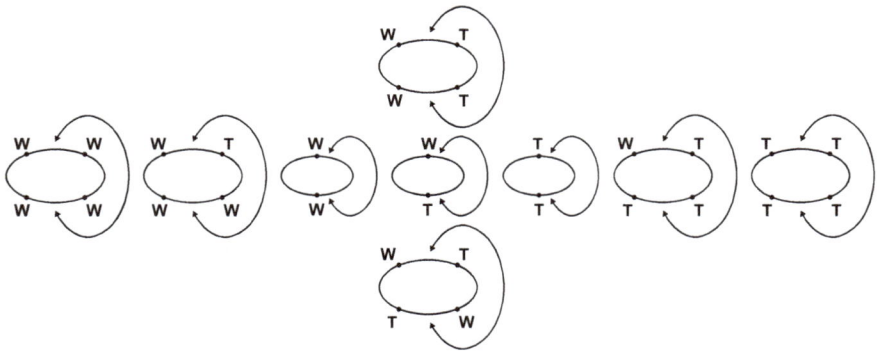

Fig. 10 This diagram shows all the possible transition pair cancellation and generation bifurcations in a 3D linear tensor field for one degenerate curve

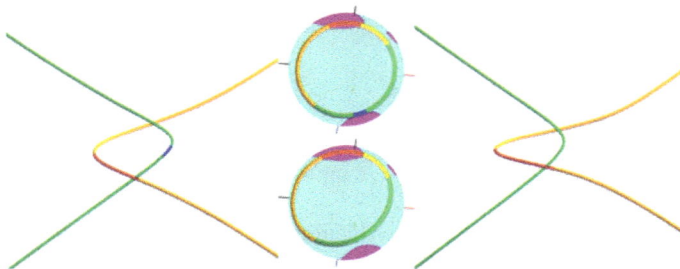

Fig. 11 A transition point pair cancellation bifurcation that removes two transition points and one blue segment from one of the two degenerate curves (left), resulting in the same number of degenerate curves with two fewer transitions points (right)

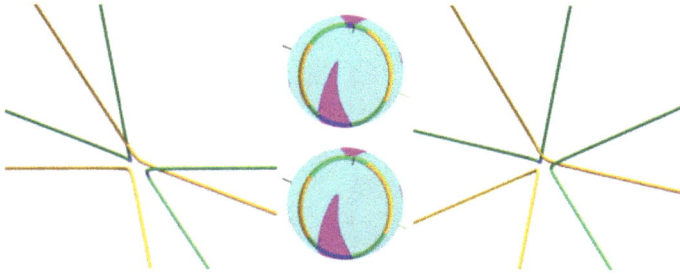

Fig. 12 A transition point pair cancellation operation (from to right) occurs in the case of four degenerate curves

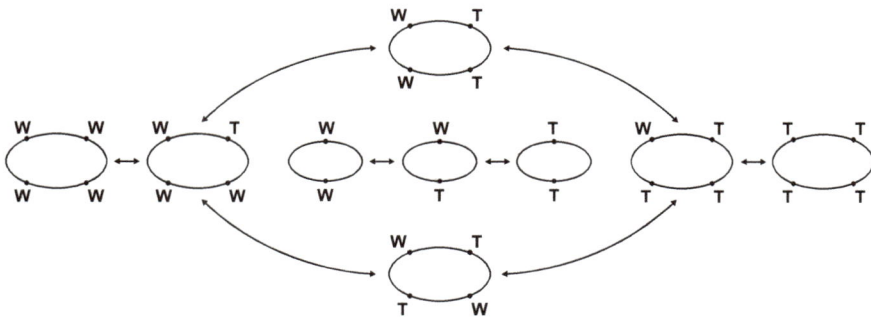

Fig. 13 This diagram shows all the possible transition point relocation bifurcations in a 3D linear tensor field. Note that these bifurcations have the effect of switching one W symbol to one T symbol, or vice versa. Consequently, there are only eight possible such bifurcations, all observed in our experiment

4.4 Transition Point Relocation

Transition points, along with ∞ points, divide the degenerate curves into segments of pure L/P and W/T types. While the total number of transition points in a 3D linear tensor field is eight, its distribution is not uniform among the degenerate curves.

Transition point relocation refers to moving one transition point from one degenerate curve to another degenerate curve of the opposite L/P type. From the viewpoint of the parameterization, this is the swap of the positions of the transition point with one adjacent infinity point on the ellipse in the counterclockwise order. This results in one fewer transition point on the original degenerate curve and one more transition point on the new degenerate curve but keeps the total number of transition points constant. Moreover, the W/T type of the segment also changes. Consequently, the segment between the transition point and infinity point will change its L/P type but maintains its W/T type, thus will change colors between red and green, or between yellow and blue. Figures 14 and 15 provide two examples of this type of bifurcations that involve respectively two degenerate curves and four degenerate curves.

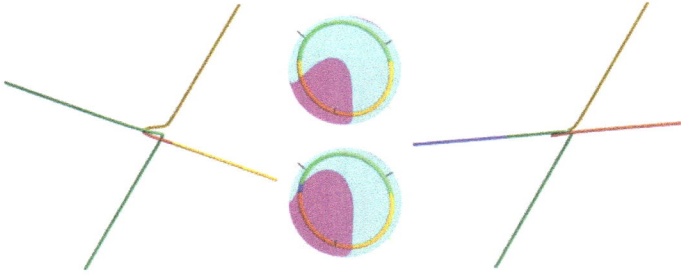

Fig. 14 This figure shows a transition point relocation bifurcation from the case of two WW curves (left) to two WT curves (right)

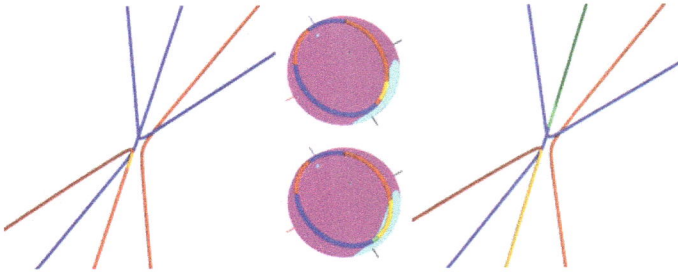

Fig. 15 Another transition point relation bifurcation takes the case of four TT curves (left) to two TT and two WT curves (right)

Figure 13 shows the possible transition point relation bifurcations. Note that these bifurcations result in a switch of a W infinity point to a T type of infinity point, or vice versa. Note that each edge in this diagram corresponds to two transition point relocation bifurcations. As there are eight edges in the diagram, there is a total of 16 bifurcations, all of which have been observed during experiment.

5 Conclusion

Degenerate curves and transition points play an important role in describing the behavior of a 3D tensor field. In this book chapter, we study possible bifurcations for any 3D linear tensor field.

This is done based on the realization that there is a total of nine scenarios and four types bifurcations. Theoretically, we have identified ten degenerate curve removal bifurcations, ten degenerate curve generation bifurcations, seven degenerate curve reconnection bifurcation, nine transition point pair cancellation bifurcations, nine transition point pair generation bifurcations, and 16 transition point relation bifurcations. This leads to a total of 61 topological bifurcations for 3D linear tensor fields, all of which have been observed.

There are a number of future research directions.

First, the number of transition points can further divide each scenario into sub-scenarios. For example, for the scenario of two WW curves, there can be zero, two, four, six, or eight transition points. Furthermore, transition points are not evenly distributed on degenerate curves. This leads to multiple cases per scenario. Enumerating of all bifurcations from any sub-scenario to any other sub-scenario can lead to deeper insight about the topology of 3D symmetric tensor fields.

Second, the bifurcations among 3D tensor fields can be used to generate a multiscale framework for the topological analysis of these objects, such as degenerate curve clustering and removal. Such a framework has the potential of enabling users to inspect the topology of their tensor field data at various scales. We plan to investigate this possibility in our future research.

Third, simulation data sets usually involve *piecewise linear tensor fields*, which can have potentially more types of bifurcations than linear tensor fields that we investigate in this paper. This is a future direction that can lead to more practical applications for tensor field topology.

Fourth, neutral surfaces are the other important constituent of 3D tensor field topology. Enumerating the scenarios of neutral surfaces and bifurcations of these different scenarios can provide insight into the structure of a tensor field. We will explore this direction.

Acknowledgments We wish to thank our anonymous reviewers for their valuable feedback and suggestions. This research is partially supported by NSF awards (# 1566236) and (#1619383).

References

1. Delmarcelle, T., Hesselink, L.: Visualizing second-order tensor fields with hyperstream lines. IEEE Comput. Graph. Appl. **13**(4), 25–33 (1993)
2. Delmarcelle, T., Hesselink, L.: The topology of symmetric, second-order tensor fields. In: Proceedings IEEE Visualization 1994 (1994)
3. Hesselink, L., Levy, Y., Lavin, Y.: The topology of symmetric, second-order 3D tensor fields. IEEE Trans. Visual. Comput. Graph. **3**(1), 1–11 (1997)
4. Kratz, A., Auer, C., Stommel, M., Hotz, I.: Visualization and analysis of second-order tensors: Moving beyond the symmetric positive-definite case. Comput. Graph. Forum **32**(1), 49–74 (2013). http://dblp.uni-trier.de/db/journals/cgf/cgf32.html#KratzASH13
5. Palacios, J., Roy, L., Kumar, P., Hsu, C.Y., Chen, W., Ma, C., Wei, L.Y., Zhang, E.: Tensor field design in volumes. ACM Trans. Graph. **36**(6), 188:1–188:15 (2017). https://doi.org/10.1145/3130800.3130844
6. Palacios, J., Yeh, H., Wang, W., Zhang, Y., Laramee, R.S., Sharma, R., Schultz, T., Zhang, E.: Feature surfaces in symmetric tensor fields based on eigenvalue manifold. IEEE Trans. Visual. Comput. Graph. **22**(3), 1248–1260 (2016). https://doi.org/10.1109/TVCG.2015.2484343
7. Raith, F., Blecha, C., Nagel, T., Parisio, F., Kolditz, O., Günther, F., Stommel, M., Scheuermann, G.: Tensor field visualization using fiber surfaces of invariant space. IEEE Trans. Visual. Comput. Graph. **25**(1), 1122–1131 (2019). https://doi.org/10.1109/TVCG.2018.2864846
8. Roy, L., Kumar, P., Zhang, Y., Zhang, E.: Robust and fast extraction of 3D symmetric tensor field topology. IEEE Trans. Visual. Comput. Graph. **25**(1), 1102–1111 (2019), (IEEE VisWeek 2018)

9. Tricoche, X., Kindlmann, G., Westin, C.F.: Invariant crease lines for topological and structural analysis of tensor fields. IEEE Trans. Visual. Comput. Graph. **14**(6), 1627–1634 (2008). https://doi.org/10.1109/TVCG.2008.148
10. Zhang, E., Hays, J., Turk, G.: Interactive tensor field design and visualization on surfaces. IEEE Trans. Visual. Comput. Graph. **13**(1), 94–107 (2007)
11. Zhang, S., Kindlmann, G., Laidlaw, D.H.: Diffusion tensor MRI visualization. In: Visualization Handbook. Academic Press (2004). http://www.cs.brown.edu/research/vis/docs/pdf/Zhang-2004-DTM.pdf
12. Zhang, Y., Gao, X., Zhang, E.: Applying 2d tensor field topology to solid mechanics simulations. In: Schultz, T., Özarslan, E., Hotz, I. (eds.) Modeling, Analysis, and Visualization of Anisotropy, pp. 29–41. Springer International Publishing, Cham (2017)
13. Zhang, Y., Palacios, J., Zhang, E.: Topology of 3d linear symmetric tensor fields. In: Hotz, I., Schultz, T. (eds.) Visualization and Processing of Higher Order Descriptors for Multi-Valued Data, pp. 73–91. Springer International Publishing, Cham (2015)
14. Zhang, Y., Roy, L., Sharma, R., Zhang, E.: Maximum number of transition points in 3d linear tensor fields. In: Carr, H., Fujishiro, I., Sadlo, F., Takahashi, S. (eds.) Topological Methods in Data Analysis and Visualization V, pp. 237–250. Springer International Publishing (2020)
15. Zhang, Y., Tzeng, Y.J., Zhang, E.: Maximum number of degenerate curves in 3d linear tensor fields. In: Carr, H., Garth, C., Weinkauf, T. (eds.) Topological Methods in Data Analysis and Visualization IV, pp. 221–234. Springer International Publishing, Cham (2017)
16. Zheng, X., Pang, A.: Topological lines in 3d tensor fields. In: Proceedings IEEE Visualization 2004, VIS 2004, pp. 313–320. IEEE Computer Society, Washington, DC, USA (2004). https://doi.org/10.1109/VISUAL.2004.105
17. Zheng, X., Parlett, B., Pang, A.: Topological structures of 3D tensor fields. Proc. IEEE Visual. **2005**, 551–558 (2005)
18. Zheng, X., Parlett, B.N., Pang, A.: Topological lines in 3d tensor fields and discriminant hessian factorization. IEEE Trans. Visual. Comput. Graph. **11**(4), 395–407 (2005)

Image Processing and Visualization

4

Uncertainty in the DTI Visualization Pipeline

Faizan Siddiqui, Thomas Höllt, and Anna Vilanova

Abstract Diffusion-Weighted Magnetic Resonance Imaging (DWI) enables the in-vivo visualization of fibrous tissues such as white matter in the brain. Diffusion-Tensor Imaging (DTI) specifically models the DWI diffusion measurements as a second order-tensor. The processing pipeline to visualize this data, from image acquisition to the final rendering, is rather complex. It involves a considerable amount of measurements, parameters and model assumptions, all of which generate uncertainties in the final result which typically are not shown to the analyst in the visualization. In recent years, there has been a considerable amount of work on the visualization of uncertainty in DWI, and specifically DTI. In this chapter, we primarily focus on DTI given its simplicity and applicability, however, several aspects presented are valid for DWI as a whole. We explore the various sources of uncertainties involved, approaches for modeling those uncertainties, and, finally, we survey different strategies to visually represent them. We also look at several related methods of uncertainty visualization that have been applied outside DTI and discuss how these techniques can be adopted to the DTI domain. We conclude our discussion with an overview of potential research directions.

F. Siddiqui (✉) · T. Höllt · A. Vilanova
Delft University of Technology, Delft, The Netherlands
e-mail: F.P.Siddiqui@tudelft.nl

T. Höllt
e-mail: T.Hollt-1@tudelft.nl

A. Vilanova
e-mail: A.Vilanova@tue.nl

T. Höllt
Leiden University Medical Center, Leiden, The Netherlands

A. Vilanova
Eindhoven University of Technology, Eindhoven, The Netherlands

1   Introduction

Recent advancements in magnetic resonance imaging (MRI) technology have led to the development of various remarkable techniques for the interpretation of brain anatomy. The most promising one is diffusion-weighted imaging (DWI), the only non-invasive technique for the assessment of brain white matter connectivity. This approach relies on the measurement of anisotropic diffusion of water molecules. The imaging and the interpretation of the molecular diffusion have improved with the development of techniques like diffusion tensor imaging (DTI) and high angular resolution diffusion imaging (HARDI). In this chapter, we will discuss the visualization pipeline of DTI, given its clinical applicability. However, several visualization strategies and sources of uncertainties associated are valid for more advanced models like HARDI models.

DTI allows direct in-vivo examination of the fibrous structure in the brain at a relatively low acquisition cost. By analyzing the three-dimensional shape of the diffusion tensor, it provides valuable information about the microstructure of brain tissues. Despite many advantages of this technique, some downsides limit its widespread use. The main reason is that the complexity in the data makes it notoriously difficult to infer and analyze.

The DTI visualization pipeline consists of four main stages, from data acquisition to the final visual representation of the results, as shown in Fig. 1. Each stage is based on assumptions, parameters, and estimations subject to considerable uncertainties. The uncertainties involved in each of the pipelines' stage can lead to unpredictable variations in the final output.

Several state-of-the-art reports exist on DWI visualization [62, 92, 93, 99]. However, none of them give an overview of uncertainty, or they focus on some specific aspects. Most of the visualization literature about uncertainty in DTI focuses on issues related to the visual representation rather than sources of error involved in the pipeline [36, 47, 92]. In this chapter, we discuss the DTI visualization pipeline and analyze the sources of uncertainties present at each stage. We briefly cover the approaches used for quantification of uncertainties, which are often omitted in other studies [36, 47]. We review state-of-the-art strategies for uncertainty visualization in DTI and compare their main characteristics and drawbacks. We further investigate several methodologies for uncertainty visualization in other domains that have not been explored in DWI and discuss how these techniques can be adopted in this domain. DWI models beyond DTI share a similar pipeline as the one shown in Fig. 1. However, some parameters, error sources, and visual representations would differ from the tensor model. Specifically, diffusion modeling and fiber tracking would be based on different parameters and algorithms. In this chapter, we will indicate the methods from the DTI that are valid for the more general DWI pipeline.

In Sect. 2, we discuss the background and review the visualization techniques for DTI. In Sect. 3, we discuss the sources of uncertainties involved in the visualization

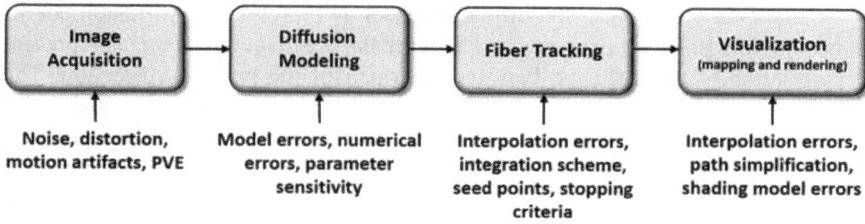

Fig. 1 The DTI visualization pipeline with sources of uncertainties at each step

pipeline and proceed with the uncertainty modeling techniques in Sect. 4. We review uncertainty visualization methods in Sect. 5 and conclude with open issues and research direction in Sect. 6.

2 Background

Diffusion refers to the constant rapid movement of microscopic particles due to the presence of thermal energy, i.e., 'Brownian motion'. DWI deals with the diffusion of water molecules present in biological tissues where the diffusion is usually restricted due to the hindrance by many obstacles such as axonal membranes, macromolecules, and myelin. This kind of restricted diffusion is known as anisotropic diffusion. Stejskal and Tanner [94] observed the anisotropic diffusion of water molecules in tissues and investigated the related modeling of the diffusion effects for MRI. The clinical application of this technique was first presented by Bihan et al. [61] with the introduction of diffusion MRI along with the concept of apparent diffusion coefficient (ADC). In some neurological conditions, the amount of diffusion is disturbed in the affected area. Through studying these changes in diffusion, the abnormalities can be detected. In the following section, we will summarize how these measurements have been used to visualize white matter tracts in the brain.

2.1 Diffusion Tensor

The pattern of diffusion anisotropy of white matter tracts in 3D space can be mathematically modeled by a second order tensor, called the diffusion tensor D, introduced by Basser et al. [5]. The tensor D is a symmetric, positive definite tensor represented by a 3×3 matrix with six unique elements, denoted by D_{ij} as follows:

$$D = \begin{pmatrix} D_{11} & D_{12} & D_{13} \\ D_{21} & D_{22} & D_{23} \\ D_{31} & D_{32} & D_{33} \end{pmatrix} \tag{1}$$

The diagonal elements in the diffusion tensor D represent the diffusion coefficients along the principle axes x, y and z, while the off-diagonal elements represent the correlation of the diffusion between each pair of the principal direction. The diffusion tensor D is symmetric about the diagonal axis ($D_{ij} = D_{ji}$). By analysis of the eigenvectors and eigenvalues $\lambda_1, \lambda_2, \lambda_3$ of the diffusion matrix, the length and the direction of the principal axes of the diffusion tensor can be determined.

The six unique values in the tensor D provide the intervoxel diffusion information and the microstructure of a particular voxel. However, the six-dimensional diffusion tensor is hard to infer and present to a user. For this reason, several scalar quantities have been introduced to simplify the tensor to a single value. Fractional Anisotropy (FA), the most widely used scalar measure in diffusion tensor imaging [9], represents the extent of the diffusion anisotropicity. A lower value of FA indicates that the diffusion is free (FA $= 0$; isotropic), while a high value of FA implies that the diffusion is restricted to a single direction (FA $= 1$; anisotropic). Another popular scalar measure in DTI is mean diffusivity (MD), which represents the overall amount of diffusion. Many other scalar measures have been proposed based on the more complex behaviour of molecular diffusion and are explained in detail in surveys by Novikov et al. [75], Rajagopalan et al. [83] and Vilanova et al. [99].

Visualization strategies: A Glyph is a general term for geometrically plotted specifier that represent multidimensional data values. Data information is mapped to glyph characteristics such as shape and color. Glyphs provide a way to represent the full six-dimensional information of a diffusion tensor by mapping the eigenvectors and eigenvalues to the orientation and shape of a geometric primitive. The most straight forward approach to visualize the diffusion tensor are ellipsoidal glyphs [82] as shown in Fig. 2a. The orientation of the ellipsoid represents the direction of the major eigenvector, while the length represents the corresponding eigenvalue. Westin et al. [104] introduced three metrics to measure linear ($\lambda_1 > \lambda_2, \lambda_3$), planar ($\lambda_1 = \lambda_2 > \lambda_3$) and spherical diffusion ($\lambda_1 = \lambda_2 = \lambda_3$). Figure. 2 represents the barycentric space of diffusion tensor shapes in which the three extremes (linear, planar, and spherical) are at the corner of triangles. Among several other proposed techniques, the super quadratic glyph is considered state-of-the-art for glyph-based tensor visualization. Instead of interpolating between ellipsoidal shapes, Kindlmann [57] represents the diffusion by superquadrics with shape parameters defined by the barycentric coordinates, as shown in Fig. 2b.

(a) Ellipsoid glyphs   (b) Superquadratic glyphs

Fig. 2 Barycentric space of diffusion tensor shapes

Outside DTI, glyphs are also used to represent the orientation distribution function (ODF) of the molecular diffusion, which is commonly estimated by models that go beyond the diffusion tensor. ODF specifies the overall diffusion in a given direction, integrated over displacement magnitudes [103]. Spherical polar plots [98] parametrized surfaces [77], the HARDI glyph [81] and the HOME glyph [89] are some of the common glyph based visualization techniques for representing ODFs.

2.2 Fiber Tracking

The diffusion tensor provides per-voxel information about the orientation of the underlying neural tracts by analyzing the derived eigenvectors. By combining this information with other scalar measure, e.g. FA, one can estimate trajectories of white matter bundles in 3D space. The process of virtual reconstruction of the neural fiber tract on the basis of the diffusion tensor field is named Fiber Tracking or Tractography [8, 71]. These techniques of generating brain anatomical connectivity from the diffusion information have been summarized in review articles [42, 72, 102]. Fiber

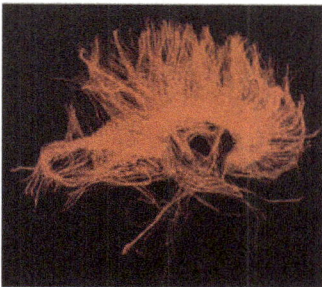

(a) Polylines

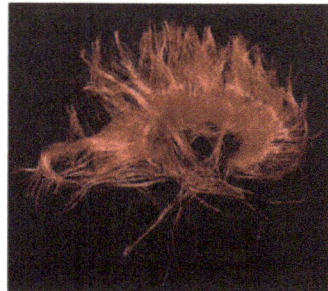

(b) Illuminated streamlines

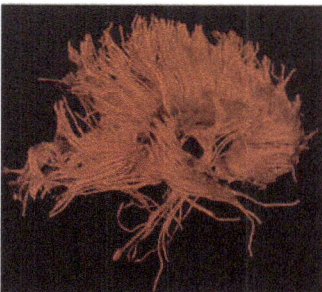

(c) Cylindrical tubes

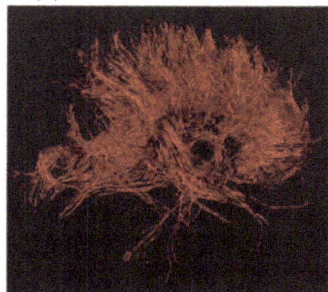

(d) Streamtubes

Fig. 3 Visualization techniques for deterministic tractography. Images are generated using vIST/e [1]

tracking methods have found its way in many neurological applications [18, 31, 45, 73, 74].

The fiber tracking strategies can be mainly divided into deterministic, probabilistic, and global geometric techniques. Deterministic techniques always produce the same output with the same set of inputs. Probabilistic techniques, however, add randomness in the tracking process to incorporate the inherent uncertainty. We defer an extensive discussion of probabilistic methods to Sect. 4. Global geometric methods deduce connectivity in the white matter by globally optimizing a certain cost function based on the diffusion tensor information and are outside of the scope of this chapter.

Streamline tracing is the most commonly used algorithm for tractography. It is a deterministic technique that generates trajectories by integrating the vector field defined by the main eigenvector at each voxel position. The tracing process ends when the stopping criteria are met. Several constraints can be used as stopping criteria, such as maximum turning angle or FA, to limit tracts to the region where tensors realistically represent the fiber tracts.

Visualization strategies: Line-based approaches are the most straightforward technique to represent deterministic fiber tracts. Numerous strategies have been introduced for the visualization of the white matter tracts, such as thin polylines [70] illuminated streamlines [114] or cylindrical tubes [8]. Zhang et al. [112] introduced streamtubes to encode the local diffusion tensor information along the cross-section of the fiber tracts at each voxel. This technique has previously been used to represent the tensor field in fluid dynamics, where they were called Hyper-streamlines [25]. Figure 3 shows the most commonly used representations for deterministic tractography.

So far, we have discussed the visualization pipeline methods used in DTI without the involvement of uncertainties. In the following sections, we will review the sources of uncertainty present in the pipeline, the modeling techniques and the strategies used to visualize them.

3   Sources of Uncertainty

Understanding the sources of uncertainties is essential to provide effective visualization. The DTI visualization pipeline involves complex stages of mathematical modeling, analysis, mapping, and rendering strategies, therefore, it is prone to uncertainty from various sources. Noise, patient movement, modeling residuals, and distortion from imaging artifacts produce uncertainty in the orientation of the diffusion tensor and are detrimental to fiber tracking algorithms. These uncertainties hamper the link between the data being measured and visualized. The sources of uncertainty involved at each stage of the DTI visualization pipeline are shown in Fig. 1. In this section, we will go through this pipeline and discuss the sources of error present at each stage. Even though we focus on the DTI modality, several of the sources of uncertainty are present in DWI pipelines that go beyond DTI.

3.1 Image Acquisition

MRI-based techniques usually suffer from various acquisition errors such as noise, motion artifacts, partial volume effects, etc. Signal to noise ratio in DWI sequences is relatively high given that signal attenuation is being measured. The effect of noise on the fiber tracking output has been widely studied in literature [3, 46, 59]. There has been a growing trend of increasing the gradient direction in DTI acquisition to improve the tractography quality. However, this further increases the acquisition time. In HARDI, the gradient directions for acquisition are much higher than that of DTI and, therefore, it needs more time. With higher acquisition time, it is more likely that the subject move during the scan, which in turn, introduces misalignment in the acquired image. These kinds of artifacts are known as motion artifacts. Providentially, these misalignments can be corrected during the registration process. Several automated techniques have been introduced to remove this artifact [113]. The finite resolution of the results also affects the output of the process. The resolution of a clinical DWI acquisition is typically in the order of millimeter (mm) in each direction, which is much lower than that of actual axons. Therefore, the signal values have to be averaged to be able to fit in a single image voxel. This loss of information is called the partial-volume effect (PVE). Several studies have been conducted in neurological literature to investigate the PVE in DTI [84, 85, 101]. Other sources of error during image acquisition involve magnetic distortion, scanner setting and others [14].

3.2 Diffusion Tensor Calculation

In DTI, the diffusion of a water molecule is mathematically represented by a second-order tensor, known as the diffusion tensor. Numerous measurements are performed along various gradient directions to determine the molecular diffusion at each voxel. The least-squares method is the most commonly used fitting technique to calculate the diffusion tensor, but other more accurate regression procedures can also be used [5, 6]. This fitting procedure introduces a fitting error and involves a model choice. Therefore it adds variation in the outcome of the DTI procedure. DTI technique can only estimate one dominant diffusion direction per voxel, and thus, is incapable of determining the structure where the multi-fiber direction is present and, therefore, results in unreliable outcomes. HARDI models emerge to overcome this limitation and able to model complex fibrous regions of the brain. It provides a way to estimate the multi-fiber populations that can then be used for robust tractography. HARDI models are more complex and usually introduce more parameters and choices to be determined than DTI.

3.3 Fiber Tracking

Fiber tracking involves the reconstruction of the fibrous structure of the brain white matter by gradually following the local fiber orientation estimated from the diffusion tensor. There are several parameters in fiber tracking to control the tracking process, however, these parameters add variability to the fiber tracking results. There are four major sources of uncertainties in the fiber tracking algorithm:

1. Region definition and filtering
2. Numerical approximation
3. Interpolation
4. Stopping criteria

Region definition and filtering: Regions are usually defined by the user to start, end, or control the fiber pathways. The seeding region refers to the starting point of the tracking process and defines the initial conditions for numerical integration. Regions are also used to extract a specific bundle of interest and filter out others to avoid visual clutter. The region definition in the fiber tracking process also adds variation in the outcome. Usually, these regions are defined manually, and therefore introduce an implicit user bias. A minor variation in the definition can result in largely different pathways. Recently, several techniques have been proposed to minimize the effect of seed region in the fiber tracking algorithms [21, 46, 102].

Numerical approximation: Different types of numerical approximation schemes can be implemented in the fiber tracking algorithm. Euler integration is the most straight forward technique [71] but it is a strong approximation. Higher-order methods, such as 2nd or 4th order Runge-Kutta methods [8], are typically less sensitive to noise and can be used for accurate fiber tracking. The integration step, or step size, further affect the quality of these integration schemes [97].

Interpolation: During the numerical approximation process, most of the time, the sample position after each integration step lies between volume grid points, hence, interpolation is needed to estimate values, based on the neighboring grid points. Several studies have been conducted to address the effect of interpolation in fiber tracking [32, 109]. Various kinds of interpolation schemes are present, each result in different pathways, and therefore, add variability in the results.

Stopping criteria: Different scalar measures, such as FA, MD, or curve angle, can be used as stopping criteria in the fiber tracking process. Fiber tracking algorithms are often highly sensitive to these values, meaning that a very small variation in the stopping criteria can lead to a very large change in the resulting fiber [97]. Brecheisen et al. [16] propose a visual exploration tool that allows users to investigate the behavior and sensitivity of DTI fiber tracking to stopping criteria.

In fiber tracking algorithms for HARDI models the principal directions are extracted from a multifiber representation which adds another layer of complexity to the algorithms.

3.4   Visualization

The visualization stage involves the mapping of the data into a geometric representation or visual primitives that are finally rendered on to the screen. This process can be another source the uncertainty. Various photo-realistic rendering techniques are used to simulate real world lighting as exact as possible, but this further complexity adds uncertainty in the outcome. Lighting models and shadows enhance the structural perception of the fibers and as such improve the recognition of the spatial relations between tracts; however, the controlling parameters can add further variability in the final results.

4   Uncertainty Modeling

As discussed in the previous section, many sources of uncertainties are present at each stage of the DTI visualization pipeline that affect the outcome of the process. These uncertainties propagate through the pipeline adding uncertainty in the derived quantities including diffusion tensor and fiber orientations. Estimating the error distribution of different sources is not a straight forward task. Different approaches have been used to model the uncertainty, however, each with pros and cons. We have classified the methods used for the uncertainty quantification into two categories:

1. Analytical methods
2. Stochastic methods

4.1   Analytical Methods

Analytical methods refer to approaches that provide an explicit mathematical formulation of the error distribution. These modeling techniques are based on the Bayes theorem [56] and were first introduced by Behrens et al. [12] in DWI. They estimated the probability distribution function (PDF) of the fiber orientation by a Bayesian model. The main disadvantage of this modeling technique is that they rely on the assumption of prior and noise present in the data. These techniques are computationally inexpensive, however, their dependence on the prior assumption limit their widespread use. Most of the Bayesian model-based techniques are often combined with random sampling methods, such as Markov Chain Monte Carlo (MCMC), to determine the distribution of model parameters [11, 12, 33]. The application of Bayesian model based methods in DTI and HARDI has been reported several times [48, 54, 64].

Shortest path algorithms are another useful approach for quantifying structural brain connectivity and were first introduced by O'Donnell et al. [78]. This approach relies on computing the connections between regions of interest rather than connections from a seed. Schober et al. [88] presented the distribution of the shortest path as a Gaussian process over the solution to an ordinary differential equation (ODE). This strategy offers novel ways to quantify and visualize uncertainty arising from the numerical computation and allow marginalization over a space of feasible solutions. Hauberg et al. [40] extended this work and incorporated data uncertainty in DTI by sub-sampling the diffusion gradients and solving the noisy ODE. Several other studies using the shortest path algorithms in fiber tracking can be found in the literature [39, 63].

4.2   Stochastic Methods

Describing the probability distribution analytically and propagating it through the visualization pipeline is extremely difficult and often not feasible. The alternative and the most straight forward way to estimate the probability distribution function is to repeat the acquisition multiple times, this approach is called the bootstrap method [27]. However, for robust estimation of the PDFs, hundreds of data sets are required, which is not practical in a clinical setting. Several stochastic techniques were proposed to overcome this limitation [20]. Among them, the most widely used techniques are residual bootstrapping and wild bootstrapping [106]. These techniques rely on a single scan and estimate the probability distribution from the residuals that remain after fitting diffusion tensor to the data. In residual bootstrapping [24], the distribution is estimated by randomly assigning the residuals among gradient directions. Another possibility is to resample the data based on randomly flipping the sign of the residuals by assuming symmetry in the distribution. The latter approach is called wild bootstrapping [23]. A detailed comparison of bootstrap methods has been presented by Chung et al. [20]. Stochastic bootstrapping has been widely used for DTI [50, 60, 79, 100]. These techniques generate multiple DTI volumes through stochastic simulations for estimating the probability distribution, however, they are computationally very expensive.

Various stochastic algorithms were introduced to incorporate uncertainty in tractography by adding randomness in the tracking process. These techniques are called probabilistic tractography [11, 87, 95]. These algorithms estimate the probability density function of the fiber orientation at each voxel and determine the propagation direction by drawing random samples from the distribution. Probabilistic fiber tracking is preferable in most cases as it takes uncertainty into account and can estimate the confidence interval for each reconstructed pathways, however, they are computationally expensive [12, 26]. Koch et al. [58] propose to use Monte Carlo random walks for the estimation of the fiber connectivity. The fiber tracking algorithm proceeds through each randomly selected neighboring voxel depending on the angle between the voxel's main eigenvector and its connecting angle with the neighboring voxels. A similar approach has been used in other studies to establish a connectivity map in a probabilistic sense [10, 13, 34, 80]. Monte Carlo methods have also been used to generate fiber tracks based on random particle movement [38]. The PDF obtained from the analytical methods can be used to perform tractography with these stochastic techniques [33, 48]. These studies are based on DTI, however, the concept is extendable to HARDI as well, but they are not used much in this context [64].

5 Uncertainty Visualization

So far, we have discussed the sources of uncertainty present in the visualization pipeline and the methods used for their quantification. Visualization provides a way to communicate data effectively and efficiently, however, uncertainty is often omitted in the process. Visualizing uncertainty information in DWI can help assess the accuracy of the acquisition and modeling, which ultimately guide the users in making critical decisions. However, the visualization of complex data in itself is not straightforward, adding uncertainty representation to it further complicates the process. Issues of visual cluttering and loss of anatomical context are some of the few complications when visualizing uncertainties.

In this section, we will survey the strategies used for the visualization of the uncertainties in DTI and also discuss some related techniques used in the HARDI model. We also summarize these strategies in Table 1. The modeling column refers to the uncertainty quantification techniques, such as stochastic, bootstrapping, or analytical methods. Domain indicates the application area of the study and ensemble column categorizes the method into the local or global level. The representation specifies the measure used for the aggregations of the ensemble, and finally, the visualization column indicates the technique used to display the uncertainties. The visualization of uncertainty in DTI can roughly be divided into two categories.

1. Local uncertainty visualization
2. Global uncertainty visualization

5.1 *Local Uncertainty Visualization*

Local representations of the uncertainty depict variation per voxel inside the vector or tensor fields. Glyphs are typically used to depict the voxel-wise information of the data. Several glyph-based techniques have been proposed to visualize the inherent local uncertainty in DTI. Jones et al. [50] proposed a method to represent the confidence interval of the main fiber direction by rendering an uncertainty cone, as shown in Fig. 4a. Basser et al. [4] used a similar technique to represent the main eigenvector and their associated uncertainties. This visualization approach allows the representation of the main diffusion direction and the confidence interval concurrently, also described in Table 1. Schultz et al. [90] demonstrate a new glyph design, called HiFiVE, that provides a more detailed impression of the uncertainty. It represents the variation corresponding to the main eigenvector by rendering a double cone (blue

Table 1 Summary of uncertainty representation and visualization strategies

References	Modeling	Domain	Ensemble	Representation	Visualization
Jones et al. [50]	Stochastic	DWI	Local	Direction interval	Interval glyph
Basser et al. [4]	Stochastic	DWI	Local	Direction interval	Interval glyph
Schultz et al. [90]	Stochastic	DWI	Local	Probability distribution	HIfive glyph
Jones et al. [52]	Stochastic	DWI	Local	Mean and median	Overlay glyph
Zhang et al. [110]	–	DWI	Local	Mean and variance	Halo and texture
Zhang et al. [111]	–	DWI	Local	Difference encoding	Overlay glyph
Tournier et al. [96]	Stochastic	DWI	Local	ODF mean and variance	Semi-transparent glyph
Jiao et al. [49]	Stochastic	DWI	Local	ODF SIP	Volume rendered glyph
Basser et al. [7]	Analytical	DWI	Local	Mean and covariance	Superimpose glyph
Abbasloo et al. [2]	Analytical	DWI	Local	Mean and covariance	Overlay/Animation glyph
Gerrits et al. [35]	Analytical	Both	Local	Mean and covariance	Superimpose glyph
Wittenbrink et al. [108]	Bootstrap	Non-DWI	Local	Mean and variance	Flow-field glyph
Zuk et al. [115]	Bootstrap	Non-DWI	Local	Probability distribution	Flow-field glyph
Hlawatsch et al. [43]	Bootstrap	Non-DWI	Local	Mean and variance	Flow-field glyph
Lodha et al. [65]	Bootstrap	Non-DWI	Local	Interval	Flow-field glyph
Otten et al. [76]	–	DWI	Global	Line and interval	Illustrative
Hermosilla et al. [41]	–	DWI	Global	Line and interval	Illustrative
Brecheisen et al. [15]	Stochastic	DWI	Global	Line and interval	Illustrative
Corouge et al. [22]	Bootstrap	DWI	Global	Ensembles	Spaghetti plot
Bjornemo et al. [13]	Stochastic	DWI	Global	Ensembles	Spaghetti plot
Jones et al. [51]	Stochastic	DWI	Global	Ensembles	Spaghetti plot
Hangmann et al. [38]	Stochastic	DWI	Global	Ensembles	Color coded spaghetti plot
Ehricke et al. [28]	Stochastic	DWI	Global	Ensembles	Color coded spaghetti plot
Enders et al. [29]	–	DWI	Global	Fiber clusters	Wrapped geometrical hull
Chen et al. [19]	–	DWI	Global	Fiber clusters	Wrapped geometrical hull

(continued)

Table 1 (continued)

References	Modeling	Domain	Ensemble	Representation	Visualization
Merhof et al. [68]	–	DWI	Global	Fiber clusters	Wrapped geometrical hull
Jones et al. [53]	Bootstrap	DWI	Global	Ensemble/local estimates	Streamtubes
Wiens et al. [107]	Stochastic	DWI	Global	Ensemble/local estimates	Streamtubes
Goldau et al. [37]	Stochastic	DWI	Global	Fiber density	Stipples glyphs
Hlawitschka et al. [44]	Stochastic	DWI	Global	Fiber density	Stipples glyphs
Goldau et al. [36]	Stochastic	DWI	Global	Fiber density	Stipples glyphs
Brown et al. [17]	Stochastic	DWI	Global	Fiber density	Confidence region
Schultz et al. [91]	Stochastic	DWI	Global	Connectivity Probability	Confidence region
Kapri et al. [55]	–	DWI	Global	Connectivty Probability	Volume rendering
McGraw et al. [67]	Stochastic	DWI	Global	Connectivity Probability	Volume rendering
Koch et al. [58]	Stochastic	DWI	Global	Connectivity Probability	Density map
Parker et al. [80]	Stochastic	DWI	Global	Connectivity Probability	Density map
Kaden et al. [54]	Analytical	DWI	Global	Connectivity Probability	Density map
Schober et al. [88]	Analytical	DWI	Global	Ensembles	Wobbly Spaghetti plot
Hauberg et al. [40]	Analytical	DWI	Global	Ensembles	Wobbly spaghetti plot
Mirzargar et al. [69]	Bootstrap	Non-DWI	Global	Band Depth	Wrapped geometrical hull
Whitaker et al. [105]	Bootstrap	Non-DWI	Global	Band Depth	Contour lines
Ferstl et al. [30]	Bootstrap	Non-DWI	Global	Line and interval	Wrapped geometrical hull
Sanyal et al. [86]	Bootstrap	Non-DWI	Global	Mean and std. deviation	Ribbon

color) and the density estimation of the uncertainty around it (represented as a gray surface), as shown in Fig. 4b.

Another way to represent the uncertainty in multivariate data is to estimate its covariance. It does not only express the variance in each coefficient but also indicates their linear dependencies. Since the diffusion tensor is a second-order tensor, its covariance is represented by a fourth-order tensor, however, the visualization of the fourth-order tensor is rather difficult in this context. Basser et al. [7] presented a novel technique for the spectral decomposition of the fourth-order covariance tensor

(a) Uncertainity cones [50]

(b) HiFiVE Glyphs [90]

(c) Decomposed ensemble representation [110]

(d) ODF glyphs [96]

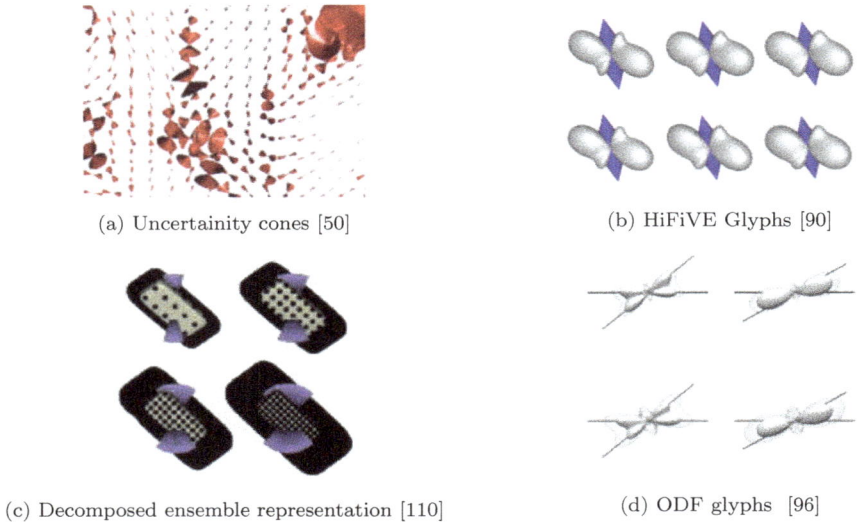

Fig. 4 Glyphs with uncertainty encoding

and introduced the concept of tensorial normal distribution. They proposed a glyph representation, called radial glyphs , which depicts the overall variance and a composite glyph for representing the eigentensor of the fourth-order covariance. They visualized the expected mean tensor and its standard deviation as three isosurfaces. Abbasloo et al. [2] highlight that the radial glyph does not convey the correlation with the mean tensor and also suffers from high visual complexity in the tensor field. They proposed a more intuitive approach for the visualization of the covariance by using multiple levels of detail. Unlike Basser et al., Abbasloo et al. visualize the confidence interval at each eigenmode separately by glyph overlays and used animation to visualize the differences in each mode. Gerrits et al. [35] pointed out the shortcoming in both of these visualization techniques and proposed a generic approach that incorporates all the coefficients of the mean tensor and covariance in a single glyph.

Various studies have been published concerning the representation of the tensor ensemble directly. Jones et al. [52] visualize the ensemble data simply by overlaying several glyphs. Although the superposition depicts the overall picture of the data, it adds visual clutter and occlusion during display. To remedy this, Zhang et al. [111] used transparency to minimize the occlusion. Abbasloo et al. [2] tried to minimize this problem by rendering the superimposed glyphs in complementary colors. Zhang et al. [110] proposed an approach to decompose the tensor data into three properties (i.e., scale, shape, and orientation), representing the structure of the underlying fibers, and measure the variation per property. A glyph based representation has been presented in this study to visualize the ensemble effectively. The variation in the ensemble is represented by Halo and texture over the surface as shown in Fig. 4c.

The orientation distribution function (ODF), associated with HARDI, specifies the overall amount of diffusion in a given direction. Unlike the diffusion tensor model, ODFs can have multiple maxima, and therefore are capable of modeling complex fibrous structure more accurately than DTI. However, this technique is computationally expensive. The representation of the ODF itself is a challenging task and adding uncertainty information only increases the complexity. Jiao et al. [49] proposed a technique to visualize uncertainty over polar ODF glyphs by using a volume rendering technique. They introduced shape inclusion probability (SIP) function to represent the orientation uncertainty of the tensor. Tournier et al. [96] presented a method to visualize uncertainties associated with ODFs by using semitransparent glyphs. They represent the mean ODF by the opaque surface and the mean + standard deviation by the transparent surface, as shown in Fig. 4d.

The visualization of uncertainty in a diffusion tensor is similar to the uncertainty representation in a vector field where orientation is considered important. Several glyph-based techniques exist in this scope. Wittenbrink et al. [108] presented a glyphs based representation of the uncertainty for atmospheric and oceanographic data. Likewise, Hlawatsch [43] and Lodha et al. [65] visualize the local uncertainty in a fluid flow field using glyphs. Zuk et al. [115] proposed a glyph design to provide uncertainty information in a bidirectional vector field. These techniques rely on the representation of the vector direction and magnitude with encoded uncertainties to depict the local uncertainty present in the field.

5.2 Global Uncertainty Visualization

In contrast to the local strategies, global uncertainty visualization in DTI aims at providing information on how accurate fiber tract information is throughout the complete tensor field, and how the inherent uncertainties accumulate during the tracking process. In DWI independently of DTI or HARDI models being used, probabilistic tractography is often used to incorporate these uncertainties. The most widely used approach to visualize fibers obtained through probabilistic tractography is to superimpose the resulting fibers in a so-called spaghetti plot [13, 22, 51], see Fig. 5a. This visualization technique, however, does not depict a clear view of the region-wise fiber connections and its uncertainty and suffers from strong cluttering. Color coding the fiber tracts according to their seed points [22] does not suffice to minimize the complexity of the visualization. Schober et al. [88] and Hauberg et al. [40] used wobbly spaghetti plot that emphasize the fact that the individual resulting paths cannot be considered as real fibers in the brain which is a common misinterpretation in spaghetti plot. Instead, they are uncertain estimates of fibers.

To overcome the complexity and clutter caused by the multiple superimposed tracts, Enders et al. [29] presented a technique to group the fibers related to a certain nerve tract and generate a surface that wraps the resulting fibers. Similarly, Mehrof et al. [68] and Chen et al. [19] presented a method to cluster the fiber with a proximity-based algorithm and generate hulls encompassing the fiber bundles, as

(a) Spaghetti plot [51]

(b) Wrapped streamlines [19]

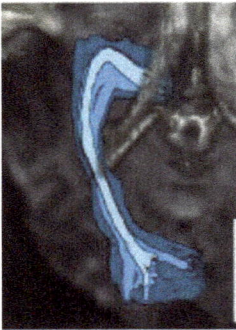

(c) Illustrative visualization [15]

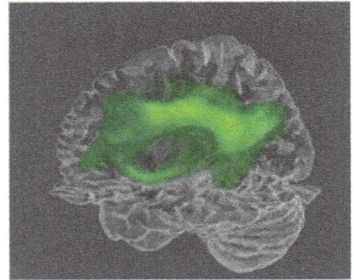

(d) Connectivity mapping [55]

Fig. 5 Global uncertainty visualization strategies

shown in Fig. 5b. The anatomical grouping helps the user to understand the under-lying fibrous structure. Outside of DTI, Frest et al. [30] used a similar technique to visualize uncertainty in flow field ensembles. They performed principal component analysis to cluster the streamlines in a low dimensional space and determine the mean and confidence interval in an ensemble. These representations are visualized with a line enclosed by a transparent surface. The geometrical hulls and enclosed surfaces reduce clutter, however, they cannot resolve complex cluster shapes. To alleviate these problems, Illustrative techniques have been proposed to represent the confi-dence interval of the fiber bundle by creating silhouette, outline, and contours [15, 76], as shown in Fig. 5c.

To improve the understanding of ensembles of curves, it has been proposed to visualize the statistical information such as mean or confidence intervals rather than the direct ensemble visualization as spaghetti plot. Table 1 indicates the various representations used by the studies. These representations, e.g, mean and confidence interval, are the summarization of the raw samples. Unlike scalar values, the statistical measures are not well defined for curves, and therefore, several approaches have been proposed for the estimation of these terms. Brecheisen et al. [15] proposed to compute median and confidence intervals based on pre-selected distance measures between curves. In the field of fluid dynamics, a band-depth concept [66] has been introduced to analyze curve ensembles in two-dimensions [105] and three-dimensions [69]. This concept provides a way to determine centrality within the present curves and estimate the variations. Sanyal et al. [86] visualize the uncertainty in the wind trajectories by creating a ribbon along the ensemble mean. The width of a ribbon represents the variability at each point.

A widely used approach for the visualization of the global uncertainty is to represent and visualize measures derived from the probabilistic tractography. Voxel-wise fiber density computes the probability that a fiber tract traverses a voxel for a given seed region [17]. Voxel-wise fiber density [28, 38] helps to infer the anatomical connections. Another measure is the connectivity probability, which represents the probability of a fiber tract crossing a given voxel while connecting two fixed anatomical regions [91]. Von Kapri et al. [55] and McGraw et al. [67] used volume rendering for the visualization of density maps, as shown in Fig. 5d. The global visualization of the fiber tracts does not provide the local tensor information. To visualize the local uncertainty along with the probabilistic tracts, a stream tube technique has been proposed [53, 107], which maps the local uncertainty measure onto the cross-section of the tube.

A common problem with the three-dimensional approaches is that the geometrical representation often occludes the underlying information, hampering its interpretation. Various slice-based methods have been proposed for the visualization of probabilistic fibers [58, 80]. These techniques have been used in neuroscience as they provide a way to directly visualize the anatomical information, making it easy to interpret anatomical context. Goldou et al. [36, 37] presented a novel slice based approach for visualizing the probability by rendering fiber stipples. The number of stipples, present at a particular region depicts the fiber density. Hlawitschka et al. [44] proposes to use poisson-disk sampling for the generation of the fiber stipples.

Table 1 summarizes the survey indicating the domain, representation, and visualization strategies used to display the uncertainty. The table covers the approaches used for local and global uncertainty visualization in both the DWI and non-DWI domain.

6 Conclusion

Diffusion-weighted imaging relies on complex stages of signal acquisition, mathematical modeling, model assumptions, and hence, is exposed to many sources of uncertainty. Excluding this uncertainty from the visualization does not only affect the result but also cripples the user to make an effective decision. However, the efficient visualization of the uncertainty in DWI is nontrivial, as the data itself has high visual complexity and adding uncertainty to it only adds further complexity. In this chapter, we explored uncertainty in the various stages of the DTI visualization pipeline. Several of the problems and solutions discussed throughout this chapter are also valid for other models beyond Diffusion tensor, such as HARDI models. Even though we have not covered the technical background, where applicable, we have discussed the applicability of the strategies beyond DTI. Further, we have reviewed applicable uncertainty visualization techniques beyond the DWI domain.

DWI is still a growing field, considering the recent advancements and the frequent development of new techniques, this survey should not be considered complete, it rather should be enhanced in the future. Studies on uncertainty visualization so far are mostly focused on the research aspect, however, no uncertainty visualization solution exist to specifically support clinical tractography. Visual analytics, an emerging field in visualization, can be helpful in enabling detailed analysis of uncertainty present in DWI data, building on the top of the studies present in this survey. Most of the presented studies deal with uncertainty on the noise and modeling level, dealing with other sources of uncertainty and visualizing them as a whole part of the exploration is another open research direction.

We focus on the visualization pipeline and techniques mostly related to DTI, a study including other modeling techniques would be beneficial. In summary, even though the uncertainty visualization in DWI has evolved considerably in the last few years, we believe, a lot of work still needs to be done for the effective visualization and exploration of DWI data.

References

1. Vist/e: Interactive visualization of dti, hardi and other complex imaging data https://sourceforge.net/projects/viste/
2. Abbasloo, A., Wiens, V., Hermann, M., Schultz, T.: Visualizing tensor normal distributions at multiple levels of detail. IEEE Trans. Visual. Comput. Graph. **22**(1), 975–984 (2015)
3. Anderson, A.W.: Theoretical analysis of the effects of noise on diffusion tensor imaging. Magn. Reson. Med. Official J. Int. Soc. Magn. Reson. Med. **46**(6), 1174–1188 (2001)
4. Basser, P.J.: Quantifying errors in fiber-tract direction and diffusion tensor field maps resulting from mr noise. In: Fifth Annual meeting of ISMRM, Vancouver, p. 1740 (2007)
5. Basser, P.J., Mattiello, J., LeBihan, D.: Estimation of the effective self-diffusion tensor from the nmr spin echo. J. Magn. Reson. Ser. B **103**(3), 247–254 (1994)
6. Basser, P.J., Mattiello, J., LeBihan, D.: Mr diffusion tensor spectroscopy and imaging. Biophys. J. **66**(1), 259–267 (1994)
7. Basser, P.J., Pajevic, S.: Spectral decomposition of a 4th-order covariance tensor: Applications to diffusion tensor mri. Signal Process. **87**(2), 220–236 (2007)
8. Basser, P.J., Pajevic, S., Pierpaoli, C., Duda, J., Aldroubi, A.: In vivo fiber tractography using dt-mri data. Magn. Reson. Med. **44**(4), 625–632 (2000)
9. Basser, P.J., Pierpaoli, C.: Microstructural and physiological features of tissues elucidated by quantitative-diffusion-tensor mri. J. Magn. Reson. **213**(2), 560–570 (2011)
10. Batchelor, P.G., Hill, D.L., Atkinson, D., Calamante, F.: Study of connectivity in the brain using the full diffusion tensor from mri. In: Biennial International Conference on Information Processing in Medical Imaging, pp. 121–133. Springer, Berlin (2001)
11. Behrens, T.E., Berg, H.J., Jbabdi, S., Rushworth, M.F., Woolrich, M.W.: Probabilistic diffusion tractography with multiple fibre orientations: what can we gain? Neuroimage **34**(1), 144–155 (2007)

12. Behrens, T.E., Woolrich, M.W., Jenkinson, M., Johansen-Berg, H., Nunes, R.G., Clare, S., Matthews, P.M., Brady, J.M., Smith, S.M.: Characterization and propagation of uncertainty in diffusion-weighted mr imaging. Magn. Reson. Med. Official J. Int. Soc. Magn. Reson. Med. **50**(5), 1077–1088 (2003)
13. Björnemo, M., Brun, A., Kikinis, R., Westin, C.F.: Regularized stochastic white matter tractography using diffusion tensor mri. In: International Conference on Medical Image Computing and Computer-Assisted Intervention, pp. 435–442. Springer, Berlin (2002)
14. Brecheisen, R.: Visualization of uncertainty in fiber tracking based on diffusion tensor imaging. Ph.D. thesis, Technische Universiteit Eindhoven, Department of Biomedical Engineering (2012)
15. Brecheisen, R., Platel, B., ter Haar Romeny, B.M., Vilanova, A.: Illustrative uncertainty visualization of dti fiber pathways. Vis. Comput. **29**(4), 297–309 (2013)
16. Brecheisen, R., Vilanova, A., Platel, B., ter Haar Romeny, B.: Parameter sensitivity visualization for dti fiber tracking. IEEE Trans. Visual. Comput. Graph. **15**(6), 1441–1448 (2009)
17. Brown, C.J., Booth, B.G., Hamarneh, G.: Uncertainty in tractography via tract confidence regions. In: Computational Diffusion MRI and Brain Connectivity, pp. 129–138 (2014)
18. Catani, M.: Diffusion tensor magnetic resonance imaging tractography in cognitive disorders. Curr. Opin. Neurol. **19**(6), 599–606 (2006)
19. Chen, W., Zhang, S., Correia, S., Ebert, D.S.: Abstractive representation and exploration of hierarchically clustered diffusion tensor fiber tracts. In: Computer Graphics Forum, vol. 27, pp. 1071–1078. Wiley Online Library (2008)
20. Chung, S., Lu, Y., Henry, R.G.: Comparison of bootstrap approaches for estimation of uncertainties of dti parameters. NeuroImage **33**(2), 531–541 (2006)
21. Ciccarelli, O., Parker, G., Toosy, A., Wheeler-Kingshott, C., Barker, G., Boulby, P., Miller, D., Thompson, A.: From diffusion tractography to quantitative white matter tract measures: a reproducibility study. Neuroimage **18**(2), 348–359 (2003)
22. Corouge, I., Fletcher, P.T., Joshi, S., Gouttard, S., Gerig, G.: Fiber tract-oriented statistics for quantitative diffusion tensor mri analysis. Med. Image Anal. **10**(5), 786–798 (2006)
23. Davidson, R., Flachaire, E.: The wild bootstrap, tamed at last. J. Econometr. **146**(1), 162–169 (2008)
24. Davison, A.C., Hinkley, D.V.: Bootstrap methods and their application **1**, (1997)
25. Delmarcelle, T., Hesselink, L.: Visualizing second-order tensor fields with hyperstreamlines. IEEE Comput. Graph. Appl. **13**(4), 25–33 (1993)
26. Descoteaux, M., Deriche, R., Knosche, T.R., Anwander, A.: Deterministic and probabilistic tractography based on complex fibre orientation distributions. IEEE Trans. Med. Imaging **28**(2), 269–286 (2008)
27. Efron, B., Tibshirani, R.J.: An Introduction to the Bootstrap (1994)
28. Ehricke, H.H., Klose, U., Grodd, W.: Visualizing mr diffusion tensor fields by dynamic fiber tracking and uncertainty mapping. Comput. Graph. **30**(2), 255–264 (2006)
29. Enders, F., Sauber, N., Merhof, D., Hastreiter, P., Nimsky, C., Stamminger, M.: Visualization of white matter tracts with wrapped streamlines (2005)
30. Ferstl, F., Bürger, K., Westermann, R.: Streamline variability plots for characterizing the uncertainty in vector field ensembles. IEEE Trans. Visual. Comput. Graph. **22**(1), 767–776 (2015)
31. ffytche, D.H., Catani, M.: Beyond localization: from hodology to function. Philoso. Trans. R. Soc. B: Biol. Sci. **360**(1456), 767–779 (2005)
32. Florack, L., Haije, T.D., Fuster, A.: Direction-controlled dti interpolation. In: Visualization and Processing of Higher Order Descriptors for Multi-Valued Data, pp. 149–162 (2015)
33. Friman, O., Farneback, G., Westin, C.F.: A bayesian approach for stochastic white matter tractography. IEEE Trans. Med. Imaging **25**(8), 965–978 (2006)
34. Gembris, D., Schumacher, H., Suter, D.: Solving the diffusion equation for fiber tracking in the living human brain. In: Proceedings of the International Society for Magnetic Resonance Medicine (ISMRM), vol. 9, p. 1529 (2001)

35. Gerrits, T., Rössl, C., Theisel, H.: Towards glyphs for uncertain symmetric second-order tensors. In: Computer Graphics Forum, vol. 38, pp. 325–336. Wiley Online Library (2019)
36. Goldau, M., Hlawitschka, M.: Multi-modal visualization of probabilistic tractography. Visual. Med. Life Sci. **III**, 195–218 (2016)
37. Goldau, M., Wiebel, A., Gorbach, N.S., Melzer, C., Hlawitschka, M., Scheuermann, G., Tittgemeyer, M.: Fiber stippling: An illustrative rendering for probabilistic diffusion tractography. In: 2011 IEEE Symposium on Biological Data Visualization (BioVis), pp. 23–30. IEEE (2011)
38. Hagmann, P., Thiran, J.P., Jonasson, L., Vandergheynst, P., Clarke, S., Maeder, P., Meuli, R.: Dti mapping of human brain connectivity: statistical fibre tracking and virtual dissection. Neuroimage **19**(3), 545–554 (2003)
39. Hao, X., Whitaker, R.T., Fletcher, P.T.: Adaptive riemannian metrics for improved geodesic tracking of white matter. In: Biennial International Conference on Information Processing in Medical Imaging, pp. 13–24. Springer, Berlin (2011)
40. Hauberg, S., Schober, M., Liptrot, M., Hennig, P., Feragen, A.: A random riemannian metric for probabilistic shortest-path tractography. In: International Conference on Medical Image Computing and Computer-Assisted Intervention, pp. 597–604. Springer, Berlin (2015)
41. Hermosilla, P., Brecheisen, R., Vázquez, P.P., Vilanova, A.: Uncertainty Visualization of Brain Fibers (2012)
42. Hess, C.P., Mukherjee, P.: Visualizing white matter pathways in the living human brain: diffusion tensor imaging and beyond. Neuroimaging Clin. North Am. **17**(4), 407–426 (2007)
43. Hlawatsch, M., Leube, P., Nowak, W., Weiskopf, D.: Flow radar glyphs-static visualization of unsteady flow with uncertainty. IEEE Trans. Visual. Comput. Graph. **17**(12), 1949–1958 (2011)
44. Hlawitschka, M., Goldau, M., Wiebel, A., Heine, C., Scheuermann, G.: Hierarchical poisson-disk sampling for fiber stipples (2013)
45. Holodny, A.I., Schwartz, T.H., Ollenschleger, M., Liu, W.C., Schulder, M.: Tumor involvement of the corticospinal tract: diffusion magnetic resonancetractography with intraoperative correlation: case illustration. J. Neurosurg. **95**(6), 1082–1082 (2001)
46. Huang, H., Zhang, J., Van Zijl, P.C., Mori, S.: Analysis of noise effects on dti-based tractography using the brute-force and multi-roi approach. Magn. Reson. Med. Official J. Int. Soc. Magn. Reson. Med. **52**(3), 559–565 (2004)
47. Isenberg, T.: A survey of illustrative visualization techniques for diffusion-weighted mri tractography. In: Visualization and Processing of Higher Order Descriptors for Multi-Valued Data, pp. 235–256 (2015)
48. Jbabdi, S., Woolrich, M.W., Andersson, J.L., Behrens, T.: A bayesian framework for global tractography. Neuroimage **37**(1), 116–129 (2007)
49. Jiao, F., Phillips, J.M., Gur, Y., Johnson, C.R.: Uncertainty visualization in hardi based on ensembles of odfs. In: 2012 IEEE Pacific Visualization Symposium, pp. 193–200. IEEE (2012)
50. Jones, D.K.: Determining and visualizing uncertainty in estimates of fiber orientation from diffusion tensor mri. Magn. Reson. Med. Official J. Int. Soc. Magn. Reson. Med. **49**(1), 7–12 (2003)
51. Jones, D.K.: Tractography gone wild: probabilistic fibre tracking using the wild bootstrap with diffusion tensor mri. IEEE Transa. Med. Imaging **27**(9), 1268–1274 (2008)
52. Jones, D.K., Griffin, L.D., Alexander, D.C., Catani, M., Horsfield, M.A., Howard, R., Williams, S.C.: Spatial normalization and averaging of diffusion tensor mri data sets. Neuroimage **17**(2), 592–617 (2002)
53. Jones, D.K., Travis, A.R., Eden, G., Pierpaoli, C., Basser, P.J.: Pasta: pointwise assessment of streamline tractography attributes. Magn. Reson. Med. Official J. Int. Soc. Mag. Reson. Med. **53**(6), 1462–1467 (2005)
54. Kaden, E., Knösche, T.R., Anwander, A.: Parametric spherical deconvolution: inferring anatomical connectivity using diffusion mr imaging. NeuroImage **37**(2), 474–488 (2007)

55. von Kapri, A., Rick, T., Caspers, S., Eickhoff, S.B., Zilles, K., Kuhlen, T.: Evaluating a visualization of uncertainty in probabilistic tractography. In: Medical Imaging 2010: Visualization, Image-Guided Procedures, and Modeling, vol. 7625, p. 762534. International Society for Optics and Photonics (2010)

56. Kendall, M.G., et al.: The advanced theory of statistics. vols. 1. The advanced theory of statistics, vol. 1. 1(Ed. 4) (1948)

57. Kindlmann, G.: Superquadric tensor glyphs. In: Proceedings of the Sixth Joint Eurographics-IEEE TCVG conference on Visualization, pp. 147–154. Eurographics Association (2004)

58. Koch, M.A., Norris, D.G., Hund-Georgiadis, M.: An investigation of functional and anatomical connectivity using magnetic resonance imaging. Neuroimage 16(1), 241–250 (2002)

59. Lazar, M., Alexander, A.L.: An error analysis of white matter tractography methods: synthetic diffusion tensor field simulations. Neuroimage 20(2), 1140–1153 (2003)

60. Lazar, M., Alexander, A.L.: Bootstrap white matter tractography (boot-trac). NeuroImage 24(2), 524–532 (2005)

61. Le Bihan, D., Breton, E., Lallemand, D., Grenier, P., Cabanis, E., Laval-Jeantet, M.: Mr imaging of intravoxel incoherent motions: application to diffusion and perfusion in neurologic disorders. Radiology 161(2), 401–407 (1986)

62. Leemans, A.: Visualization of diffusion mri data. In: Diffusion MRI, pp. 354–379 (2010)

63. Lenglet, C., Deriche, R., Faugeras, O.: Inferring white matter geometry from diffusion tensor mri: application to connectivity mapping. In: European Conference on Computer Vision, pp. 127–140. Springer, Berlin (2004)

64. Liang, R., Wang, Z., Zhang, S., Feng, Y., Jiang, L., Ma, X., Chen, W., Tate, D.F.: Visual exploration of hardi fibers with probabilistic tracking. Inf. Sci. 330, 483–494 (2016)

65. Lodha, S.K., Pang, A., Sheehan, R.E., Wittenbrink, C.M.: Uflow: Visualizing uncertainty in fluid flow. In: Proceedings of Seventh Annual IEEE Visualization'96, pp. 249–254. IEEE (1996)

66. López-Pintado, S., Sun, Y., Lin, J.K., Genton, M.G.: Simplicial band depth for multivariate functional data. Adv. Data Anal. Classif. 8(3), 321–338 (2014)

67. McGraw, T., Nadar, M.: Stochastic dt-mri connectivity mapping on the gpu. IEEE Trans. Visual. Comput. Graph. 13(6), 1504–1511 (2007)

68. Merhof, D., Meister, M., Bingöl, E., Nimsky, C., Greiner, G.: Isosurface-based generation of hulls encompassing neuronal pathways. Stereotact. Funct. Neurosurg. 87(1), 50–60 (2009)

69. Mirzargar, M., Whitaker, R.T., Kirby, R.M.: Curve boxplot: Generalization of boxplot for ensembles of curves. IEEE Trans. Visual. Comput. Graph. 20(12), 2654–2663 (2014)

70. Mori, S., Crain, B.J., Chacko, V.P., Van Zijl, P.C.: Three-dimensional tracking of axonal projections in the brain by magnetic resonance imaging. Ann. Neurol. Official J. Am. Neurol. Assoc. Child Neurol. Soc. 45(2), 265–269 (1999)

71. Mori, S., Van Zijl, P.C.: Fiber tracking: principles and strategies-a technical review. NMR Biomed. Int. J. Devoted Develop. Appl. Magn. Reson. Vivo 15(7–8), 468–480 (2002)

72. Mori, S., Wakana, S., Van Zijl, P.C., Nagae-Poetscher, L.: MRI Atlas of Human White Matter (2005)

73. Mori, S., Zhang, J.: Principles of diffusion tensor imaging and its applications to basic neuroscience research. Neuron 51(5), 527–539 (2006)

74. Nimsky, C., Ganslandt, O., Hastreiter, P., Wang, R., Benner, T., Sorensen, A.G., Fahlbusch, R.: Preoperative and intraoperative diffusion tensor imaging-based fiber tracking in glioma surgery. Neurosurgery 56(1), 130–138 (2005)

75. Novikov, D.S., Fieremans, E., Jespersen, S.N., Kiselev, V.G.: Quantifying brain microstructure with diffusion mri: Theory and parameter estimation. NMR Biomed. 32(4), e3998 (2019)

76. Otten, R., Vilanova, A., Van De Wetering, H.: Illustrative white matter fiber bundles. In: Computer Graphics Forum, vol. 29, pp. 1013–1022. Wiley Online Library (2010)

77. Özarslan, E., Mareci, T.H.: Generalized diffusion tensor imaging and analytical relationships between diffusion tensor imaging and high angular resolution diffusion imaging. Magn. Reson. Med. Official J. Int. Soc. Magn. Reson. Med. 50(5), 955–965 (2003)

78. O'Donnell, L., Haker, S., Westin, C.F.: New approaches to estimation of white matter connectivity in diffusion tensor mri: Elliptic pdes and geodesics in a tensor-warped space. In: International Conference on Medical Image Computing and Computer-Assisted Intervention, pp. 459–466. Springer, Berlin (2002)

79. Pajevic, S., Basser, P.J.: Parametric and non-parametric statistical analysis of dt-mri data. J. Magn. Reson. **161**(1), 1–14 (2003)

80. Parker, G.J., Haroon, H.A., Wheeler-Kingshott, C.A.: A framework for a streamline-based probabilistic index of connectivity (pico) using a structural interpretation of mri diffusion measurements. J. Magn. Reson. Imaging Official J. Int. Soc. Magn. Reson. Med. **18**(2), 242–254 (2003)

81. Peeters, T.H., Prckovska, V., van Almsick, M., Vilanova, A., ter Haar Romeny, B.M.: Fast and sleek glyph rendering for interactive hardi data exploration. In: 2009 IEEE Pacific Visualization Symposium, pp. 153–160. IEEE (2009)

82. Pierpaoli, C., Basser, P.J.: Toward a quantitative assessment of diffusion anisotropy. Magn. Reson. Med. **36**(6), 893–906 (1996)

83. Rajagopalan, V., Jiang, Z., Stojanovic-Radic, J., Yue, G., Pioro, E., WYLIE, G., Das, A.: Ea basic introduction to diffusion tensor imaging mathematics and image processing steps. Brain Disord. Ther. **6**(229), 2 (2017)

84. Roine, T., Jeurissen, B., Perrone, D., Aelterman, J., Leemans, A., Philips, W., Sijbers, J.: Isotropic non-white matter partial volume effects in constrained spherical deconvolution. Front. Neuroinform. **8**, 28 (2014)

85. Salminen, L.E., Conturo, T.E., Bolzenius, J.D., Cabeen, R.P., Akbudak, E., Paul, R.H.: Reducing csf partial volume effects to enhance diffusion tensor imaging metrics of brain microstructure. Technol. Innovation **18**(1), 5 (2016)

86. Sanyal, J., Zhang, S., Dyer, J., Mercer, A., Amburn, P., Moorhead, R.: Noodles: a tool for visualization of numerical weather model ensemble uncertainty. IEEE Trans. Visual. Comput. Graph. **16**(6), 1421–1430 (2010)

87. Sarwar, T., Ramamohanarao, K., Zalesky, A.: Mapping connectomes with diffusion mri: deterministic or probabilistic tractography? Magn. Reson. Med. **81**(2), 1368–1384 (2019)

88. Schober, M., Kasenburg, N., Feragen, A., Hennig, P., Hauberg, S.: Probabilistic shortest path tractography in dti using gaussian process ode solvers. In: International Conference on Medical Image Computing and Computer-Assisted Intervention, pp. 265–272. Springer, Berlin (2014)

89. Schultz, T., Kindlmann, G.: A maximum enhancing higher-order tensor glyph. In: Computer Graphics Forum, vol. 29, pp. 1143–1152. Wiley Online Library (2010)

90. Schultz, T., Schlaffke, L., Schölkopf, B., Schmidt-Wilcke, T.: Hifive: a hilbert space embedding of fiber variability estimates for uncertainty modeling and visualization. In: Computer Graphics Forum, vol. 32, pp. 121–130. Wiley Online Library (2013)

91. Schultz, T., Theisel, H., Seidel, H.P.: Topological visualization of brain diffusion mri data. IEEE Trans. Visual. Comput. Graph. **13**(6), 1496–1503 (2007)

92. Schultz, T., Vilanova, A.: Diffusion mri visualization. NMR Biomed. **32**(4), e3902 (2019)

93. Schultz, T., Vilanova, A., Brecheisen, R., Kindlmann, G.: Fuzzy fibers: Uncertainty in dmri tractography. In: Scientific Visualization, pp. 79–92 (2014)

94. Stejskal, E.O., Tanner, J.E.: Spin diffusion measurements: spin echoes in the presence of a time-dependent field gradient. J. Chem. Phys. **42**(1), 288–292 (1965)

95. Tournier, J.D., Calamante, F., Connelly, A.: Improved probabilistic streamlines tractography by 2nd order integration over fibre orientation distributions. In: Proceedings of the International Society for Magnetic Resonance in Medicine, vol. 18, p. 1670. Ismrm (2010)

96. Tournier, J.D., Calamante, F., Gadian, D.G., Connelly, A.: Direct estimation of the fiber orientation density function from diffusion-weighted mri data using spherical deconvolution. NeuroImage **23**(3), 1176–1185 (2004)

97. Tournier, J.D., Calamante, F., King, M., Gadian, D., Connelly, A.: Limitations and requirements of diffusion tensor fiber tracking: an assessment using simulations. Magn. Reson. Med. Official J. Int. Soc. Magn. Reson. Med. **47**(4), 701–708 (2002)

98. Tuch, D.S.: Q-ball imaging. Magn. Reson. Med. Official J. Int. Soc. Magn. Reson. Med. **52**(6), 1358–1372 (2004)
99. Vilanova, A., Zhang, S., Kindlmann, G., Laidlaw, D.: An introduction to visualization of diffusion tensor imaging and its applications. In: Visualization and Processing of Tensor Fields, pp. 121–153 (2006)
100. Vorburger, R.S., Habeck, C.G., Narkhede, A., Guzman, V.A., Manly, J.J., Brickman, A.M.: Insight from uncertainty: bootstrap-derived diffusion metrics differentially predict memory function among older adults. Brain Struct. Funct. **221**(1), 507–514 (2016)
101. Vos, S.B., Jones, D.K., Viergever, M.A., Leemans, A.: Partial volume effect as a hidden covariate in dti analyses. Neuroimage **55**(4), 1566–1576 (2011)
102. Wakana, S., Jiang, H., Nagae-Poetscher, L.M., Van Zijl, P.C., Mori, S.: Fiber tract-based atlas of human white matter anatomy. Radiology **230**(1), 77–87 (2004)
103. Wedeen, V.J., Hagmann, P., Tseng, W.Y.I., Reese, T.G., Weisskoff, R.M.: Mapping complex tissue architecture with diffusion spectrum magnetic resonance imaging. Magn. Reson. Med. **54**(6), 1377–1386 (2005)
104. Westin, C.F.: Geometrical diffusion measures for mri from tensor basis analysis. In: Proceedings ISMRM 1997 (1997)
105. Whitaker, R.T., Mirzargar, M., Kirby, R.M.: Contour boxplots: a method for characterizing uncertainty in feature sets from simulation ensembles. IEEE Trans. Visual. Comput. Graph. **19**(12), 2713–2722 (2013)
106. Whitcher, B., Tuch, D.S., Wisco, J.J., Sorensen, A.G., Wang, L.: Using the wild bootstrap to quantify uncertainty in diffusion tensor imaging. Hum. Brain Mapp. **29**(3), 346–362 (2008)
107. Wiens, V., Schlaffke, L., Schmidt-Wilcke, T., Schultz, T.: Visualizing uncertainty in hardi tractography using superquadric streamtubes. In: EuroVis (Short Papers) (2014)
108. Wittenbrink, C.M., Pang, A.T., Lodha, S.K.: Glyphs for visualizing uncertainty in vector fields. IEEE Trans. Visual. Comput. Graph. **2**(3), 266–279 (1996)
109. Yang, F., Zhu, Y.M., Luo, J.H., Robini, M., Liu, J., Croisille, P.: A comparative study of different level interpolations for improving spatial resolution in diffusion tensor imaging. IEEE J. Biomed. Health Inf. **18**(4), 1317–1327 (2014)
110. Zhang, C., Caan, M.W., Höllt, T., Eisemann, E., Vilanova, A.: Overview+ detail visualization for ensembles of diffusion tensors. In: Computer Graphics Forum, vol. 36, pp. 121–132. Wiley Online Library (2017)
111. Zhang, C., Schultz, T., Lawonn, K., Eisemann, E., Vilanova, A.: Glyph-based comparative visualization for diffusion tensor fields. IEEE Trans. Visual. Comput. Graph. **22**(1), 797–806 (2015)
112. Zhang, S., Demiralp, C., Laidlaw, D.H.: Visualizing diffusion tensor mr images using streamtubes and streamsurfaces. IEEE Trans. Visual. Comput. Graph. **9**(4), 454–462 (2003)
113. Zhou, Z., Liu, W., Cui, J., Wang, X., Arias, D., Wen, Y., Bansal, R., Hao, X., Wang, Z., Peterson, B.S., et al.: Automated artifact detection and removal for improved tensor estimation in motion-corrupted dti data sets using the combination of local binary patterns and 2d partial least squares. Magn. Reson. Imaging **29**(2), 230–242 (2011)
114. Zockler, M., Stalling, D., Hege, H.C.: Interactive visualization of 3d-vector fields using illuminated stream lines. In: Proceedings of Seventh Annual IEEE Visualization 1996, pp. 107–113. IEEE (1996)
115. Zuk, T., Downton, J., Gray, D., Carpendale, S., Liang, J.: Exploration of uncertainty in bidirectional vector fields. In: Visualization and Data Analysis 2008, vol. 6809, p. 68090B. International Society for Optics and Photonics (2008)

5

Tensor Approximation for Multidimensional and Multivariate Data

Renato Pajarola, Susanne K. Suter, Rafael Ballester-Ripoll, and Haiyan Yang

Abstract Tensor decomposition methods and multilinear algebra are powerful tools to cope with challenges around multidimensional and multivariate data in computer graphics, image processing and data visualization, in particular with respect to compact representation and processing of increasingly large-scale data sets. Initially proposed as an extension of the concept of matrix rank for 3 and more dimensions, tensor decomposition methods have found applications in a remarkably wide range of disciplines. We briefly review the main concepts of tensor decompositions and their application to multidimensional visual data. Furthermore, we will include a first outlook on porting these techniques to multivariate data such as vector and tensor fields.

1 Introduction

Data approximation is widely used in the fields of computer graphics and scientific visualizations. One way to achieve it is to decompose the data into a more compact and compressed representation. The general idea is to express a dataset by a set of bases, which are used to reconstruct the dataset to its approximation when needed (see Fig. 1). More specifically, such a representation usually consists of the actual bases and the corresponding coefficients describing the relationship between the original data and the actual bases. Typically, such compact representations consist of less data than the original dataset, capture the most significant features, and, moreover, describe the data in a format that is convenient for adaptive data loading and access.

R. Pajarola (✉) · H. Yang
University of Zürich, Zürich, Switzerland
e-mail: pajarola@ifi.uzh.ch

S. K. Suter
Zurich University of Applied Sciences, Zürich, Switzerland
e-mail: sues@zhaw.ch

R. Ballester-Ripoll
IE University, Madrid, Spain
e-mail: rafael.ballester@ie.edu

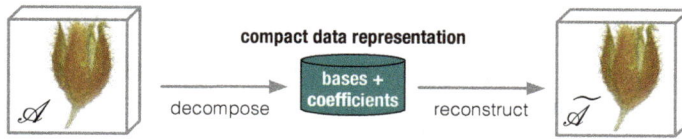

Fig. 1 Compact data representation for a 3rd-order tensor $\mathcal{A}$ (a volume) by bases and coefficients that can be used to reconstruct the data to its approximation $\widetilde{\mathcal{A}}$

Bases for compact data representation can be classified into two types: *pre-defined* and *learned* bases. Pre-defined bases comprise a given function or filter, which is applied to the dataset without any a-priori knowledge of the correlations in the dataset. In contrast, learned bases are generated from the dataset itself. Established examples of the former are the Fourier transform (FT) and the wavelet transform (WT). Well-known examples of the latter are the principal component analysis (PCA), the singular value decomposition (SVD) and vector quantization. Using predefined bases is often computationally cheaper, but learned bases potentially remove more redundancy from a dataset.

Generally, PCA-like methods are able to extract the main data direction of the dataset and represent the data in another coordinate system such that it makes it easier for the user to find the major contributions within the dataset. To exploit this, PCAs higher-order extension—*tensor approximation* (TA)—can be used for multidimensional datasets. The first occurrence of TA was in [21]. The idea of multi-way analysis, however, is generally attributed to Catell in 1944 [11]. It took a few decades until tensor approximations received more widespread attention, namely by several authors in the field of psychometrics [10, 19, 52].

1.1 Higher-Order Data Decompositions

The matrix SVD works on 2D data arrays and exploits the fact that the dataset can be represented with a few highly significant coefficients and corresponding recon-struction vectors based on the matrix rank reduction concept. The SVD, being a multilinear algebra tool, computes (a) a rank-R decomposition and (b) orthonormal row and column vector matrices. Unlike for 2D, the extension to higher-orders is not unique and these two main properties are captured by two different models that are both called tensor approximations : the Tucker model [14, 15, 25, 49, 52] preserves the orthonormal factor matrices while the CP model (from CANDECOMP [10] and PARAFAC [19]) preserves the rank-R decomposition.

Generally speaking, a tensor is a term for a higher-order generalization of a vector or a multidimensional array. In TA approaches, a multidimensional input dataset in array form, i.e., a tensor, is factorized into a sum of rank-one tensors or into a product of a core tensor (coefficients that describe the relationship to input data) and matrices (bases), i.e., one for each dimension. This factorization process is generally known as *tensor decomposition*, while the reverse process of the decomposition is the *tensor reconstruction*.

Tensor decompositions have been widely studied in other fields and were reviewed [13, 25, 34] and summarized [27, 44]. Since TA was emerging from various disciplines, it was developed under multiple names. The CP model was independently developed under the terms CANDECOMP and PARAFAC, therefore, it is sometimes referenced with a single name. The Tucker model takes its name from Tucker, who initially worked on the *three-mode factor analysis* (3MFA), which is sometimes referred to as the Tucker3 model, also called the *three mode PCA* (3MPCA) [27, 28, 49]. Similarly the model was referenced to as *n-mode PCA* [23] since it is equivalent to applying PCA *n* times, each time along a different mode of the tensor. In [14] all these previous works were captured and written down as generalization of the SVD as *multilinear singular SVD*, which is usually termed as higher-order SVD or HOSVD. Furthermore, it was also called *n-mode SVD* in [54, 55].

Given an input tensor, different decompositions capture different types of structures and result in varying numbers of coefficients. For a given accuracy, the number of CP ranks required to decompose a tensor is usually much larger to that of Tucker ranks. On the other hand, CP's storage cost grows only linearly with respect to its ranks, whereas that relationship becomes exponential in the case of Tucker. In sum, there is no silver bullet: CP is more suitable than Tucker for certain types of data, and vice versa. As a rule of thumb:

- Dense tensors of moderate dimensionality *n* over continuous variables, for example $n = 3$ or $n = 4$ including spatial and temporal axes, can often be compressed more compactly via the Tucker model.
- Tensors with categorical variables, sparse tensors, and tensors of higher dimensionality (say, $n \geq 5$) often benefit more from the CP model.

The data sets addressed in this chapter fit in the first category, and so we restrict our experiments to the Tucker model. To further illustrate the usual advantage of Tucker over CP for spatial visual data, see Fig. 2, we compare the number of coefficients and root mean squared error (RMSE) obtained with CP vs. Tucker using different numbers of ranks for a $256 \times 256 \times 256$ CT scan of a bonsai.

1.2   TA Applications in Graphics and Visualization

TA approaches have been applied to a wide range of application domains. Starting from psychometrics, in recent years, TA has been applied to visual data. A highly studied area is TA used for image ensembles [20, 35, 42, 43, 54, 58–60, 63] and/or TA used for pattern recognition, e.g., [17, 32, 40, 41, 43, 59]. In (real-time) rendering, tensor decompositions have been used as a method for global illumination models, e.g., for bidirectional reflectance distribution functions (BRDFs) [9, 45] or precomputed radiance transfer (PRT) [45, 50, 51]. Furthermore, TAs have successfully been used in graphics in the context of bidirectional texture functions (BTFs) [3,

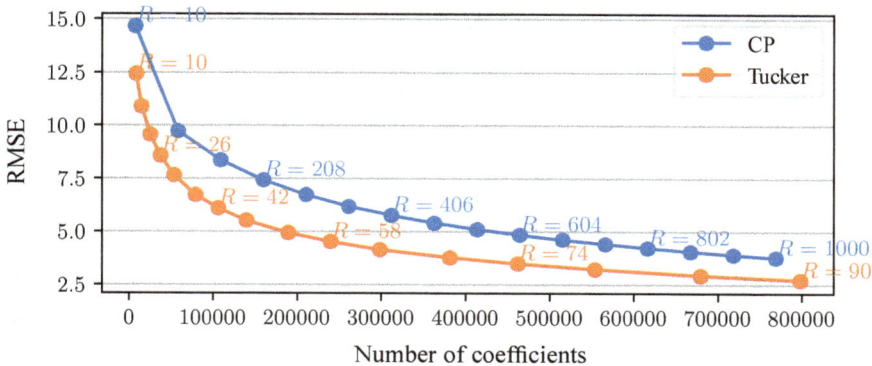

Fig. 2 RMSE achieved after CP vs. Tucker compression at different ranks R. For Tucker we take $R_1 = R_2 = R_3 = R$. The dataset used is a bonsai CT scan of size 256^3

18, 38, 39, 51, 55, 61, 62], texture synthesis [62], time-varying visual data [61, 62], 3D face scanning [56] and compression in animation [29, 31, 33, 36, 37, 53, 57].

In scientific visualization, TA methods have first been introduced for interactive multiresolution and multiscale direct volume rendering [6–8, 46–48]. Additionally, their compact representation power has been exploited for 3D volume data compression [4, 6] with notable advantages over other approaches at extreme compression ratios [2]. In this work, we explore the multiscale feature expressiveness of TA methods for the first time on vector fields, i.e. multidimensional multivariate data . Hence we interpret the vector field as a 4D array, or as a 4th-order data tensor. However, the rank-reduction of TA is only applied to the three spatial dimension.

1.3 Motivation and Contributions

In the results section (Sect. 9) we demonstrate that the feature sensitive approximation power of TA methods carries over from scalar to vector fields. These first results are promising and encourage the extension of tensor compression techniques to multivariate data fields, e.g. for compact storage and quick visualization at variable feature scales of large vector data in the fields of computational fluid dynamics or biomedicine. Moreover, based on the compression-domain data filtering and processing capabilities demonstrated in [5], we expect that important vector-field operators such as divergence or vorticity as well as other feature extraction operations can be analyzed and performed directly in the compressed TA format.

2   Singular Value Decomposition

The SVD is a widely used matrix factorization method to solve linear least-square problems. The SVD can be applied to any square or rectangular matrix $\mathbf{A} \in \mathbb{R}^{M \times N}$. Hence, the decomposition is always possible. The aim of the SVD is to produce a diagonalization of the input matrix $\mathbf{A}$. Since the input matrix $\mathbf{A}$ is not symmetric, two bases (matrices) are needed to diagonalize $\mathbf{A}$. Therefore, the SVD produces a matrix factorization into two orthogonal bases $\mathbf{U} \in \mathbb{R}^{M \times M}$ and $\mathbf{V} \in \mathbb{R}^{N \times N}$ and a diagonal matrix $\Sigma \in \mathbb{R}^{M \times N}$, as expressed in Eq. (1) (matrix form) or Eq. (2) (summation form).

$$\mathbf{A} = \mathbf{U \Sigma V^T} \tag{1}$$

$$a_{mn} = \sum_{r=1}^{R} u_{mr} \sigma_r v_{nr} \tag{2}$$

The bases $\mathbf{U}$ and $\mathbf{V}$ contain orthogonal unit length vectors $\mathbf{u}_j$ and $\mathbf{v}_j$, respectively, and represent a r-dimensional column space ($\mathbb{R}^M$) and a r-dimensional row space ($\mathbb{R}^N$). Hence, the bases $\mathbf{U}$ and $\mathbf{V}$ are even orthonormal. The diagonal matrix Σ contains the *singular values* σ_r, where $\sigma_1 \geq \sigma_2 \geq \cdots \geq \sigma_R \geq 0$, where $R = \min(M, N)$. A *singular value* and a pair of *singular vectors* of a square or rectangular matrix $\mathbf{A}$ correspond to a non-negative scalar σ and two non-zero vectors $\mathbf{u}_j$ and $\mathbf{v}_j$, respectively. The vectors $\mathbf{u}_j$ are the *left singular vectors*, and the vectors $\mathbf{v}_j$ are the *right singular vectors* (see Fig. 3). The number of non-zero singular values determines the rank R of the matrix $\mathbf{A}$.

In applications truncated versions of the SVD are frequently desired. That is, only the first K singular values $\sigma_1 \ldots \sigma_K$ and the corresponding K singular vectors $\mathbf{u}_1 \ldots \mathbf{u}_K$ and $\mathbf{v}_1 \ldots \mathbf{v}_K$ are used for the reconstruction. This approach is referred to as low-rank approximation of a truncated SVD. Basically, each weighted outer vector-product term $\sigma_j \cdot \mathbf{u}_j \circ \mathbf{v}_j$ corresponds to a rank-one component (see also Fig. 4), and the SVD of matrices or images consequently represents a 2D data array eventually as a sum of such rank-one components.

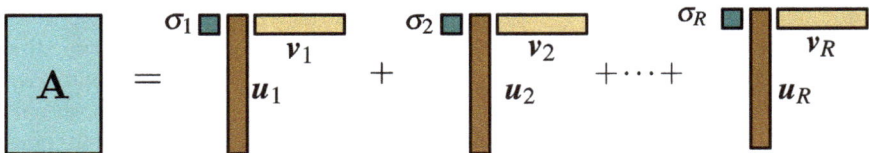

Fig. 3 Visualization of the summed form of the SVD as shown in Eq. (2)—illustrating the singular values with the corresponding left and right singular vector pairs

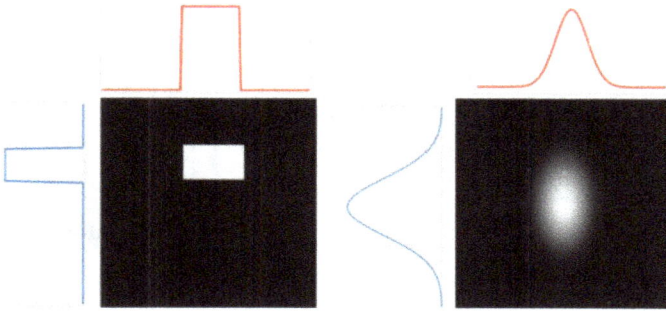

Fig. 4 Simple 2D functions can be represented as an outer product of two 1D functions, and more complex ones as a weighted sum of several such components

3 Tensor Approximation Notation and Definitions

The notation taken here is mostly taken from De Lathauwer et al. [13, 14], Smilde et al. [44], as well as Kolda and Bader [25], who follow the notation proposed by Kiers [24]. To illustrate higher-order extensions we mostly make examples of order three.

3.1 General Notation

A tensor is a multidimensional array (or an N-way data array): a 0th-order tensor (tensor0) is a scalar, a 1st-order tensor (tensor1) is a vector, a 2nd-order tensor is a matrix, and a 3rd-order (tensor3) is a volume, see Fig. 5. We consistently use the letter 'a' to represent the data, following the notation of, e.g., [14, 15, 51, 61, 62]. We use lower case letters for a scalar a, lower case boldface letters for a vector $\mathbf{a}$ in $\mathbb{R}^{I_1}$, capital boldface letters for a matrix $\mathbf{A}$ in $\mathbb{R}^{I_1 \times I_2}$, and calligraphic letters for a 3rd-order tensor $\mathcal{A}$ in $\mathbb{R}^{I_1 \times I_2 \times I_3}$.

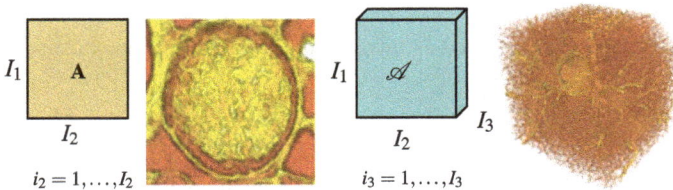

Fig. 5 A tensor is a multidimensional array: a 2nd-order tensor is a matrix $\mathbf{A}$ and a 3rd-order tensor is a volume $\mathcal{A}$

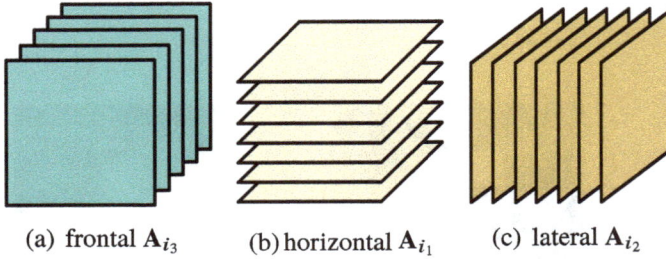

(a) frontal $\mathbf{A}_{i_3}$ (b) horizontal $\mathbf{A}_{i_1}$ (c) lateral $\mathbf{A}_{i_2}$

Fig. 6 Slices of a 3rd-order tensor $\mathcal{A}$

The *order* of a tensor is the number of data directions, also referred as *ways* or *modes*. Along a mode n, the index i_n runs from 1 to I_N. By using lower script indices for the modes, we can extend the index scheme to any order, i.e., $I_1, I_2, I_3, I_4, \ldots$. The ith entry of a vector $\mathbf{a}$ is denoted by a_i, an element (i_1, i_2) of a matrix $\mathbf{A}$ is denoted by $a_{i_1 i_2}$, and an element (i_1, i_2, i_3) of a 3rd-order tensor $\mathcal{A}$ is denoted by $a_{i_1 i_2 i_3}$.

The general term *fibers* is used as a generalization for vectors taken along different modes in a tensor. A fiber is defined by fixing every index but one. A matrix column is thus a mode-1 fiber and a matrix row is a mode-2 fiber. 3rd-order tensors have column, row, and *tube* fibers, denoted by $\mathbf{a}_{i_1}$, $\mathbf{a}_{i_2}$, and $\mathbf{a}_{i_3}$, respectively. Sometimes, fibers are also called mode-n vectors.

Slices are two-dimensional sub-sections of a tensor (e.g., one fixed index in a 3rd-order tensor). For a 3rd-order tensor $\mathcal{A}$, there are, for example, *frontal*, *horizontal*, and *lateral* slices, denoted by $\mathbf{A}_{i_3}$, $\mathbf{A}_{i_1}$, and $\mathbf{A}_{i_2}$, respectively as illustrated in Fig. 6.

For computations, a tensor is often reorganized into a matrix what we denote as *tensor unfolding* (sometimes called *matricization*). There are two main unfolding strategies, *backward cyclic unfolding* [14] and *forward cyclic unfolding* [24] as shown in Fig. 7. An unfolded tensor in matrix shape is denoted with a subscript in parentheses, e.g., $\mathbf{A}_{(n)}$.

3.2 Computing with Tensors

While many more operations on tensors exist, here we only outline the most common products used within the scope of this work.

- An Nth-order tensor is defined as $\mathcal{A} \in \mathbb{R}^{I_1 \times I_2 \times \cdots \times I_N}$.
- The *tensor product* is denoted here by $\otimes$: however, other symbols are used in the literature, too. For rank-one tensors, the tensor product corresponds to the *vector outer product* ($\circ$) of N vectors $\mathbf{b}^{(n)} \in \mathbb{R}^{I_n}$ and results in an Nth-order tensor $\mathcal{A}$. The tensor product or vector outer product for a 3rd-order rank-one tensor is illustrated

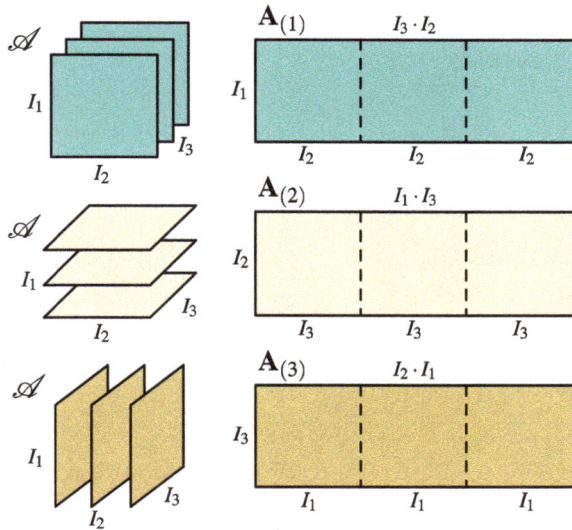

Fig. 7 Forward cyclic unfolding [24] of a 3rd-order tensor

in Fig. 8: $\mathcal{A} = \mathbf{b}^{(1)} \circ \mathbf{b}^{(2)} \circ \mathbf{b}^{(3)}$, where an element (i_1, i_2, i_3) of $\mathcal{A}$ is given by $a_{i_1 i_2 i_3} = b_{i_1}^{(1)} b_{i_1}^{(2)} b_{i_3}^{(3)}$.

- The *inner product* of two same-sized tensors $\mathcal{A}, \mathcal{B} \in \mathbb{R}^{I_1 \times I_2 \times \cdots \times I_N}$ is the sum of the products of their entries, i.e., Eq. (3).

$$(\mathcal{A}, \mathcal{B}) = \sum_{i_1=1}^{I_1} \sum_{i_2=1}^{I_2} \cdots \sum_{i_N=1}^{I_N} a_{i_1 i_2 \ldots i_N} b_{i_1 i_2 \ldots i_N} \tag{3}$$

- The *n-mode product* [14] multiplies a tensor by a matrix (or vector) in mode n. The n-mode product of a tensor $\mathcal{B} \in \mathbb{R}^{I_1 \times I_2 \times \cdots \times I_N}$ with a matrix $\mathbf{C} \in \mathbb{R}^{J_n \times I_n}$ is denoted by $\mathcal{B} \times_n \mathbf{C}$ and is of size $I_1 \times \cdots \times I_{n-1} \times J_n \times I_{n+1} \times \cdots \times I_N$. That is, element-wise we have Eq. (4).

$$(\mathcal{B} \times_n \mathbf{C})_{i_1 \ldots i_{n-1} j_n i_{n+1} \ldots i_N} = \sum_{i_n=1}^{I_n} b_{i_1 i_2 \ldots i_N} \cdot c_{j_n i_n} \tag{4}$$

Each mode-n fiber is multiplied by the matrix $\mathbf{C}$. The idea can also be expressed in terms of unfolded tensors (reorganization of a tensor into a matrix as described in Sect. 3.1).

The n-mode product of a tensor with a matrix is related to a change of basis in the case when a tensor defines a multilinear operator [25]. The n-mode product is the generalized operand to compute *tensor times matrix* (TTM) multiplications, and can best be illustrated using unfolded tensors as in Fig. 9.

Fig. 8 Three-way outer
product for a 3rd-order
rank-one tensor
$\mathcal{A} = \mathbf{b}^{(1)} \circ \mathbf{b}^{(2)} \circ \mathbf{b}^{(3)}$

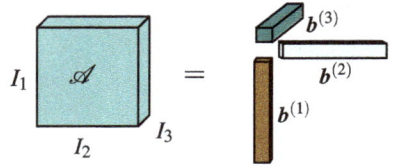

Fig. 9 TTM multiplication
$\mathbf{C} \cdot \mathbf{B}_{(n)}$, multiplying the
unfolded 3rd-order tensor
$\mathbf{B}_{(n)}$ with the matrix $\mathbf{C}$

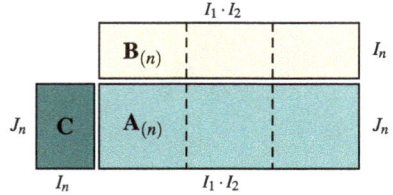

$$\mathcal{A} = \mathcal{B} \times_n \mathbf{C} \Leftrightarrow \mathbf{A}_{(n)} = \mathbf{C}\mathbf{B}_{(n)} \tag{5}$$

- The *norm of a tensor* $\mathcal{A} \in \mathbb{R}^{I_1 \times I_2 \times \cdots \times I_N}$ is defined analogously to the matrix Frobenius norm $\|\mathbf{A}\|_F$ and is the square root of the sum squares of all its elements, i.e., Eq. (6).

$$\|\mathcal{A}\|_F = \sqrt{\sum_{i_1=1}^{I_1} \sum_{i_2=1}^{I_2} \cdots \sum_{i_N=1}^{I_N} a_{i_1 i_2 \ldots i_N}^2} \tag{6}$$

3.3 Rank of a Tensor

In order to describe the definitions of the tensor rank, the definition for the matrix rank is recaptured. The *matrix rank* of a matrix $\mathbf{A}$ is defined over its column and row ranks, i.e., the column and row matrix rank of a matrix $\mathbf{A}$ is the maximal number of linearly independent columns and rows of $\mathbf{A}$ that can be chosen, respectively. For matrices, the column rank and the row rank are always equal and, a matrix rank is therefore simply denoted as rank$(\mathbf{A})$. A *tensor rank* is defined similarly to the matrix rank, however, there are differences. In fact, the extension of the rank concept is not uniquely defined in higher orders and we review the definitions for the tensor ranks from [14] here.

- The *n-rank* of a tensor $\mathcal{A}$, denoted by $R_n = \text{rank}_n(\mathcal{A})$, is the dimension of the vector space spanned by mode-n vectors, where the mode-n vectors of $\mathcal{A}$ are the column vectors of the unfolding $\mathbf{A}_{(n)}$, and $\text{rank}_n(\mathcal{A}) = \text{rank}(\mathbf{A}_{(n)})$. Unlike matrices, the different n-ranks of a tensor are not necessarily the same.

- A higher-order tensor has a so called *multilinear rank* $(R_1, R_2, \ldots, R_N)$ [21] if its mode-1 rank (row vectors), mode-2 rank (column vectors) until its mode-n rank are equal to $R_1, R_2, \ldots, R_N$, e.g., giving rise to a multilinear rank-(R_1, R_2, R_3) for a 3rd-order tensor.
- A *rank-one tensor* is an N-way tensor $\mathcal{A} \in \mathbb{R}^{I_1 \times I_2 \times \cdots \times I_N}$ under the condition that it can be expressed as the outer product of N vectors, as in Eq. (7) (see also [12, 30]). A rank-one tensor is also known under the term *Kruskal tensor*.

$$\mathcal{A} = \mathbf{b}^{(1)} \circ \mathbf{b}^{(2)} \circ \cdots \circ \mathbf{b}^{(N)} \tag{7}$$

- The *tensor rank* $R = \text{rank}(\mathcal{A})$ is the minimal number of rank-one tensors that yield $\mathcal{A}$ in a linear combination (see [12, 14, 25, 30]). Except for the special case of matrices, the tensor rank is not necessarily equal to any of its n-ranks, but it always holds that $R_n \leq R$.

4 Tensor Decompositions

In tensor decompositions an input tensor $\mathcal{A} \in \mathbb{R}^{I_1 \times I_2 \times \cdots \times I_N}$ is decomposed into a set of factor matrices $\mathbf{U}^{(n)} \in \mathbb{R}^{I_n \times R_n}$ and coefficients $\mathcal{B} \in \mathbb{R}^{R_1 \times R_2 \times \cdots \times R_N}$ that describe the relationship/interactivity between $\mathcal{A}$ and the set of $\mathbf{U}^{(n)}$.

Historically, as seen earlier, tensor decompositions are a higher-order extension of the matrix SVD. The nice properties of the matrix SVD, i.e., rank-R decomposition and orthonormal row-space vectors and column-space vectors, do not extend uniquely to higher orders. The rank-R decomposition can be achieved with the so-called CP model, while the orthonormal row and column vectors are preserved in the so-called Tucker model. An extensive review of the two models and further hybrid models can be found in [25]. Here, we only outline the Tucker model that we apply in our experiments.

4.1 Tucker Model

The Tucker model is a widely used approach for tensor decompositions. As given in Eq. (8), any higher-order tensor is approximated by a product of a core tensor $\mathcal{B} \in \mathbb{R}^{R_1 \times R_2 \times \cdots \times R_N}$ and its factor matrices $\mathbf{U}^{(n)} \in \mathbb{R}^{I_n \times R_n}$, where the products $\times_n$ denote the n-mode product as outlined in Sect. 3.2. This decomposition can then again be reconstructed to its approximation $\widetilde{\mathcal{A}}$. The missing information of the input tensor $\mathcal{A}$ that cannot be captured by $\widetilde{\mathcal{A}}$ is denoted with the error e. The Tucker decomposition is visualized for a 3rd-order tensor in Fig. 10. Equivalently, a Tucker decomposition can also be represented as a sum of rank-one tensors as in Eq. (9) and illustrated in Fig. 11.

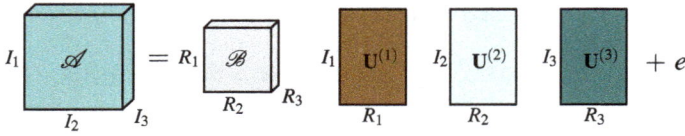

Fig. 10 Tucker 3rd-order tensor: $\mathcal{A} = \mathcal{B} \times_1 \mathbf{U}^{(1)} \times_2 \mathbf{U}^{(2)} \times_3 \mathbf{U}^{(3)} + e$

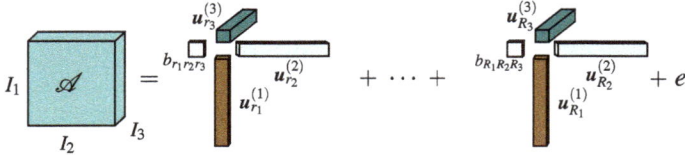

Fig. 11 Tucker 3rd-order tensor as a sum of rank-one tensors: $\mathcal{A} = \sum_{r_1=1}^{R_1} \sum_{r_2=1}^{R_2} \sum_{r_3=1}^{R_3} b_{r_1 r_2 r_3} \cdot \mathbf{u}_{r_1}^{(1)} \circ \mathbf{u}_{r_2}^{(2)} \circ \mathbf{u}_{r_3}^{(3)} + e$

$$\mathcal{A} = \mathcal{B} \times_1 \mathbf{U}^{(1)} \times_2 \mathbf{U}^{(2)} \times_3 \cdots \times_N \mathbf{U}^{(N)} + e \tag{8}$$

$$\mathcal{A} = \sum_{r_1=1}^{R_1} \sum_{r_2=1}^{R_2} \cdots \sum_{r_N=1}^{R_N} b_{r_1 r_2 \ldots r_N} \cdot \mathbf{u}_{r_1}^{(1)} \circ \mathbf{u}_{r_2}^{(2)} \circ \cdots \circ \mathbf{u}_{r_N}^{(N)} + e \tag{9}$$

The column vectors $\mathbf{u}_{r_n}^{(n)}$ of the factor matrices $\mathbf{U}^{(n)} \in \mathbb{R}^{I_n \times R_n}$ are usually orthonormal and can be thought of as principal components R_n in each mode n [25]. The core tensor $\mathcal{B} \in \mathbb{R}^{R_1 \times R_2 \times \cdots \times R_N}$ represents a projection of the original data $\mathcal{A} \in \mathbb{R}^{I_1 \times I_2 \times \cdots \times I_N}$ onto its factor matrices and is always of the same order as the input data. The core tensor is computed in general, as shown in Eq. (10), and for orthogonal factor matrices as in Eq. (11). The element-wise core tensor computation is denoted in Eq. (12). In other words, the core tensor coefficients $b_{r_1 r_2 \ldots r_N}$ represent the relationship between the Tucker model and the original data.

$$\mathcal{B} = \mathcal{A} \times_1 \mathbf{U}^{(1)(-1)} \times_2 \mathbf{U}^{(2)(-1)} \times_3 \cdots \times_N \mathbf{U}^{(N)(-1)} \tag{10}$$

$$\mathcal{B} = \mathcal{A} \times_1 \mathbf{U}^{(1)\top} \times_2 \mathbf{U}^{(2)\top} \times_3 \cdots \times_N \mathbf{U}^{(N)\top} \tag{11}$$

$$\mathcal{B} = \sum_{i_1=1}^{I_1} \sum_{i_2=1}^{I_2} \cdots \sum_{i_N=1}^{I_N} a_{i_1 i_2 \ldots i_N} \cdot \mathbf{u}_{i_1}^{(1)\top} \circ \mathbf{u}_{i_2}^{(2)\top} \circ \cdots \circ \mathbf{u}_{i_N}^{(N)\top} \tag{12}$$

The Tucker decomposition is not unique, which means that we can modify the core tensor $\mathcal{B}$ without affecting the model fit as long as we apply the same changes to the factor matrices (so-called core tensor rotations), for more details see [25].

Often, we are interested in compact models, which enable a compression of the input data. For example, after computing a (full) Tucker decomposition the core tensor $\mathcal{B}$ has the same size as the original input $\mathcal{A}$ and all the factor matrices are square. However, we are more interested in reduced-size, approximative Tucker decompositions, where $\mathcal{B}$ is an element of $\mathbb{R}^{R_1 \times R_2 \times R_3}$ with $R_1 < I_1$, $R_2 < I_2$ and $R_3 < I_3$. Using so-called *rank-reduced tensor decompositions* or *truncated tensor decompositions* one can directly obtain more compact decompositions.

5 Tensor Rank Reduction

As seen in Sect. 3.3, the extension of the matrix rank concept to higher orders is not unique and we will mostly follow the rank-$(R_1, R_2, \ldots, R_N)$ tensor decomposition and reduced-rank approximation of the Tucker model here.

5.1 Rank-R and Rank-$(R_1, R_2, \ldots, R_N)$ Approximations

A simple rank-one approximation is defined as $\widetilde{\mathcal{A}} = \lambda \cdot \mathbf{u}^{(1)} \circ \mathbf{u}^{(2)} \cdots \circ \mathbf{u}^{(N)}$ from the rank-one tensor (vector) product ($\circ$) of its N basis vectors $\mathbf{u}^{(n)} \in \mathbb{R}^{I_n}$ and a weight factor λ. Hence a tensor $\mathcal{A}$ could be approximated by a linear combination of many rank-one approximations as in Eq. (13). This approximation, also known as a CP model, is called a *rank-R approximation*.

$$\widetilde{\mathcal{A}} = \sum_{r=1}^{R} \lambda_r \cdot \mathbf{u}_r^{(1)} \circ \mathbf{u}_r^{(2)} \circ \cdots \circ \mathbf{u}_r^{(N)} \tag{13}$$

Alternatively, if we allow all weighted tensor (vector) products $\mathbf{u}_{i_1}^{(1)} \circ \mathbf{u}_{i_2}^{(2)} \circ \cdots \circ \mathbf{u}_{i_N}^{(N)}$ of any arbitrary index combinations $i_1 i_2 \ldots i_N$, we end up with the Tucker model of Sect. 4.1 and Eq. 12 where the weight factors for all index combinations form the core tensor $\mathcal{B}$. Choosing $R_{1,\ldots,N} < I_{1,\ldots,N}$ we end up with a *rank-$(R_1, R_2, \ldots, R_N)$ approximation* of $\mathcal{A}$, which is given by a decomposition into a lower-rank tensor $\widetilde{\mathcal{A}} \in \mathbb{R}^{I_1 \times I_2 \times \cdots \times I_N}$ with $rank_n(\widetilde{\mathcal{A}}) = R_n \leq rank_n(\mathcal{A})$. The approximated tensor $\widetilde{\mathcal{A}}$ is the n-mode product $\times_n$ of factor matrices $\mathbf{U}^{(n)} \in \mathbb{R}^{I_n \times R_n}$ and a core tensor $\mathcal{B} \in \mathbb{R}^{R_1 \times R_2 \times \cdots \times R_N}$ in a given reduced rank space (Eq. (14)). This rank-$(R_1, R_2, \ldots, R_N)$ approximation was previously introduced as the *Tucker* model.

$$\widetilde{\mathcal{A}} = \mathcal{B} \times_1 \mathbf{U}^{(1)} \times_2 \mathbf{U}^{(2)} \times_3 \cdots \times_N \mathbf{U}^{(N)} \tag{14}$$

In general, a rank-reduced approximation is sought such that the least-squares cost function of Eq. (15) is minimized.

$$\widetilde{\mathcal{A}} = \arg \min_{\widetilde{\mathcal{A}}} \|\mathcal{A} - \widetilde{\mathcal{A}}\|^2 \tag{15}$$

Given that $(R_1, R_2, \ldots, R_N)$ are sufficiently smaller than the initial dimensions $(I_1, I_2, \ldots, I_N)$, the core coefficients $\mathcal{B} \in \mathbb{R}^{R_1 \times R_2 \times \cdots \times R_N}$ and the factor matrices $\mathbf{U}^{(n)} \in \mathbb{R}^{I_n \times R_n}$ can lead to a compact approximation of $\widetilde{\mathcal{A}}$ of the original tensor $\mathcal{A}$. In particular, the multilinear rank-$(R_1, R_2, \ldots, R_N)$ is typically explicitly chosen to be smaller than the initial ranks in order to achieve a compression of the input data (see also [2, 4, 6]).

5.2 Truncated Tensor Decomposition

Similar to matrix SVD, tensor rank reduction can be used to generate lower-rank reconstructions $\widetilde{\mathcal{A}}$ of the input $\mathcal{A}$. The tensor rank parameters R_n are chiefly responsible for the number of TA bases and coefficients that are used for the reconstruction and hence are responsible for the approximation level. In higher orders, the CP decomposition is not directly rank-reducible, however, the truncation of the Tucker decomposition is possible due to the *all-orthogonality* property of the core tensor.

For a 3rd-order tensor, all-orthogonality means that the different horizontal matrix slices of the core $\mathcal{B}$ (the first index i_1 is kept fixed, while the two other indices, i_2 and i_3, are free) are mutually orthogonal with respect to the scalar product of matrices (i.e., the sum of the products of the corresponding entries vanishes). The same holds for the other slices with fixed indices i_2 and i_3 (see [14]). Therefore, given an initial sufficiently accurate rank-(R_1, R_2, R_3) Tucker model, we can progressively choose lower ranks $K_n < R_n$ for reduced quality reconstructions. As indicated in Fig. 12, the ranks K_n indicate how many factor matrix columns and corresponding core tensor entries are used for the reconstruction.

Note that the ordering of the coefficients in the Tucker core tensor $\mathcal{B}$ is not strictly decreasing in contrast to the decreasing singular values in the matrix SVD case. However, in practice it can be shown that progressive tensor rank reduction in the Tucker model works very well for adaptive reconstruction of the data at different accuracy levels.

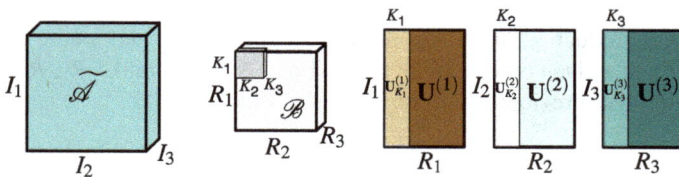

Fig. 12 Illustration of a rank reduced Tucker tensor reconstruction: A reduced range of factor matrix columns with corresponding fewer core tensor entries reconstructs a lower quality approximation but at full resolution

6   Tucker Decomposition Algorithms

There are various strategies for how to compute and generate a tensor decomposition. The most popular and widely used group of algorithms belongs to the *alternating least squares* (ALS) algorithms, the other group of algorithms uses various Newton methods. The respective algorithms differ for the computation of the different tensor models, and we will mainly focus on the Tucker model in our review.

For the Tucker model, the first decomposition algorithms used were a simple higher-order SVD (HOSVD) (see [14]), the so-called *Tucker1* [52], a three-mode SVD. However, the truncated decompositions of higher orders are not optimal in terms of best fit, which is measured by the Frobenius norm of the difference. Starting from a HOSVD algorithm, tensor approximation ALS algorithms [26, 28] were developed, where one of the first Tucker ALS was the so-called *TUCKALS* [49]. Later various improvements accelerated [1] or optimized the basic TUCKALS method. The *higher-order orthogonal iteration* (HOOI) algorithm [15] is an iterative algorithm that performs a better fit for a truncated HOSVD version.

Newton methods are also used for the Tucker decomposition or rank-$(R_1, R_2, \ldots, R_N)$ approximation. They typically start with a HOOI initialization and then converge faster to the final point. [16] developed a *Newton-Grassman optimization* approach, which takes much fewer iterations than the basic HOOI - even though one single iteration is more expensive due to the computation of the Hessian. While the HOOI is not guaranteed to converge, the Newton-Grassmann Tucker decomposition is guaranteed to converge to a stationary point. Another Newton method was proposed by [22], who developed a *differential-geometric Newton* algorithm with a fast quadratic convergence of the algorithm in a neighborhood of the solution. Since this method is not guaranteed to converge to a global maximum, they support the method by starting with an initial guess of several HOOI iterations, which increases the chances of converging to a solution.

7   Tensor Reconstruction

The *tensor reconstruction* from a reduced-rank tensor decomposition can be achieved in multiple ways. One alternative is a progressive reconstruction: Each entry in the core tensor $\mathcal{B}$ is considered as weight for the outer product between the corresponding column vectors in the factor matrices $\mathbf{U}^{(n)}$. This gives rise to Eq. (16) for the Tucker reconstruction.

$$\widetilde{\mathcal{A}} = \sum_{r_1=1}^{R_1} \sum_{r_2=1}^{R_2} \cdots \sum_{r_N=1}^{R_N} b_{r_1 r_2 \ldots r_N} \cdot \mathbf{u}_{r_1}^{(1)} \circ \mathbf{u}_{r_2}^{(2)} \circ \cdots \circ \mathbf{u}_{r_N}^{(N)} \tag{16}$$

This reconstruction strategy corresponds to forming rank-one tensors and cumulatively summing them up. The accumulated weighted *subtensors* then form the approximation $\widetilde{\mathcal{A}}$ of the original data $\mathcal{A}$. In particular for the Tucker model, this is an expensive reconstruction strategy since it involves multiple for-loops to run over all the summations, for a total cost of $O(R^N \cdot I^N)$ operations.

7.1 Element-Wise Reconstruction

A simple approach is to reconstruct each required element of the approximated dataset individually, which we call *element-wise reconstruction*. Each element $\widetilde{a}_{i_1 i_2 i_3}$ is reconstructed, as shown in Eq. (17) for the Tucker reconstruction. That is, all core coefficients multiplied with the corresponding coefficients in the factor matrices are summed up (weighted sum).

$$\widetilde{a}_{i_1 i_2 \ldots i_N} = \sum_{r_1 r_2 \ldots r_N} b_{r_1 r_2 \ldots r_N} \cdot u_{i_1 r_1}^{(1)} \cdot u_{i_2 r_2}^{(2)} \cdot \ldots \cdot u_{i_N r_N}^{(N)} \tag{17}$$

Element-wise reconstruction requires $O(R^N)$ operations on average. It can be beneficial for applications where only a sparse set of reconstructed elements are needed.

7.2 Optimized Tucker Reconstruction

A third reconstruction approach—applying only to the Tucker reconstruction—is to build the n-mode products along every mode, which leads to a TTM multiplication for each mode, e.g., TTM1 along mode 1, (see also Eq. (5)). This is analogous to the Tucker model given by Eq. (14). The intermediate results are then multiplied along the next modes, e.g., TTM2 and finally TTM3. In Fig. 13 we visualize the TTM reconstruction, and the intermediate results $\mathcal{B}'$ and $\mathcal{B}''$ as well as the final approximation $\widetilde{\mathcal{A}}$, applied to a 3rd-order tensor using n-mode products.

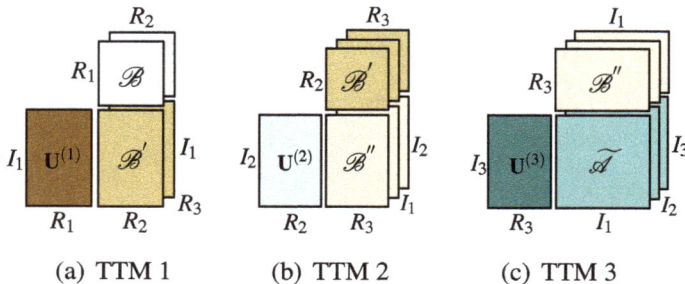

(a) TTM 1   (b) TTM 2   (c) TTM 3

Fig. 13 Forward cyclic TTM multiplications after [24] along the three modes (n-mode products)

Since it exploits matrix multiplications, this optimized algorithm is much faster than progressive reconstruction (Eq. (16)). Its cost is dominated by $O(R_1 \cdot I^N)$. In 3D, in particular, it takes $O(R_1 \cdot I_1 \cdot I_2 \cdot I_3 + R_1 \cdot R_2 \cdot I_2 \cdot I_3 + R_1 \cdot R_2 \cdot R_3 \cdot I_3)$ operations.

8 Useful TA Properties for Scientific Visualization

As stated in the introduction, TA is the higher-order generalization of the matrix SVD, which can offer either properties of (a) rank-R decomposition or (b) orthonormal row-space and column-space vectors. In higher orders, the orthonormal row and column vectors are preserved in the Tucker model which thus supports progressive rank-reduced approximations and reconstructions.

8.1 Spatial Selectivity and Subsampling

The Tucker model (Sect. 4.1) consists of one factor matrix per mode (data direction) $\mathbf{U}^{(\mathbf{n})} \in \mathbb{R}^{I_n \times R_n}$ and one core tensor $\mathcal{B} \in \mathbb{R}^{R_1 \times R_2 \times \cdots \times R_N}$. The core tensor $\mathcal{B}$ is in effect a projection of the original data $\mathcal{A}$ onto the basis of the factor matrices $\mathbf{U}^{(\mathbf{n})}$. In case of a volume, the Tucker model has three modes, as illustrated in Fig. 10, and defines an approximation $\widetilde{\mathcal{A}} = \mathcal{B} \times_1 \mathbf{U}^{(1)} \times_2 \mathbf{U}^{(2)} \times_3 \mathbf{U}^{(3)}$ of the original volume $\mathcal{A}$ (using n-mode products $\times_n$).

The row and column axes of the factor matrices represent two different spaces: (1) the rows correspond to the spatial dimension in the corresponding mode, and (2) the columns to the approximation quality. These two properties can be exploited for multiresolution modeling (spatial selection and subsampling of rows) and multiscale approximation (rank reduction on the columns) (see also Fig. 14). In [5, 47] we demonstrated how these features can be exploited for multiresolution and multiscale reconstruction as well as filtering of compressed volume data in the context of interactive visualization.

Fig. 14 Factor matrix properties along the vertical axis supporting: (1.1) spatial selectivity, (1.2) spatial subsampling, and (2) low-rank approximation

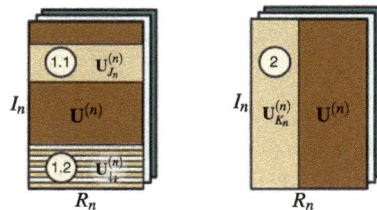

8.2 Approximation and Rank Reduction

As described earlier, the Tucker model defines a rank-(R_1, R_2, R_3) approximation, where a small R_n corresponds to a low-rank approximation (with details removed) and a large R_n corresponds to a more accurate approximation of the original. The highest rank R_n for the initial Tucker decomposition has to be given explicitly. However, rank reductions can be applied after the initial decomposition (similar to the rank reduction in matrix SVD). Even though the core tensor coefficients are not guaranteed to be in decreasing order, as in matrix SVD, in practice it can be shown that progressive tensor rank reduction in the Tucker model works well for adaptive visualization of the data at different feature scales [2, 4–6, 47].

As illustrated in Fig. 12 in Sect. 5.2, the ranks indicate how many factor matrix columns and corresponding core tensor entries are used for a desired reconstruction. Thus, given a rank-(R_1, R_2, R_3) Tucker model, we can specifically or progressively choose lower ranks $K_n < R_n$ for a reduced quality reconstruction, at the original spatial output resolution given by I_n (or also subsampled).

Figure 15 shows the progressive rank reduction from an initial rank-$(256, 256, 256)$ Tucker decomposition of an original 512^3 example volume. Shown are the visual results and the data reduction of the approximation at variable reduced-rank reconstructions. The numbers of coefficients include all core tensor and factor matrix entries that are used, e.g. a rank-$(32, 32, 32)$ reconstruction corresponds to $32^3 + 3 \cdot 512 \cdot 32 = 81'920$ coefficients. The data reduction ratio can be derived by dividing the number of coefficients by 512^3, which for $R = 32$ results in using only 0.06% of the original amount of data. In particular, Fig. 15 demonstrates the power of low-rank tensor approximations that can be used for multiscale feature visualization or progressive image refinement in volume rendering.

Fig. 15 Multiscale volume visualization by tensor rank reduction, corresponding to 0.02, 0.06, 0.27, 1.71 and 12.79% of the original amount of data used for $R = 16, 32, 64, 128, 256$

9 Application to Multivariate Data

Encouraged by the data reduction power and the approximation quality of tensor approximations as shown in Fig. 16, we have extended the Tucker decomposition and reduced rank reconstruction to vector field data. Compared to scalar 3D volumes (e.g. from MRI or CT scans), a 3D vector field $\mathcal{V}$ (e.g. of velocities from a weather or fluid simulation) can be interpreted as a multivariate data field with D-dimensional vector-valued entries at each position in a $I_1 \times I_2 \times I_3$ grid. Thus we can interpret $\mathcal{V}$ as a 4th-order tensor $\mathcal{V} \in \mathbb{R}^{I_1 \times I_2 \times I_3 \times D}$. Note that the TA rank-reduction (Sect. 5) is only applied to the three spatial dimension as we are not interested in a dimensionality reduction of the vector-valued field value itself.

9.1 Dataset

For our first experiments we used a data set from the *Johns Hopkins Turbulence Database* http://turbulence.pha.jhu.edu/ representing a direct numerical simulation of incompressible magnetohydrodynamic (MHD) equations (see also Fig. 17). The data set contains, among other output variables, 3 velocity components that we used, hence $D = 3$, covers a 3D grid of 256^3 cells (downsampled subset from the original 1024^3), hence $I_1 = I_2 = I_3 = 256$, and is the first time step from the output of the simulation.

The vector field therefore covers a 3D cube which we visualize using direct volume and streamline rendering techniques with color-coding of velocity, divergence, vorticity or error magnitudes in ParaView. Transparency is used to reduce clutter and opacity such as to focus the visualizations on the high-magnitude value regions. Note that this cubic vector field is dense and rendered over a black background, thus always appearing as a cube like object in the images.

(a) Original   (b) 10:1   (c) 300:1

Fig. 16 **a** A 512^3 isotropic turbulence volume of 512MB, **b** visually identical compression result using 51.2MB and **c** result after extreme compression down to only 1.71MB using the TTHRESH method [2]. All the visualizations are generated using ParaView

9.2 Vector Field Magnitude and Angle

As can be seen in Fig. 17, the velocity magnitude and structure of the flow directions is very well maintained down to fairly course reconstructions using $R = 32$. Especially the (important) regions with larger velocity magnitudes are very well preserved with respect to their flow orientation as visible from the streamline visualization. The rendering applies an opacity transfer function setting which is almost linear, slight curve below the diagonal, to the *vector field magnitude*, highlighting the important high-velocity regions.

To see how many details are lost after rank reduction of the velocity field data set, we calculated and visualized the error information for both magnitude and angle deviation in Figs. 18 and 19, respectively. For the magnitude, we normalized the error to be the percentage relative to the local vector field magnitude value. The rendering uses an opacity mapping which is almost linear, slight curve below the diagonal, to the *vector field magnitude*. This makes low magnitude and low opacity areas, and their errors, transparent or dark and highlights any errors in the more critical high-velocity regions. The completely dark images in Fig. 18e–h demonstrate that the magnitude error after rank reduction using Tucker decomposition is close to zero even with for $R = 16$, which is $1/16$ of the original full rank of 256. The barely noticeable dark blue regions in Fig. 18a–d correspond to low errors in high-velocity regions, with errors in the range of less than 10% down to 10^{-8}% (with 60% being white).

Compared to the almost negligible relative magnitude errors of the velocity field, however, the angular error of the same data set after rank reduction is more prominent as shown in Fig. 19. The maximum angular error $\pi = 180°$ is shown in red while small errors are shown in dark blue (0°). The rendering applies an opacity setting almost linear to the *angle offset*, hence in Fig. 19 large angular errors are highlighted.

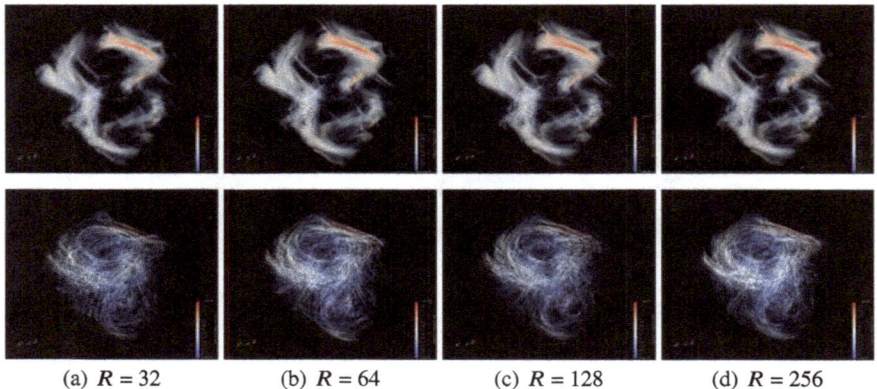

| (a) $R = 32$ | (b) $R = 64$ | (c) $R = 128$ | (d) $R = 256$ |

Fig. 17 Scalar velocity magnitude (top) and vector field streamline (bottom) visualization of the forced MHD turbulence simulation at variable reduced rank reconstructions $R = 32, 64, 128, 256$

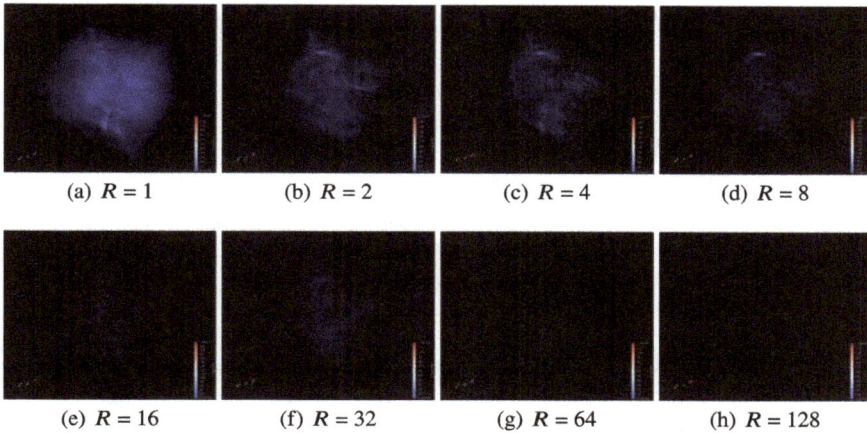

Fig. 18 Relative magnitude error (in %) of the velocity field with reduced rank reconstructions $R = 1, 2, 4, 8, 16, 32, 64, 128$

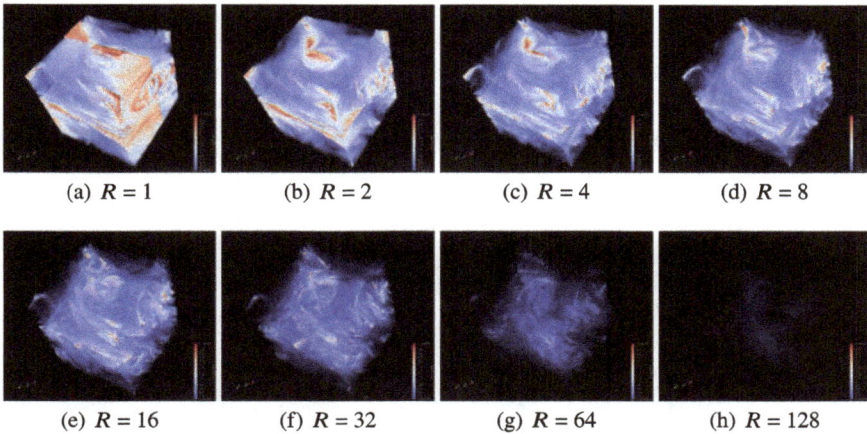

Fig. 19 Absolute angular error $[0°, 180°]$ of the velocity field with reduced rank reconstructions $R = 1, 2, 4, 8, 16, 32, 64, 128$

However, this does not relate to the local strength of the vector field since large angle differences in low-magnitude areas are not that important. Directly comparing Figs 17 and 19, which have the same viewing configuration, one can observe that large angular errors occur mostly in very low-velocity areas and may thus not be that relevant. In particular, since for very short vectors small errors can cause large angular changes.

9.3 Vorticity and Divergence

We also inspected the influence of low-rank tensor compression on the two most fundamental features of vector field data: divergence and vorticity. As can be seen in Fig. 20, the global structure of the magnitude from both the divergence field and the vorticity field are still very well preserved with rank $R = 64$, which is $1/4$ of the original full rank of 256. The values are given in their absolute ranges of $[0.36, -0.43]$ and $[0.52, 10^{-5}]$ for the divergence and vorticity respectively. The rendering applies an opacity setting which is almost linear, slight curve below the diagonal, to the *magnitude values*, highlighting the important regions.

For the vorticity vector field, we also calculated the relative magnitude error as for the velocity vector field data shown in Fig. 18, thus as percentage of the vorticity magnitude, and similar conclusions can also be drawn here for the results shown in Fig. 21. All errors are very low, and barely noticeable for rank-reductions down to $R = 16$, which corresponds to $1/16$ of the original data volume.

Additionally, we also computed the absolute angular error for the vorticity vector field, in this case applying an opacity proportional to the *magnitude of the raw vector field*, and the results are shown in Fig. 22. We can observe that the angular error almost everywhere is in the color range of dark to light blue which maps to errors from 0 up to 45° in this plot, and thus the directional vorticity information seems to be well preserved.

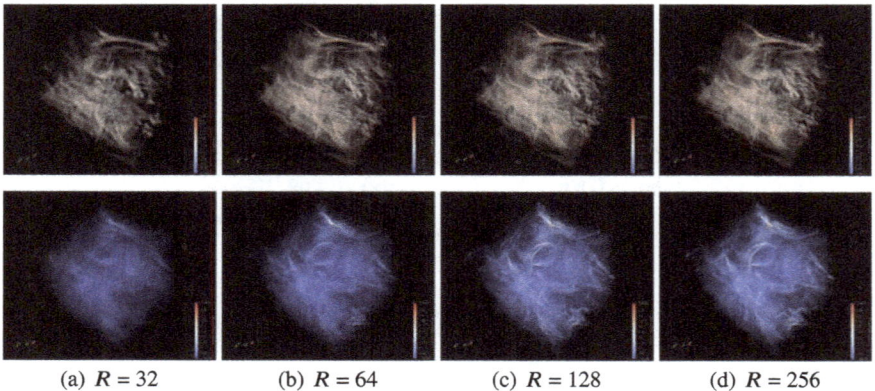

(a) $R = 32$ (b) $R = 64$ (c) $R = 128$ (d) $R = 256$

Fig. 20 Magnitude of the vector field's divergence (top) and vorticity (bottom) at variable reduced rank reconstructions $R = 32, 64, 128, 256$

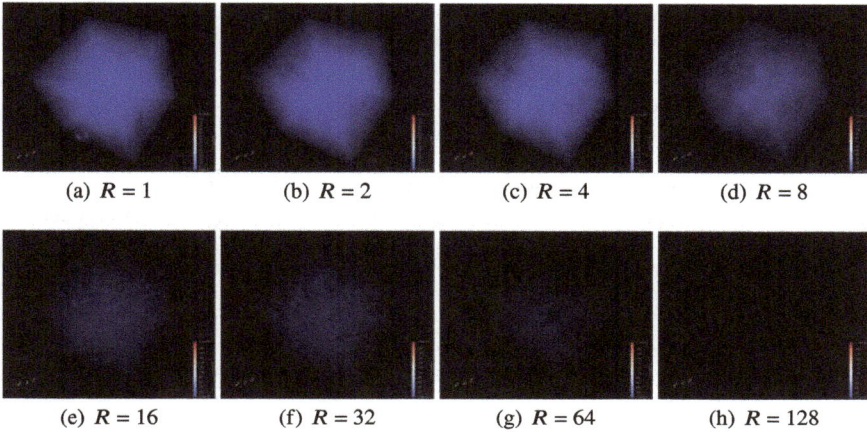

Fig. 21 Relative magnitude error (in %) of the vorticity field with reduced rank reconstructions $R = 1, 2, 4, 8, 16, 32, 64, 128$

(a) $R = 1$ (b) $R = 2$ (c) $R = 4$ (d) $R = 8$

(e) $R = 16$ (f) $R = 32$ (g) $R = 64$ (h) $R = 128$

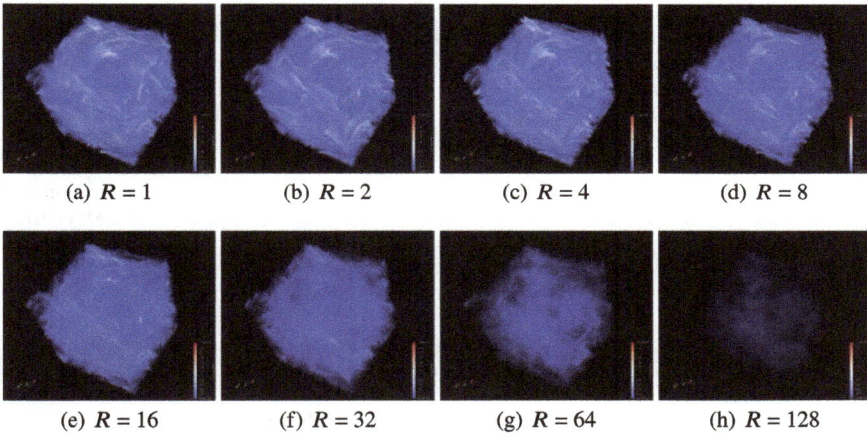

(a) $R = 1$ (b) $R = 2$ (c) $R = 4$ (d) $R = 8$

(e) $R = 16$ (f) $R = 32$ (g) $R = 64$ (h) $R = 128$

Fig. 22 Absolute angular error $[0°, 130°]$ of the vorticity field with reduced rank reconstructions $R = 1, 2, 4, 8, 16, 32, 64, 128$

10 Conclusions

In the analysis conducted in this first study on vector fields, we have shown that tensor approximation methods are not only very useful for multidimensional scalar fields but can also be applied to multivariate data, thus extending to vector and possibly tensor fields. In general we can observe that for the important high-velocity vector field regions, the low-rank reconstructions maintain the important overall structures of the flow features, in particular also the vorticity. Furthermore, we note that the MHD simulation model in theory should be divergence-free and thus the numerical results report very low divergence numbers. Therefore, we are very satisfied with the result that the low-rank tensor reconstructions do not result in an unexpected and uncontrolled enlargement of these divergence values, keeping them within the numerical range of the simulation.

Acknowledgements This work was partially supported by the University of Zurich's Forschungskredit "Candoc" (grant numbers FK-16-012 and 53511401), a Swiss National Science Foundation grant (SNF) (project n°200021_132521), a Hasler Foundation grant (project number 12097), and the EU FP7 People Programme (Marie Curie Actions) under REA Grant Agreement n°290227. Furthermore, we would like to acknowledge the *Computer-Assisted Paleoanthropology* group and the *Visualization and MultiMedia Lab* at University of Zürich for the acquisition of the Hazelnut dataset (https://www.ifi.uzh.ch/en/vmml/research/datasets.html in Fig. 15. Also we acknowledge the Johns Hopkins Turbulence Database http://turbulence.pha.jhu.edu/ for the data used in Fig. 16 as well as their forced MHD simulation data http://turbulence.pha.jhu.edu/Forced_MHD_turbulence.aspx used in our experiments.

References

1. Andersson, C.A., Bro, R.: Improving the speed of multi-way algorithms: Part I. Tucker3. Chemometr. Intell. Lab. Syst. **42**, 93–103 (1998)
2. Ballester-Ripoll, R., Lindstrom, P., Pajarola, R.: TTHRESH: Tensor compression for multidimensional visual data. IEEE Trans. Visual. Comput. Graph. to appear (2020). https://doi.org/10.1109/TVCG.2019.2904063
3. Ballester-Ripoll, R., Pajarola, R.: Compressing bidirectional texture functions via tensor train decomposition. In: Proceedings Pacific Graphics Short Papers (2016). https://doi.org/10.2312/pg.20161329
4. Ballester-Ripoll, R., Pajarola, R.: Lossy volume compression using Tucker truncation and thresholding. Visual Comput. **32**(11), 1433–1446 (2016). https://doi.org/10.1007/00371-015-1130-y
5. Ballester-Ripoll, R., Steiner, D., Pajarola, R.: Multiresolution volume filtering in the tensor compressed domain. IEEE Trans. Visual. Comput. Graph. **24**(10), 2714–2727 (2018). https://doi.org/10.1109/TVCG.2017.2771282
6. Ballester-Ripoll, R., Suter, S.K., Pajarola, R.: Analysis of tensor approximation for compression-domain volume visualization. Comput. Graph. **47**, 34–47 (2015). https://doi.org/10.1016/j.cag.2014.10.002
7. Balsa Rodríguez, M., Gobbetti, E., Iglesias Guitián, J.A., Makhinya, M., Marton, F., Pajarola, R., Suter, S.K.: A survey of compressed GPU direct volume rendering. In: Eurographics State of The Art Reports (STAR), pp. 117–136 (2013). https://doi.org/10.2312/conf/EG2013/stars/117-136
8. Balsa Rodríguez, M., Gobbetti, E., Iglesias Guitián, J.A., Makhinya, M., Marton, F., Pajarola, R., Suter, S.K.: State-of-the-art in compressed GPU-based direct volume rendering. Comput. Graph. Forum **33**(6), 77–100 (2014). https://doi.org/10.1111/cgf.12280
9. Bilgili, A., Öztürk, A., Kurt, M.: A general BRDF representation based on tensor decomposition. Comput. Graph. Forum **30**(8), 2427–2439 (2011)
10. Caroll, J.D., Chang, J.J.: Analysis of individual differences in multidimensional scaling via an *n*-way generalization of "Eckart-Young" decompositions. Psychometrika **35**, 283–319 (1970)
11. Cattell, R.B.: Parallel proportional profiles and other principles for determining the choice of factors by rotation. Psychometrika **9**(4), 267–283 (1944)
12. Comon, P., Mourrain, B.: Decomposition of quantics in sums of powers of linear forms. Signal Process. Special Issue Higher Order Stat. **53**, 93–108 (1996)
13. de Lathauwer, L.: A survey of tensor methods. In: Proceedings IEEE International Symposium on Circuits and Systems, pp. 2773–2776 (2009)

14. de Lathauwer, L., de Moor, B., Vandewalle, J.: A multilinear singular value decomposition. SIAM J. Matrix Anal. Appl. **21**(4), 1253–1278 (2000)
15. de Lathauwer, L., de Moor, B., Vandewalle, J.: On the best rank-1 and rank-$(R_1, R_2, ..., R_N)$ approximation of higher-order tensors. SIAM J. Matrix Anal. Appl. **21**(4), 1324–1342 (2000)
16. Elden, L., Savas, B.: A Newton-Grassmann method for computing the best multilinear rank-(r_1, r_2, r_3) approximation of a tensor. SIAM J. Matrix Anal. Appl. **31**(2), 248–271 (2009)
17. Ergin, S., Çakir, S., Gerek, O.N., Gülmezoğlu, M.B.: A new implementation of common matrix approach using third-order tensors for face recognition. Exp. Syst. Appl. **38**(4), 3246–3251 (2011)
18. Furukawa, R., Kawasaki, H., Ikeuchi, K., Sakauchi, M.: Appearance based object modeling using texture database: acquisition, compression and rendering. In: Proceedings Eurographics Workshop on Rendering, pp. 257–266 (2002)
19. Harshman, R.A.: Foundations of the PARAFAC procedure: models and conditions for an "explanatory" multi–modal factor analysis. UCLA Working Papers Phonetics **16**, 1–84 (1970)
20. He, X., Cai, D., Liu, H., Han, J.: Image clustering with tensor representation. In: Proceedings ACM Multimedia Conference, pp. 132–140 (2005)
21. Hitchcock, F.L.: The expression of a tensor or a polyadic as a sum of products. J. Math. Phys. **6**, 164–169 (1927)
22. Ishteva, M., De Lathauwer, L., Absil, P.A., Van Huffel, S.: Differential-geometric Newton method for the best rank-(R_1, R_2, R_3) approximation of tensors. Numer. Algorithms **51**(2), 179–194 (2009)
23. Kapteyn, A., Neudecker, H., Wansbeek, T.: An approach to n-mode components analysis. Psychometrika **51**(2), 269–275 (1986)
24. Kiers, H.A.: Towards a standardized notation and terminology in multiway analysis. J. Chemometr. **14**(3), 105–122 (2000)
25. Kolda, T.G., Bader, B.W.: Tensor decompositions and applications. SIAM Rev. **51**(3), 455–500 (2009)
26. Kroonenberg, P.M.: Three-mode Principal Component Analysis: Theory and Applications. DSWO Press, Leiden (1983)
27. Kroonenberg, P.M.: Applied Multiway Data Analysis. Wiley, New York (2008)
28. Kroonenberg, P.M., De Leeuw, J.: Principal component analysis of three-mode data by means of alternating least squares algorithms. Psychometrika **45**, 69–97 (1980)
29. Krüger, B., Tautges, J., Müller, M., Weber, A.: Multi-mode tensor representation of motion data. J. Virtual Reality Broadcast. **5**(5), (2008)
30. Kruskal, J.B.: Rank, decomposition, and uniqueness for 3-way and n-way arrays. In: Multiway Data Analysis, pp. 7–18. North-Holland Publishing Co. (1989)
31. Liu, G., Xu, M., Pan, Z., El Rhalibi, A.: Human motion generation with multifactor models. J. Visual. Comput. Anim. **22**(351–359), 4 (2011)
32. Liu, J., Liu, J., Wonka, P., Ye, J.: Sparse non-negative tensor factorization using columnwise coordinate descent. Pattern Recognit. **45**(1), 649–656 (2012)
33. Min, J., Liu, H., Chai, J.: Synthesis and editing of personalized stylistic human motion. In: Proceedings ACM SIGGRAPH Symposium on Interactive 3D Graphics and Games, pp. 39–46 (2010)
34. Moravitz, Martin, C.D.: Tensor decompositions workshop discussion notes. American Institute of Mathematics, Palo Alto, CA (2004)
35. Morozov, O.V., Unser, M., Hunziker, P.: Reconstruction of large, irregularly sampled multidimensional images: a tensor-based approach. IEEE Trans. Med. Imaging **30**(2), 366–74 (2011)
36. Mukai, T., Kuriyama, S.: Multilinear motion synthesis with level-of-detail controls. In: Proceedings Pacific Conference on Computer Graphics and Applications, pp. 9–17 (2007)
37. Perera, M., Shiratori, T., Kudoh, S., Nakazawa, A., Ikeuchi, K.: Multilinear analysis for task recognition and person identification. In: Proceedings IEEE Conference on Intelligent Robots and Systems, pp. 1409–1415 (2007)
38. Ruiters, R., Klein, R.: BTF compression via sparse tensor decomposition. Comput. Graph. Forum **28**(4), 1181–1188 (2009)

39. Ruiters, R., Schwartz, C., Klein, R.: Data driven surface reflectance from sparse and irregular samples. Comput. Graph. Forum **31**(2), 315–324 (2012)
40. Savas, B., Eldén, L.: Handwritten digit classification using higher order singular value decomposition. Pattern Recognit. **40**(3), 993–1003 (2007)
41. Schultz, T., Seidel, H.P.: Estimating crossing fibers: a tensor decomposition approach. IEEE Trans. Visual. Comput. Graph. **14**(6), 1635–1642 (2008)
42. Shashua, A., Hazan, T.: Non-negative tensor factorization with applications to statistics and computer vision. Proc. ACM Mach. Learn. **119**, 792–799 (2005)
43. Shashua, A., Levin, A.: Linear image coding for regression and classification using the tensor-rank principle. In: Proceedings IEEE Computer Vision and Pattern Recognition Conference, pp. 42–49 (2001)
44. Smilde, A., Bro, R., Geladi, P.: Multi-Way Analysis: Applications in the Chemical Sciences. Wiley, West Sussex, England (2004)
45. Sun, X., Zhou, K., Chen, Y., Lin, S., Shi, J., Guo, B.: Interactive relighting with dynamic BRDFs. ACM Trans. Graph. **26**(3), 27 (2007). https://doi.org/10.1145/1276377.1276411
46. Suter, S.K., Iglesias Guitián, J.A., Marton, F., Agus, M., Elsener, A., Zollikofer, C.P., Gopi, M., Gobbetti, E., Pajarola, R.: Interactive multiscale tensor reconstruction for multiresolution volume visualization. IEEE Trans. Visual. Comput. Graph. **17**(12), 2135–2143 (2011)
47. Suter, S.K., Makhinya, M., Pajarola, R.: TAMRESH: Tensor approximation multiresolution hierarchy for interactive volume visualization. Comput. Graph. Forum **32**(3), 151–160 (2013). https://doi.org/10.1111/cgf.12102
48. Suter, S.K., Zollikofer, C.P., Pajarola, R.: Application of tensor approximation to multiscale volume feature representations. In: Proceedings Vision, Modeling and Visualization, pp. 203–210 (2010)
49. Ten Berge, J.M.F., De Leeuw, J., Kroonenberg, P.M.: Some additional results on principal components analysis of three-mode data by means of alternating least squares algorithms. Psychometrika **52**, 183–191 (1987)
50. Tsai, Y.T., Shih, Z.C.: All-frequency precomputed radiance transfer using spherical radial basis functions and clustered tensor approximation. ACM Trans. Graph. **25**(3), 967–976 (2006)
51. Tsai, Y.T., Shih, Z.C.: K-clustered tensor approximation: a sparse multilinear model for real-time rendering. ACM Trans. Graph. **31**(3), 19:1–19:17 (2012)
52. Tucker, L.R.: Some mathematical notes on three-mode factor analysis. Psychometrika **31**(3), 279–311 (1966)
53. Vasilescu, M.A.O.: Human motion signatures: analysis, synthesis, recognition. Proc. IEEE Conf. Pattern Recognit. **3**, 456–460 (2002)
54. Vasilescu, M.A.O., Terzopoulos, D.: Multilinear analysis of image ensembles: tensorFaces. Proc. Eur. Conf. Comput. Vis. **2350**, 447–460 (2002)
55. Vasilescu, M.A.O., Terzopoulos, D.: TensorTextures: multilinear image-based rendering. ACM Trans. Graph. **23**(3), 336–342 (2004)
56. Vlasic, D., Brand, M., Pfister, H., Popović, J.: Face transfer with multilinear models. ACM Trans. Graph. **24**(3), 426–433 (2005)
57. Wampler, K., Sasaki, D., Zhang, L., Popović, Z.: Dynamic, expressive speech animation from a single mesh. In: Proceedings SIGGRAPH/Eurographics Symposium on Computer Animation, pp. 53–62 (2007)
58. Wang, H., Ahuja, N.: Compact representation of multidimensional data using tensor rank-one decomposition. In: Proceedings Pattern Recognition Conference, pp. 44–47 (2004)
59. Wang, H., Ahuja, N.: Rank-R approximation of tensors: using image-as-matrix representation. In: Proceedings IEEE Conference on Computer Vision and Pattern Recognition, pp. 346–353 (2005)
60. Wang, H., Ahuja, N.: A tensor approximation approach to dimensionality reduction. Int. J. Comput. Vis. **76**(3), 217–229 (2008)
61. Wang, H., Wu, Q., Shi, L., Yu, Y., Ahuja, N.: Out-of-core tensor approximation of multi-dimensional matrices of visual data. ACM Trans. Graph. **24**(3), 527–535 (2005)

62. Wu, Q., Xia, T., Chen, C., Lin, H.Y.S., Wang, H., Yu, Y.: Hierarchical tensor approximation of multidimensional visual data. IEEE Trans. Visual. Comput. Graph. **14**(1), 186–199 (2008)
63. Yan, S., Wang, H., Tu, J., Tang, X., Huang, T.S.: Mode-kn factor analysis for image ensembles. IEEE Trans. Image Process. **18**(3), 670–676 (2009)

6

Challenges for Tractogram Filtering

Daniel Jörgens, Maxime Descoteaux, and Rodrigo Moreno

Abstract Tractography aims at describing the most likely neural fiber paths in white matter. A general issue of current tractography methods is their large false-positive rate. An approach to deal with this problem is tractogram filtering in which anatomically implausible streamlines are discarded as a post-processing step after tractography. In this chapter, we review the main approaches and methods from literature that are relevant for the application of tractogram filtering. Moreover, we give a perspective on the central challenges for the development of new methods, including modern machine learning techniques, in this field in the next few years.

Keywords Diffusion MRI · Tractography · Tractogram filtering

1 Introduction

Diffusion magnetic resonance imaging (dMRI) is currently the method of choice for assessing the local microstructure in the white matter (WM) of the human brain in vivo. Tractography methods use this local microstructure to generate streamlines aiming at modeling the underlying anatomy of neural fibers in the brain. These streamlines are locally aligned with the estimated tissue orientation. The set of obtained streamlines is usually referred to as a *tractogram* and is the basis for different subsequent analyses.

D. Jörgens (✉) · R. Moreno
Department of Biomedical Engineering and Health Systems, KTH Royal Institute
of Technology, Hälsovägen 11C, 14157 Huddinge, Stockholm, Sweden
e-mail: danjorg@kth.se

R. Moreno
e-mail: rodmore@kth.se

M. Descoteaux
Université de Sherbrooke, Sherbrooke Connectivity Imaging Laboratory, 2500 Boulevard de
l'Université, Sherbrooke, QC J1K 0A5, Canada
e-mail: M.Descoteaux@USherbrooke.ca

Tractography has potential for clinically-relevant applications. For instance, visualizations of tractography data are used by neurosurgeons with the goal of preserving important neural bundles during brain tumor resection [9, 16, 64]. Moreover, it is possible to group streamlines into bundles and perform analyses based on these. This so-called bundle-based or tract-based analysis (BBA/TBA) allows to compute statistics for individual bundles, either averaged or along the streamlines. This can be used to perform group comparisons [8, 10, 60, 70] or to analyse WM changes in patients over time [5]. Focusing on the tractogram as a whole, the concept of *structural connectivity* is employed for the analysis of how pairs of cortical and subcortical regions are connected through streamlines of the tractogram [49, 59]. These connections can be summarized in graphs, which can be seen as estimations of the actual anatomical connections between GM regions through neural pathways in the WM. Such graphs serve as the basis for different group comparisons. For example, this approach has been applied to different neurological diseases, including among others epilepsy (e.g., [21]), multiple sclerosis (e.g., [31]) and Alzheimer's disease (e.g., [17, 41]), as well as for assessing differences in the brain due to normal aging (e.g., [13]).

Despite its potential in the aforementioned applications, the validation of tractography is an open challenge in the field. Several recent studies have unveiled that state-of-the-art methods to construct tractograms suffer both from *false positives* (FP), i.e., streamlines that are not related to anatomical structures [32], and *false negatives* (FN), i.e., missing streamlines to accurately describe known WM anatomy in its whole extent [2]. At the same time, findings indicate that tractography results are to some extent reproducible [34]. That means that FPs are an intrinsic problem of current methods [32]. Already in 2014, Thomas et al. hypothesised that there are inherent limitations of tractography that lead to effects of both, FPs and FNs [62]. The importance of these two problems is application dependent. For example, it has been argued that for structural connectivity analysis, a high specificity is of higher importance than sensitivity [71]. This means that, for this specific application, reducing FPs is more important than reducing FNs.

Despite the efforts in recent years, new approaches have still not overcome the limitations of tractography methods [53]. In the particular case of FP, an alternative to improving tractography methods is to remove the FPs from the tractogram in a post processing step. In the following, we refer to this approach as *tractogram filtering*. Several methods from the literature can be used for this goal. Even though some methods were not designed for tractogram filtering, they can be adapted for this purpose. In this chapter, we consider tractogram filtering as a binary classification problem in which we aim at assigning either a positive (P) or a negative (N) label to each streamline. This allows us to assess the possibility to use a particular method to define the labels P and N for, ideally, separating true positive (TP) and FP streamlines in a tractogram. In the following, we review the most relevant methods in literature, point out their issues and pose key challenges on the way to improved tractogram filtering.

The chapter is organized as follows. Section 2 describes the main approaches and methods for tractogram filtering. Section 3 describes the key challenges of tractogram filtering and gives a perspective on future developments in the field. Finally, we make some concluding remarks in Sect. 4. In order to illustrate specific properties or issues of the different strategies, we run some experiments on a very limited amount of data from the Human Connectome Project (HCP).

2 Approaches for Tractogram Filtering

In the following sections, we review methods that can be used for tractogram filtering. Based on the definition of the labels P and N for each streamline, we identified four main criteria that these methods build upon, namely: *explainability of the diffusion signal*, *inclusion and exclusion regions of interest* (ROIs), *streamline geometry or shape*, and *streamline similarity and clustering*. Figure 1 shows the problem of tractogram filtering interpreted as a binary classification problem and lists the most relevant methods in this realm together with the main criteria used to group the streamlines.

Table 1 lists the most relevant tractogram filtering methods that are reviewed in the next subsections. The columns of this table describe the most relevant characteristics of these methods, namely: *criteria*, use of *dMRI* data, required *context*, main *target* and whether or not the method is *data driven*. First, *criteria* refers to the strategies from Fig. 1 that are followed when employing the method for streamline classification. Column *dMRI* in the table indicates if the method used the dMRI data or not. As shown, only a few filtering methods make use of the acquired dMRI for performing the classification. The listed methods also differ in the required streamline *context*. While some are able to perform the classification individually per streamline, others also require the streamlines in the bundles of interest or the complete tractogram as extra inputs (column *context*). Furthermore, by design, the target of some methods is to define either the positive (i.e., P) or negative (i.e., N) label but not both. In other words, some methods aim at being more specific detecting TPs than FPs, and the other way round for others. As an example, a streamline classified as negative by Recobundles can still be a TP, since it can belong to a bundle not present in the atlas. The target of the method is important to be considered, since preferring higher specificity of P or N is application dependent. This is shown in column *target* in the table. Finally, the methods can also be grouped into *rule-based* and *data-driven* ones. While the former makes use of classical approaches, the latter use machine learning techniques for performing the classification.

In the following sections, we describe the aforementioned criteria in detail and list the most representative methods that use those criteria to perform the classification of streamlines.

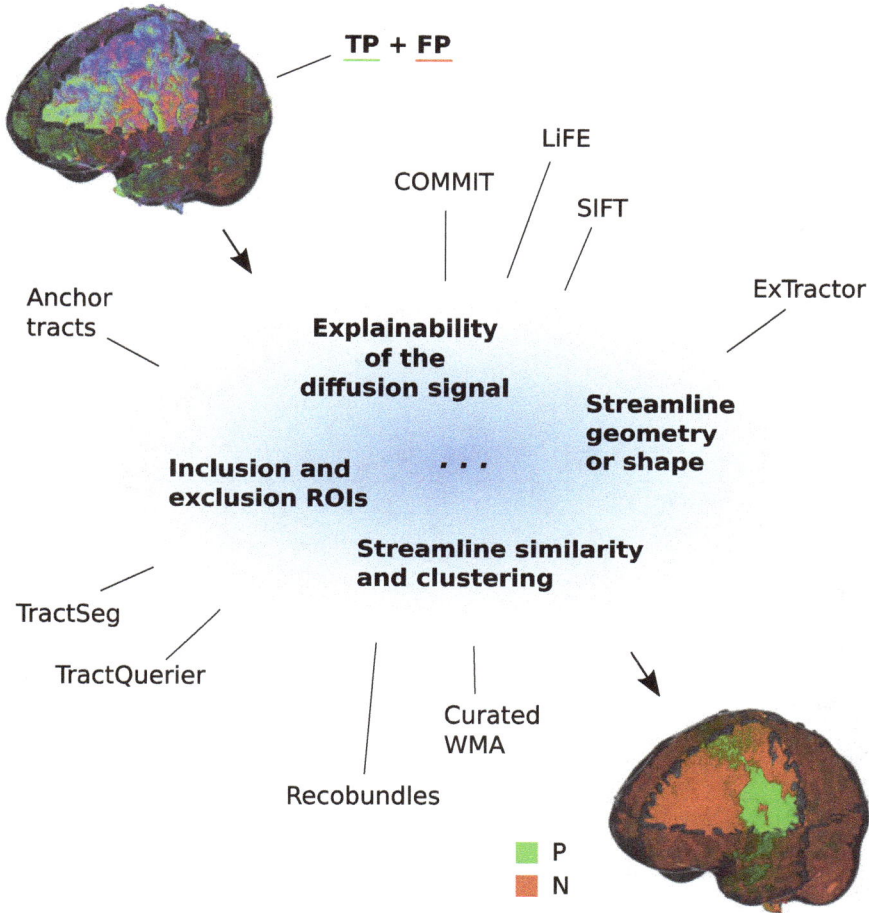

Fig. 1 Tractogram filtering can be seen as classifying each streamline in a tractogram as either positive (P) or negative (N), targeting true positives (TPs) and false positives (FPs), respectively. Relevant methods use one or more of the depicted criteria for performing this classification. Some representative methods are also depicted in the figure. For a detailed list refer to Table 1

2.1 *Explainability of the Diffusion Signal*

The idea behind this approach is that high-quality tractograms can be used to explain the acquired dMRI data. In other words, synthetic dMRI generated from tractograms should be very similar to the acquired dMRI data. Thus, these methods focus on finding a subset of streamlines which can generate data that approximates the measured signal as closely as possible. Streamlines not belonging to such a subset are likely implausible (or duplicates of other streamlines already contributing to the signal) and might be removed. This approach is shown in Fig. 2.

Table 1 Representative methods for tractogram filtering. **Criteria** shows the criteria used for streamline classification (cf. Fig. 1). Column **dMRI** indicates if the method uses diffusion data. **Context** describes the contextual information required for streamline classification. **Target** shows the target labels where the method is more specific. **Data driven** specifies if the method makes use of machine learning or not

Method	Criteria	dMRI	Context	Target	Data-driven
LiFE [42]	dMRI explainability	⊠	Tractogram	–	□
COMMIT [12]	dMRI explanability	⊠	Tractogram	–	□
SIFT [55]	dMRI explainability	⊠	Tractogram	–	□
SIFT2 [57]	dMRI explainability	⊠	Tractogram	–	□
TractQuerier [66]	ROIs	□	–	–	□
TractSeg [67]	ROIs	⊠	–	P	⊠
FiberNet [24]	Geometry	□	–	P	⊠
FiberMap [73]	Geometry	□	–	P	⊠
TRAFIC [37]	Geometry	□	–	P	⊠
DeepBundle [30]	Geometry	□	–	P	⊠
Geometric DL [1]	Geometry	□	–	N	⊠
Recobundles [19]	Streamline similarity	□	Bundles	P	□
Curated WMA [74]	Streamline similarity	□	Bundles	P	□
FS2NET [40]	Streamline similarity	□	Bundles	P	⊠
DeepFiltering [25]	dMRI pattern	⊠	–	–	⊠
BundleMAP [27]	Streamline similarity, ROIs	□	Bundles	P	⊠
ExTractor [43]	Geometry, ROIs, Streamline similarity	□	Tractogram	N	□
COMMIT2 [50]	dMRI explainability, ROIs	⊠	Tractogram	–	□
AnchorTracts [36]	dMRI explainability, ROIs, Streamline similarity	⊠	Tractogram	P	□
FiberNet2.0 [23]	ROIs, Geometry	□	–	P	⊠

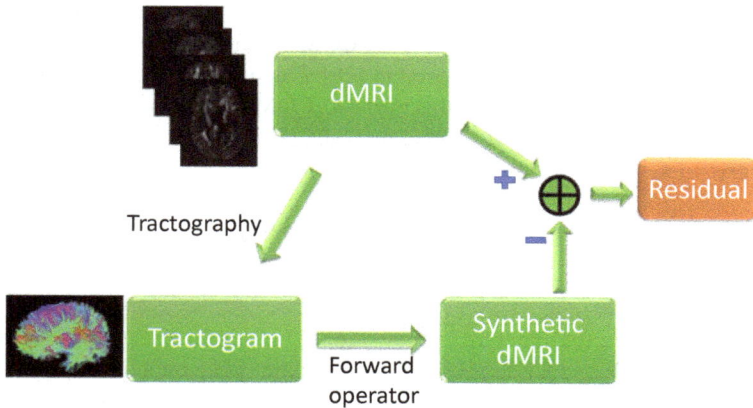

Fig. 2 dMRI signal fitting approach. Synthetic dMRI is computed based on a microstructure model and compared to the acquired dMRI. The streamlines that are not relevant to minimizing the residual might be filtered out

As shown in Table 1, these methods require the whole tractogram as an input. In the following subsections, we will describe the most commonly used methods for dMRI signal fitting and discuss their issues.

2.1.1 LiFE and COMMIT

Linear Fascicle Evaluation (LiFE) [7, 42] and Convex Optimization Modeling for Microstructure Informed Tractography (COMMIT) [12] state the problem as follows: let $\mathbf{y}$ be a vector with the acquired diffusion signals, $\mathbf{A(T)}$ be a forward operator for synthesizing diffusion data from the streamlines of the tractogram $\mathbf{T}$, $\mathbf{x}$ be a vector with weights for the contribution of every streamline to the acquired data, and η be the acquisition noise. Then $\mathbf{y}$ can be written as:

$$\mathbf{y} = \mathbf{A(T)}\,\mathbf{x} + \eta. \tag{1}$$

Since the weights $\mathbf{x}$ cannot be negative, it is possible to solve (1) through non-negative least squares:

$$\underset{\mathbf{x} \geq 0}{arg\,min} ||\mathbf{A(T)}\,\mathbf{x} - \mathbf{y}||_2^2 \tag{2}$$

Filtering is performed by discarding streamlines with low weights. This formulation allows for the use of different models to couple the information from streamlines to the measured signal or derivatives thereof. First, $\mathbf{A}$ can be chosen from a large variety of forward operators proposed in the literature [39]. As an example, in the original papers, COMMIT was based on a multi-compartment forward model while in LiFE the stick and ball model was used. Notwithstanding, both can be adapted to use any other model. Second, different solvers can be applied for solving the non-negative least-squares problem of (2). Both COMMIT and LiFE use the subspace Barzilei-Borwein (SSB) solver proposed in [28]. Due to the nature of the problem, sparsity on weights is desirable. For this purpose, COMMIT proposes a basis pursuit de-noise (BPDN) formulation of (2) that actively considers sparsity by minimizing the ℓ^1-norm of $\mathbf{x}$. Such a formulation can be written as:

$$\underset{\mathbf{x} \geq 0}{\mathrm{argmin}} ||\mathbf{x}||_1, \text{ subject to } ||\mathbf{A}(\mathbf{T})\,\mathbf{x} - \mathbf{y}||_2^2 \leq \epsilon, \tag{3}$$

where ϵ is a parameter.

In order to reduce the inherent computational burden of these strategies, $\mathbf{A}$ is implemented in LiFE and COMMIT through a lookup table on a dictionary of pre-computed estimations. Moreover, a GPU-based optimized version has recently been proposed for LiFE [29].

2.1.2   SIFT and SIFT2

Very similar to the techniques from the previous subsection is Spherical-deconvolution Informed Filtering of Tractograms (SIFT) [55]. Instead of targeting the raw dMRI data, it aims at reconstructing the fiber orientation distribution function (fODF) in each voxel. First, the fODF is obtained with constrained spherical decon-volution (CSD) [63]. Second, the contribution of every streamline to the fODF is assessed. These contributions are used to determine whether a streamline is deemed redundant/noisy or not. Third, these contributions are sorted in order to remove the least relevant streamlines. Finally, the aforementioned two steps are iterated until either a target number of streamlines or a certain residual level is reached. Unlike LiFE and COMMIT, SIFT does not generate weights per streamline. Thus, SIFT2 was proposed as a slight modification of SIFT in which an additional regularization term is added and a weight per streamline is computed [56].

In order to compare the agreement between SIFT and SIFT2, we run them with their standard parameters on a whole-brain tractogram computed with anatomically-constrained tractography (ACT) [54] from MRrtix3[1] with one million streamlines obtained from one HCP subject. In this experiment, SIFT selected 34.6% of the streamlines in the original tractogram. Figure 3 shows the histograms of weights computed with SIFT2 where the individual histograms are obtained by separating the streamlines based on whether they were accepted or discarded by SIFT. It can be seen that the two histograms have a big overlapping region. A repetition of the exper-iment with 500k streamlines showed similar results. This means that it is difficult to reproduce the results from SIFT with the weights computed with SIFT2. Thus, while SIFT2 can be useful for describing the contributions of streamlines to the acquired data, unlike SIFT, its direct use for tractogram filtering is not straightforward.

---

[1]https://www.mrtrix.org/.

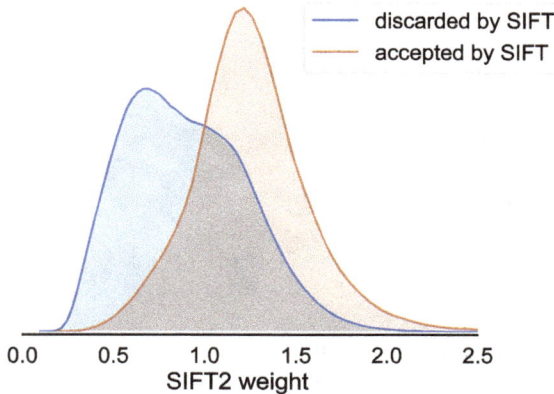

Fig. 3 Histograms showing the frequency of different SIFT2 weights computed for streamlines that are filtered out (in blue) or not (in orange) with SIFT. Both methods were computed on an HCP subject with 1 million streamlines computed with ACT in MRtrix3

2.1.3 Issues of dMRI Signal Fitting

While dMRI signal fitting is appealing, it has not been able to significantly reduce the false positive rate [50, 52]. This fact can be attributed to various reasons. Daducci et al. [11] discuss a number of issues of dMRI signal fitting. First, most of the current dMRI signal fitting methods require a whole brain tractogram, even if a single white matter bundle is of interest. This results both in an unnecessary computational burden and a higher risk for false positives when targeting specific fiber bundles. For example, if a fiber bundle is not appropriately represented in the full tractogram, which is not uncommon, there is a higher risk for implausible streamlines from other bundles to take their place in the reconstruction. Second, the computed weights of streamlines tend to be inversely proportional to the number of similar streamlines in the tractogram. This effect is shown in Fig. 4 for the case of SIFT2 (a similar behavior is expected from LiFE and COMMIT). As shown, the SIFT2 weights are lower in the centerlines of the bundles, where tractography tends to yield more streamlines. Thus, thresholding of SIFT2 weights cannot be used for progressive filtering, since that might result in discarding the most important tracts very early. Moreover, some noisy streamlines might be classified as acceptable just because of the reward they get for reaching distant regions. This issue comes from the fact that weights of SIFT2 are designed for fitting the acquired data, but not for filtering. This could potentially be solved through an extra step of weight normalization that, to our knowledge, has not been proposed so far. Third, as discussed in [11], minimizing the residual from Fig. 2 does not guarantee that the solution is plausible as the current methods are very prone to overfit due to the large amount of unknowns that must be estimated. Finally, working with incomplete streamlines can lead to biased results in the uncovered regions, which might be especially problematic for structural connectivity.

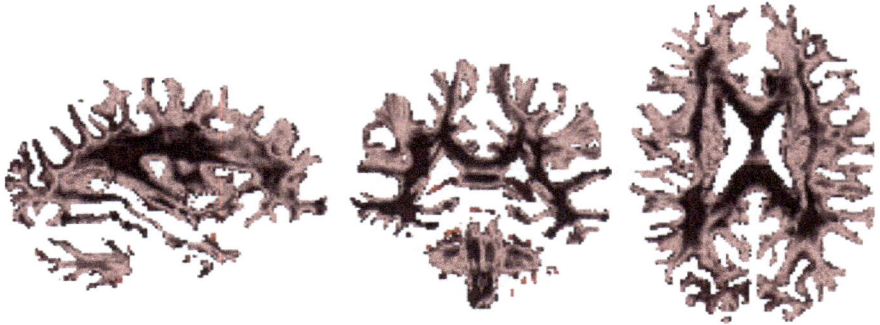

Fig. 4 Average weights computed through SIFT2 per voxel. Darker and brighter voxels correspond to lower and higher SIFT2 weights, respectively. The values are extracted from the same dataset of Fig. 3

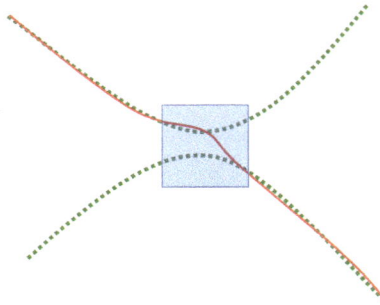

Fig. 5 The course of two close fiber bundles is depicted in dashed green. A local error (blue box) can lead to a noisy streamline (in red). Since most segments of its path are correct, it still might be classified as acceptable by most dMRI signal fitting methods

In addition to these issues, most dMRI signal fitting methods compute a single weight per streamline, which can lead to acceptance of implausible streamlines that are erroneous in a small region (cf. Fig. 5). As suggested in [42], this issue can be handled by having variable weights along the streamlines. However, this solution might come at a cost of numerical instability. Finally, an important aspect to consider is the applicability of these methods to diseased brains. For example, it was reported in [72] that using SIFT in certain types of illnesses (e.g. brain tumors) can lead to wrong conclusions of connectivity changes.

In [69], we tested the performance of SIFT2 in short synthetic streamlines. First, we generated dMRI data with Fiberfox [35] with and without noise from a set of straight and bent short streamlines. Second, we added a set of streamlines at a different angle. These added streamlines should ideally be classified as noisy as they do not comply with the generated dMRI data. Finally, SIFT2 was run in order to assess its ability to separate the 'signal-generating' streamlines from the added ones. The main result of these experiments is that SIFT2 is able to filter the implausible stream-

after a spatial normalization for bundle classification [24]. In [73], the authors extract a set of features from the shape of streamlines (named FiberMap) to train their bundle classifier. Streamlines not belonging to a known bundle are assigned to an additional class. Furthermore, TRAFIC trains a network using distances to five landmarks, curvature and torsion per tract as features for filtering [37]. Moreover, DeepBundle [30] used a graph convolutional neural network for extracting geometric features from the streamlines. Such learned features are then used to assign them to their more likely fiber bundle in an end-to-end fashion. The loss function can be designed to target false positives. Since these deep learning-based methods use streamlines of specific fiber bundles for training, their main target are the TPs in those bundles.

Similarly to the inclusion and exclusion of ROIs, geometry constraints of the streamlines are not necessarily sufficient criteria for deciding the validity of streamlines and must be combined with other priors. That does, however, not apply to the mentioned DL-approaches which—depending on the training labels—are able to learn a model of plausible streamline geometry.

2.4 Streamline Similarity and Clustering

With this strategy, streamlines are clustered in bundles before further analysis. Such clusters can be used as surrogates of the underlying structure of WM. Fibers belonging to small clusters or that do not share similar properties of bundles of interest can be removed.

The only requirement for using standard clustering algorithms for streamline clustering is to define a distance metric between streamlines. While proposing distance metrics is straightforward, it is more difficult to find the most appropriate one for streamlines. Depending on the application, a tractogram can consist of millions of streamlines. Thus, it is critical to use efficient implementations. For example, Quickbundles [18] was proposed as a tool for performing clustering very efficiently.

Once clusters of streamlines are extracted, there are different alternatives for performing filtering. For example, Recobundles uses an atlas-based strategy in which the clusters are first registered to an atlas of streamline bundles, followed by a pruning procedure of streamlines lying far away from the registered centroids [19]. Following another strategy, in [74], 800 clusters of streamlines are computed for a number of subjects that, after manual curation performed by an expert, are used for creating an atlas of streamline clusters. This curated white matter atlas (WMA) is used for filtering out streamlines far away from any cluster in the atlas. Following a different idea, BundleMAP [27] uses support vector machines on the mean and covariance of the coordinates of the streamlines in a bundle to detect FPs.

Clustering methods based on deep learning have the potential to be computationally more efficient than classical approaches. In [40], the authors proposed FS2NET, a Siamese deep neural network that uses bi-directional long short term memory (LSTM) layers for learning a distance measure between streamlines. With this distance, the method can be used to assess if two streamlines should be clustered together or not.

Implausible streamlines tend to follow more erratic paths compared to plausible ones. Thus, using clustering is appealing for filtering, since it introduces strong requirements on smoothness of streamlines. Moreover, combined with atlas-based approaches, they are able to filter both whole brain or partial tractograms. Unlike the atlas-based approaches described in Sect. 2.2, the registration is performed on tractograms, which tends to be more accurate (cf. [38] for a review of methods). Still, the inherent issues of atlas-based approaches might have an impact on the accuracy of such methods.

Since the methods of this subsection are based on bundle similarity, they target only certain bundles and, by that, only TPs.

2.5 Multiapproaches

From the previous discussion, it is natural to devise methods taking advantage of different priors for increasing accuracy. Due to the fact that the research field is relatively new, only a few multiapproach methods have been proposed. In this line, COMMIT2 [50] adds anatomy priors to the original formulation of COMMIT in order to target the issues of dMRI signal fitting methods. Another example is anchor-constrained plausibility [36], which combines streamline clustering and dMRI signal fitting for performing filtering. In [23], FiberNet2.0 has been proposed as an extension of FiberNet in which inclusion/exclusion of ROIs are added to the processing pipeline.

We have recently used deep learning for combining two methods: RecoBundles and ExTractor [25] using the dMRI signal as the only input of the neural network. From our preliminary results, it is not obvious which method should be used as gold standard, as the choice of accuracy measurement depends very much on the application. Thus, while a perfect combination of priors is not straightforward, from our experience in [25], we expect new methods that combine two or more priors to perform better on average.

3 Challenges and Perspective

Machine learning and in particular deep learning has been very successful in many medical image analysis applications in the last few years. Preliminary efforts show the potential of this approach also for the specific problem of tractogram filtering [1, 25]. However, important challenges for methods following this approach remain open.

As mentioned in [44], there are important general challenges for tractography, which also apply for tractogram filtering. Specifically, machine learning and deep learning require large amounts of training data of good quality that are difficult to obtain for tractogram filtering. Moreover, the available training data is relatively scarce and difficult to combine. Also, inter observer variability of manual annotations is particularly severe in tractography [47, 52]. Furthermore, it is in general questionable if manual annotation of whole brain tractograms would ever become a feasible goal. For this reason, the definition of adequate training labels in absence of a ground truth or a strong gold standard can be expected to remain the main challenge for machine learning-based methods in this field in the future.

One way of addressing this issue is to combine different methods as automatic annotation tools in order to define a gold standard. Following this idea, in [25], we proposed a method that builds on top of two methods, namely Recobundles and ExTractor, for defining labels. However, this combination is not straightforward. As pointed out in Table 1 and in the previous sections, it must be considered that different methods assess different characteristics of the tractogram. For example, the rejection of a streamline based on geometry priors could have a higher confidence than basing that decision on a clustering argument. The reverse is also true: a streamline close to an anatomically plausible cluster might be accepted but a streamline compliant with a finite number of geometry-based rules could still violate other unchecked rules and therefore be implausible. Filtering also depends on the application. If the goal is to obtain segmentation masks, geometrical constraints could have a lower value.

In order to investigate the process of finding a good balance in the combination of different automatic annotation tools, we run Tractquerier (TQ), which is an implementation of WMQL, Recobundles (RB), TractSeg (TS) and COMMIT (CM) in a tractogram of 10 million streamlines computed with ACT [54] for one HCP subject. A naïve approach to combine the methods would be majority voting. Figure 6 (on the left) shows the acceptance rates of streamlines for the testing dataset after performing a majority voting with different thresholds. As shown, requiring at least three methods to accept a streamline would result in a massive filtering of 95.1% of the

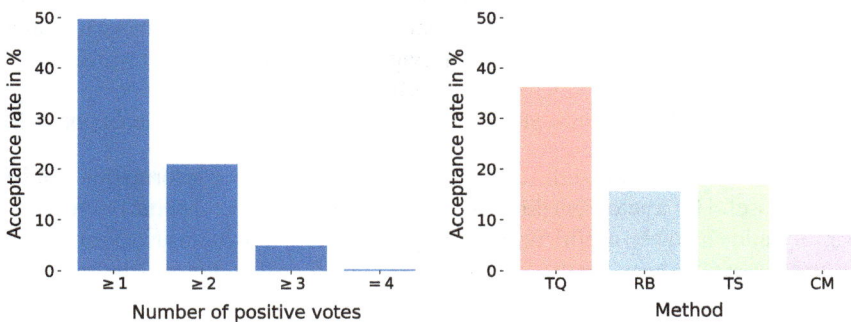

Fig. 6 Percentage of accepted streamlines obtained with Tractquerier (TQ), Recobundles (RB), TractSeg (TS) and COMMIT (CM) on a 10M tractogram of one HCP subject. **Left:** Thresholding the number of positive labels from the four methods per streamline (majority voting). **Right:** Individual methods

Fig. 7 Agreement between Tractquerier (TQ), Recobundles (RB), TractSeg (TS) and COMMIT (CM) on a 10M tractogram of one HCP subject. Every entry shows the percentage of streamlines with the same classification label obtained by the corresponding pair of methods

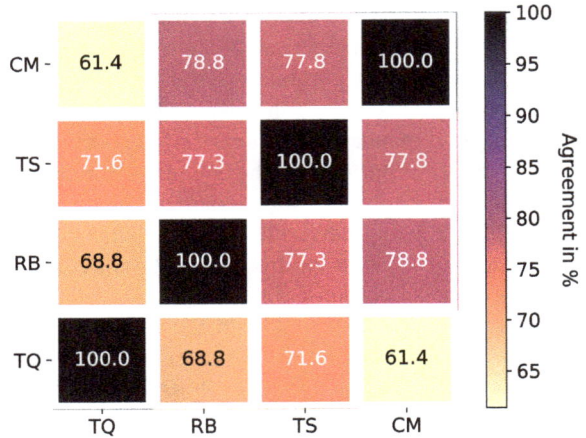

dataset. Even with a milder threshold of at least two methods, 79.1% of the dataset would be filtered out. While the amount of around 20% of accepted streamlines (i.e., two million streamlines in this example) could be enough to fill up the space of the WM with an improved TP-to-FP ratio, potential biases of such an *ad hoc* approach would need to be investigated. Also the low agreement for only 50.6% of the streamlines (0.2% for four positive votes and 50.4% for four negative votes) indicates that a more sophisticated strategy for the combination of different tools should be considered. This could maintain a higher acceptance rate of TP streamlines and by that potentially reduce the required number of streamlines in the tractogram.

Figures 6 (on the right) and 7 show the percentage of streamlines accepted by each method as well as their agreement, respectively. As shown, the most restrictive method is CM and the most relaxed one is TQ. While the other two methods are in the middle, they are also rather restrictive. Moreover, any pair of methods agrees in 60–80% of the streamlines. From these initial analyses, it is clear that more research is needed in order to find better ways of synthesising information from different methods for the purpose of tractogram filtering than just simple majority voting. Again, the final application must also be considered for assessing the ideal approach.

In addition to combining different methods, using other prior information can be potentially useful for tractogram filtering. For example, including specific microstructure information has been useful for tractography [20, 51], which can also be expected from tractogram filtering methods. Combining dMRI and functional information is promising to understand the mechanisms for brain connectivity [14]. Thus, it would be interesting to explore in the future the use of functional data for tractogram filtering.

4 Conclusion

Tractogram filtering is a relatively young but very active research area with a high potential of development. While the reviewed methods in this chapter show that there are a multitude of ways to obtain information related to the plausibility of streamlines, there is yet no holistic approach to separate the TP and FP streamlines in a tractogram in a fully satisfying way. In our opinion, machine learning-based methods have the potential to contribute substantially to tractogram filtering. However, at this moment the applicability of supervised approaches is tightly coupled to the proper definition of training labels, which is difficult to obtain in the absence of a ground truth. We see the combination of different automatic annotation tools, potentially complemented with manual annotations from neuroanatomists, as a promising avenue to address this problem, while developments are still needed in that line of research in the future.

Acknowledgments This work was partially supported by VINNOVA, through AIDA. We thank students Blanca Bastardés-Climent and Pehr Wessmark for their involvement in the project, and Jakob Wasserthal for computing and providing the tractogram we used in Sect. 3. We also thank the University of Sherbrooke Research Chair in Neuroinformatics and ComputeCanada for their invaluable computing resources. We used data from the Human Connectome Project, WU-Minn Consortium (Principal Investigators: David Van Essen and Kamil Ugurbil; 1U54MH091657) funded by the 16 NIH Institutes and Centers that support the NIH Blueprint for Neuroscience Research; and by the McDonnell Center for Systems Neuroscience at Washington University.

References

1. Astolfi, P., Verhagen, R., Petit, L., Olivetti, E., Masci, J., Boscaini, D., Avesani, P.: Tractogram filtering of anatomically non-plausible fibers with geometric deep learning (2020). arXiv:2003.11013v1
2. Aydogan, D.B., Jacobs, R., Dulawa, S., Thompson, S.L., Francois, M.C., Toga, A.W., Dong, H., Knowles, J.A., Shi, Y.: When tractography meets tracer injections: a systematic study of trends and variation sources of diffusion-based connectivity. Brain Struct. Funct. **223**(6), 2841–2858 (2018)
3. Aydogan, D.B., Shi, Y.: Track filtering via iterative correction of TDI topology. Proc. MICCAI **9349**, 20–27 (2015)
4. Aydogan, D.B., Shi, Y.: Tracking and validation techniques for topographically organized tractography. NeuroImage **181**, 64–84 (2018)
5. Boukadi, M., Marcotte, K., Bedetti, C., Houde, J.C., Desautels, A., Deslauriers-Gauthier, S., Chapleau, M., Boré, A., Descoteaux, M., Brambati, S.M.: Test-retest reliability of diffusion measures extracted along white matter language fiber bundles using HARDI-based tractography. Front. Neurosci **12**, 1055 (2019)
6. Brusini, I., Jörgens, D., Smedby, O., Moreno, R.: Voxel-wise clustering of tractography data for building atlases of local fiber geometry. Math. Visual. **226249**, 345–357 (2019)
7. Caiafa, C.F., Pestilli, F.: Multidimensional encoding of brain connectomes. Sci. Rep. **7**(1), 11491 (2017)
8. Colby, J.B., Soderberg, L., Lebel, C., Dinov, I.D., Thompson, P.M., Sowell, E.R.: Along-tract statistics allow for enhanced tractography analysis. NeuroImage **59**(4), 3227–3242 (2012)
9. Costabile, J.D., Alaswad, E., D'Souza, S., Thompson, J.A., Ormond, D.R.: Current Applications of diffusion tensor imaging and tractography in intracranial tumor resection. Front. Oncol. **9**, 426 (2019)

10. Cousineau, M., Jodoin, P.M., Garyfallidis, E., Côté, M.A., Morency, F.C., Rozanski, V., Grand'Maison, M., Bedell, B.J., Descoteaux, M.: A test-retest study on Parkinson's PPMI dataset yields statistically significant white matter fascicles. NeuroImage: Clin. **16**, 222–233 (2017)

11. Daducci, A., Dal Palú, A., Descoteaux, M., Thiran, J.P.: Microstructure informed tractography: pitfalls and open challenges. Front. Neurosci. **10**, 247 (2016)

12. Daducci, A., Dal Palu, A., Lemkaddem, A., Thiran, J.P.: COMMIT: Convex optimization modeling for microstructure informed tractography. IEEE Trans. Med. Imaging **34**(1), 246–257 (2015)

13. Damoiseaux, J.S.: Effects of aging on functional and structural brain connectivity. NeuroImage **160**, 32–40 (2017)

14. Deslauriers-Gauthier, S., Lina, J.M., Butler, R., Whittingstall, K., Gilbert, G., Bernier, P.M., Deriche, R., Descoteaux, M.: White matter information flow mapping from diffusion MRI and EEG. NeuroImage **201**, 116017 (2019)

15. Dong, X., Zhang, Z., Srivastava, A.: Bayesian tractography using geometric shape priors. Front. Neurosci. **11**, 483 (2017)

16. Essayed, W.I., Zhang, F., Unadkat, P., Cosgrove, G.R., Golby, A.J., O'Donnell, L.J.: White matter tractography for neurosurgical planning: a topography-based review of the current state of the art. NeuroImage. Clin. **15**, 659–672 (2017)

17. Frau-Pascual, A., Fogarty, M., Fischl, B., Yendiki, A., Aganj, I.: Quantification of structural brain connectivity via a conductance model. NeuroImage **189**, 485–496 (2019)

18. Garyfallidis, E., Brett, M., Correia, M.M., Williams, G.B., Nimmo-Smith, I.: QuickBundles, a method for tractography simplification. Front. Neurosci. **6**, 175 (2012)

19. Garyfallidis, E., Côté, M.A., Rheault, F., Sidhu, J., Hau, J., Petit, L., Fortin, D., Cunanne, S., Descoteaux, M.: Recognition of white matter bundles using local and global streamline-based registration and clustering. NeuroImage **170**, 283–295 (2018)

20. Girard, G., Daducci, A., Petit, L., Thiran, J.P., Whittingstall, K., Deriche, R., Wassermann, D., Descoteaux, M.: AxTract: toward microstructure informed tractography. Hum. Brain Mapp. **38**(11), 5485–5500 (2017)

21. Gleichgerrcht, E., Munsell, B., Bhatia, S., Vandergrift, W.A., Rorden, C., McDonald, C., Edwards, J., Kuzniecky, R., Bonilha, L.: Deep learning applied to whole-brain connectome to determine seizure control after epilepsy surgery. Epilepsia **59**(9), 1643–1654 (2018)

22. de Groot, M., Vernooij, M.W., Klein, S., Ikram, M.A., Vos, F.M., Smith, S.M., Niessen, W.J., Andersson, J.L.R.: Improving alignment in tract-based spatial statistics: evaluation and optimization of image registration. NeuroImage **76**, 400–11 (2013)

23. Gupta, V., Thomopoulos, S.I., Corbin, C.K., Rashid, F., Thompson, P.M.: FIBERNET 2.0: an automatic neural network based tool for clustering white matter fibers in the brain. In: Proceedings ISBI, pp. 708–711 (2018)

24. Gupta, V., Thomopoulos, S.I., Rashid, F.M., Thompson, P.M.: FiberNET: an ensemble deep learning framework for clustering white matter fibers. Proc. MICCAI **10433**, 548–555 (2017)

25. Jörgens, D., Poulin, P., Moreno, R., Jodoin, P.M., Descoteaux, M.: Towards a deep learning model for diffusion-aware tractogram filtering. In: Proceeding ISMRM, p. 3375 (2019)

26. Kasenburg, N., Liptrot, M., Reislev, N.L., Ørting, S.N., Nielsen, M., Garde, E., Feragen, A.: Training shortest-path tractography: automatic learning of spatial priors. NeuroImage **130**, 63–76 (2016)

27. Khatami, M., Schmidt-Wilcke, T., Sundgren, P.C., Abbasloo, A., Schölkopf, B., Schultz, T.: BundleMAP: anatomically localized classification, regression, and hypothesis testing in diffusion MRI. Pattern Recognit. **63**, 593–600 (2017)

28. Kim, D., Sra, S., Dhillon, I.S.: A non-monotonic method for large-scale non-negative least squares. Optim. Methods Softw. **28**(5), 1012–1039 (2013)

29. Kumar, S., Sreenivasan, V., Talukdar, P., Pestilli, F., Sridharan, D.: ReAl-LiFE: accelerating the discovery of individualized brain connectomes on GPUs. Proceedings AAAI Conference on Artificial Intelligence. **33**(1), 630–638 (2019)

30. Liu, F., Feng, J., Chen, G., Wu, Y., Hong, Y., Yap, P.T., Shen, D.: DeepBundle: fiber bundle parcellation with graph convolution neural networks. In: Proceedings International Workshop on Graph Learning in Medical Imaging, pp. 88–95 (2019)
31. Liu, Y., Duan, Y., Dong, H., Barkhof, F., Li, K., Shu, N.: Disrupted module efficiency of structural and functional brain connectomes in clinically isolated syndrome and multiple sclerosis. Front. Hum. Neurosci. **12**, 138 (2018)
32. Maier-Hein, K.H., Neher, P.F., Houde, J.C., Côté, M.A., Garyfallidis, E., Zhong, J., Chamberland, M., Yeh, F.C., Lin, Y.C., Ji, Q., Reddick, W.E., Glass, J.O., Chen, D.Q., Feng, Y., Gao, C., Wu, Y., Ma, J., Renjie, H., Li, Q., Westin, C.F., Deslauriers-Gauthier, S., González, J.O.O., Paquette, M., St-Jean, S., Girard, G., Rheault, F., Sidhu, J., Tax, C.M.W., Guo, F., Mesri, H.Y., Dávid, S., Froeling, M., Heemskerk, A.M., Leemans, A., Boré, A., Pinsard, B., Bedetti, C., Desrosiers, M., Brambati, S., Doyon, J., Sarica, A., Vasta, R., Cerasa, A., Quattrone, A., Yeatman, J., Khan, A.R., Hodges, W., Alexander, S., Romascano, D., Barakovic, M., Auría, A., Esteban, O., Lemkaddem, A., Thiran, J.P., Cetingul, H.E., Odry, B.L., Mailhe, B., Nadar, M.S., Pizzagalli, F., Prasad, G., Villalon-Reina, J.E., Galvis, J., Thompson, P.M., Requejo, F.D.S., Laguna, P.L., Lacerda, L.M., Barrett, R., Dell'Acqua, F., Catani, M., Petit, L., Caruyer, E., Daducci, A., Dyrby, T.B., Holland-Letz, T., Hilgetag, C.C., Stieltjes, B., Descoteaux, M.: The challenge of mapping the human connectome based on diffusion tractography. Nat. Commun. **8**(1), 1349 (2017)
33. Mangin, J.F., Fillard, P., Cointepas, Y., Le Bihan, D., Frouin, V., Poupon, C.: Toward global tractography. NeuroImage **80**, 290–296 (2013)
34. Nath, V., Schilling, K.G., Parvathaneni, P., Huo, Y., Blaber, J.A., Hainline, A.E., Barakovic, M., Romascano, D., Rafael-Patino, J., Frigo, M., Girard, G., Thiran, J., Daducci, A., Rowe, M., Rodrigues, P., Prčkovska, V., Aydogan, D.B., Sun, W., Shi, Y., Parker, W.A., Ould Ismail, A.A., Verma, R., Cabeen, R.P., Toga, A.W., Newton, A.T., Wasserthal, J., Neher, P., Maier-Hein, K., Savini, G., Palesi, F., Kaden, E., Wu, Y., He, J., Feng, Y., Paquette, M., Rheault, F., Sidhu, J., Lebel, C., Leemans, A., Descoteaux, M., Dyrby, T.B., Kang, H., Landman, B.A.: Tractography reproducibility challenge with empirical data (TraCED): The 2017 ISMRM diffusion study group challenge. J. Magn. Reson. Imaging **51**(1), 234–249 (2020)
35. Neher, P.F., Laun, F.B., Stieltjes, B., Maier-Hein, K.H.: Fiberfox: facilitating the creation of realistic white matter software phantoms. Magn. Reson. Med. **72**(5), 1460–1470 (2014)
36. Neher, P.F., Stieltjes, B., Maier-Hein, K.H.: Anchor-Constrained Plausibility (ACP): A Novel Concept for Assessing Tractography and Reducing False-Positives. pp. 20–27 (2018)
37. Ngattai Lam, P.D., Belhomme, G., Ferrall, J., Patterson, B., Styner, M., Prieto, J.C.: TRAFIC: fiber tract classification using deep learning. In: Proceedings of SPIE **10574** (2018)
38. O'Donnell, L.J., Daducci, A., Wassermann, D., Lenglet, C.: Advances in computational and statistical diffusion MRI. NMR Biomed. **32**(4), e3805 (2019)
39. Panagiotaki, E., Schneider, T., Siow, B., Hall, M.G., Lythgoe, M.F., Alexander, D.C.: Compartment models of the diffusion MR signal in brain white matter: A taxonomy and comparison. NeuroImage **59**(3), 2241–2254 (2012)
40. Patil, S.M., Nigam, A., Bhavsar, A., Chattopadhyay, C.: Siamese LSTM based Fiber Structural Similarity Network (FS2Net) for Rotation Invariant Brain Tractography Segmentation. Proceedings of International Conference on Computer Analysis of Images and Patterns **11679**, 459–469 (2017)
41. Pereira, J.B., van Westen, D., Stomrud, E., Strandberg, T.O., Volpe, G., Westman, E., Hansson, O.: Abnormal structural brain connectome in individuals with preclinical Alzheimer's disease. Cereb. Cortex **28**(10), 3638–3649 (2018)
42. Pestilli, F., Yeatman, J.D., Rokem, A., Kay, K.N., Wandell, B.A.: Evaluation and statistical inference for human connectomes. Nat. Methods **11**(10), 1058–1063 (2014)
43. Petit, L., Rheault, F., Descoteaux, M., Tzourio-Mazoyer, N.: Half of the streamlines built in a whole human brain tractogram is anatomically uninterpretable. In: Proceedings OHBM, p. Th785 (2019)
44. Poulin, P., Jörgens, D., Jodoin, P.M., Descoteaux, M.: Tractography and machine learning: current state and open challenges. Magn. Reson. Imaging **64**, 37–48 (2019)

45. Qi, S., Meesters, S., Nicolay, K., Ter Haar Romeny, B.M., Ossenblok, P.: Structural brain network: what is the effect of LiFE optimization of whole brain tractography? Front. Comput. Neurosci. **10**, 12 (2016)
46. Reisert, M., Mader, I., Anastasopoulos, C., Weigel, M., Schnell, S., Kiselev, V.: Global fiber reconstruction becomes practical. NeuroImage **54**(2), 955–962 (2011)
47. Rheault, F., De Benedictis, A., Daducci, A., Maffei, C., Tax, C.M.W., Romascano, D., Caverzasi, E., Morency, F.C., Corrivetti, F., Pestilli, F., Girard, G., Theaud, G., Zemmoura, I., Hau, J., Glavin, K., Jordan, K.M., Pomiecko, K., Chamberland, M., Barakovic, M., Goyette, N., Poulin, P., Chenot, Q., Panesar, S.S., Sarubbo, S., Petit, L., Descoteaux, M.: Tractostorm: The what, why, and how of tractography dissection reproducibility. In: Human Brain Mapping, p. hbm.24917 (2020)
48. Ronneberger, O., Fischer, P., Brox, T.: U-Net: Convolutional Networks for Biomedical Image Segmentation. pp. 234–241 (2015)
49. Rubinov, M., Sporns, O.: Complex network measures of brain connectivity: uses and interpretations. NeuroImage **52**(3), 1059–1069 (2010)
50. Schiavi, S., Barakovic, M., Ocampo-Pineda, M., Descoteaux, M., Thiran, J.P., Daducci, A.: Reducing false positives in tractography with microstructural and anatomical priors. bioRxiv p. 608349 (2019)
51. Schiavi, S., Pizzolato, M., Ocampo-Pineda, M., Canales-Rodriguez, E., Thiran, J., Daducci, A.: Is it feasible to directly access the bundle's specific myelin content, instead of averaging? A study with Microstructure Informed Tractography. In: Proceedings ISMRM, p. 3369 (2019)
52. Schilling, K.G., Daducci, A., Maier-Hein, K., Poupon, C., Houde, J.C., Nath, V., Anderson, A.W., Landman, B.A., Descoteaux, M.: Challenges in diffusion MRI tractography–lessons learned from international benchmark competitions. Magn. Reson. Imaging **57**, 194–209 (2019)
53. Schilling, K.G., Nath, V., Hansen, C., Parvathaneni, P., Blaber, J., Gao, Y., Neher, P., Aydogan, D.B., Shi, Y., Ocampo-Pineda, M., Schiavi, S., Daducci, A., Girard, G., Barakovic, M., Rafael-Patino, J., Romascano, D., Rensonnet, G., Pizzolato, M., Bates, A., Fischi, E., Thiran, J.P., Canales-Rodríguez, E.J., Huang, C., Zhu, H., Zhong, L., Cabeen, R., Toga, A.W., Rheault, F., Theaud, G., Houde, J.C., Sidhu, J., Chamberland, M., Westin, C.F., Dyrby, T.B., Verma, R., Rathi, Y., Irfanoglu, M.O., Thomas, C., Pierpaoli, C., Descoteaux, M., Anderson, A.W., Landman, B.A.: Limits to anatomical accuracy of diffusion tractography using modern approaches. NeuroImage **185**, 1–11 (2019)
54. Smith, R.E., Tournier, J.D., Calamante, F., Connelly, A.: Anatomically-constrained tractography: improved diffusion MRI streamlines tractography through effective use of anatomical information. NeuroImage **62**(3), 1924–1938 (2012)
55. Smith, R.E., Tournier, J.D., Calamante, F., Connelly, A.: SIFT: Spherical-deconvolution informed filtering of tractograms. NeuroImage **67**, 298–312 (2013)
56. Smith, R.E., Tournier, J.D., Calamante, F., Connelly, A.: SIFT2: enabling dense quantitative assessment of brain white matter connectivity using streamlines tractography. NeuroImage **119**, 338–351 (2015)
57. Smith, R.E., Tournier, J.D., Calamante, F., Connelly, A.: The effects of SIFT on the reproducibility and biological accuracy of the structural connectome. NeuroImage **104**, 253–265 (2015)
58. Smith, S.M., Jenkinson, M., Johansen-Berg, H., Rueckert, D., Nichols, T.E., Mackay, C.E., Watkins, K.E., Ciccarelli, O., Cader, M.Z., Matthews, P.M., Behrens, T.E.: Tract-based spatial statistics: voxelwise analysis of multi-subject diffusion data. NeuroImage **31**(4), 1487–1505 (2006)
59. Sotiropoulos, S.N., Zalesky, A.: Building connectomes using diffusion MRI: why, how and but. NMR Biomed. **32**(4), e3752 (2019)
60. Sydnor, V.J., Rivas-Grajales, A.M., Lyall, A.E., Zhang, F., Bouix, S., Karmacharya, S., Shenton, M.E., Westin, C.F., Makris, N., Wassermann, D., O'Donnell, L.J., Kubicki, M.: A comparison of three fiber tract delineation methods and their impact on white matter analysis. NeuroImage **178**, 318–331 (2018)

61. Tax, C.M., Dela Haije, T., Fuster, A., Westin, C.F., Viergever, M.A., Florack, L., Leemans, A.: Sheet Probability Index (SPI): characterizing the geometrical organization of the white matter with diffusion MRI. NeuroImage **142**, 260–279 (2016)
62. Thomas, C., Ye, F.Q., Irfanoglu, M.O., Modi, P., Saleem, K.S., Leopold, D.A., Pierpaoli, C.: Anatomical accuracy of brain connections derived from diffusion MRI tractography is inherently limited. PNAS **96**(18), 10422–10427 (2014)
63. Tournier, J.D., Calamante, F., Gadian, D.G., Connelly, A.: Direct estimation of the fiber orientation density function from diffusion-weighted MRI data using spherical deconvolution. NeuroImage **23**(3), 1176–1185 (2004)
64. Vanderweyen, D.C., Theaud, G., Sidhu, J., Sarubbo, S., Descoteaux, M., Fortin, D.: The role of diffusion tractography in refining glial tumor resection. Brain Struct. Funct. (0123456789), in press (2020)
65. Wang, J., Aydogan, D.B., Varma, R., Toga, A.W., Shi, Y.: Modeling topographic regularity in structural brain connectivity with application to tractogram filtering. NeuroImage **183**, 87–98 (2018)
66. Wassermann, D., Makris, N., Rathi, Y., Shenton, M., Kikinis, R., Kubicki, M., Westin, C.F.: The white matter query language: a novel approach for describing human white matter anatomy. Brain Struct. Funct. **221**(9), 4705–4721 (2016)
67. Wasserthal, J., Neher, P., Maier-Hein, K.H.: TractSeg–fast and accurate white matter tract segmentation. NeuroImage **183**, 239–253 (2018)
68. Wasserthal, J., Neher, P.F., Hirjak, D., Maier-Hein, K.H.: Combined tract segmentation and orientation mapping for bundle-specific tractography. Med. Image Anal. **58**, 101559 (2019)
69. Wessmark, P.: An Exploratory Approach to Generate Ground Truths of Neural Fiber Bundles. Master's thesis, KTH Royal Institute of Technology (2017)
70. Yeatman, J.D., Dougherty, R.F., Myall, N.J., Wandell, B.A., Feldman, H.M.: Tract profiles of white matter properties: automating fiber-tract quantification. PloS one **7**(11), e49790 (2012)
71. Zalesky, A., Fornito, A., Cocchi, L., Gollo, L.L., van den Heuvel, M.P., Breakspear, M.: Connectome sensitivity or specificity: which is more important? NeuroImage **142**, 407–420 (2016)
72. Zalesky, A., Sarwar, T., Ramamohanarao, K.: A cautionary note on the use of SIFT in pathological connectomes. Magn. Reson. Med. **83**(3), 791–794 (2020)
73. Zhang, F., Hoffmann, N., Karayumak, S.C., Rathi, Y., Golby, A.J., O'Donnell, L.J.: Deep White Matter Analysis: Fast, Consistent Tractography Segmentation Across Populations and dMRI Acquisitions, pp. 599–608 (2019)
74. Zhang, F., Wu, Y., Norton, I., Rigolo, L., Rathi, Y., Makris, N., O'Donnell, L.J.: An anatomically curated fiber clustering white matter atlas for consistent white matter tract parcellation across the lifespan. NeuroImage **179**, 429–447 (2018)

7

Fourth-Order Anisotropic Diffusion for Inpainting and Image Compression

Ikram Jumakulyyev and Thomas Schultz

Abstract Edge-enhancing diffusion (EED) can reconstruct a close approximation of an original image from a small subset of its pixels. This makes it an attractive foundation for PDE based image compression. In this work, we generalize second-order EED to a fourth-order counterpart. It involves a fourth-order diffusion tensor that is constructed from the regularized image gradient in a similar way as in traditional second-order EED, permitting diffusion along edges, while applying a non-linear diffusivity function across them. We show that our fourth-order diffusion tensor formalism provides a unifying framework for all previous anisotropic fourth-order diffusion based methods, and that it provides additional flexibility. We achieve an efficient implementation using a fast semi-iterative scheme. Experimental results on natural and medical images suggest that our novel fourth-order method produces more accurate reconstructions compared to the existing second-order EED.

1 Introduction

The increased availability and resolution of imaging technology, including digital cameras and medical imaging devices, along with advances in storage capacity and transfer bandwidths, have led to a proliferation of large image data. This makes image compression an important area of research. Image compression techniques can be divided into two main groups: Lossy and lossless compression. Lossless compression techniques permit restoration of the full, unmodified image data, which however limits the achievable compression rates. Our work is concerned with lossy compression, which achieves much higher compression rates by replacing the original image with an approximation that can be stored more efficiently.

I. Jumakulyyev · T. Schultz (✉)
B-IT and Department of Computer Science II, University of Bonn,
Friedrich-Hirzebruch-Allee 5, 53115 Bonn, Germany
e-mail: schultz@cs.uni-bonn.de

I. Jumakulyyev
e-mail: ijumakulyyev@cs.uni-bonn.de

We continue a line of research that has explored the use of Partial Differential Equations (PDEs) for lossy image compression [12, 13, 23, 28]. This approach is based on storing only a small subset of all pixels, and interpolating between them in order to restore the remaining ones. There is a strong similarity between that interpolation process and image inpainting, whose goal it is to reconstruct missing or corrupted parts of an image. PDE-based methods for image inpainting and compression are inspired by the physical phenomenon of heat transport. It is described by the heat diffusion equation

$$\partial_t u = \mathrm{div}(D \cdot \nabla u) , \qquad (1)$$

which relates temporal changes in a heat concentration $\partial_t u$ to the divergence of its spatial gradient ∇u. When diffusion takes place in an isotropic medium, the diffusivity D is a scalar that determines the rate of heat transfer. In an anisotropic medium, heat spreads out more rapidly in some directions than in others. In those cases, D is a diffusion tensor, i.e., a symmetric matrix that encodes this directional dependence.

When applied to image processing, the gray value at a certain location is interpreted as the heat concentration u. In diffusion-based image inpainting, Eq. (1) is used to propagate information from the known pixels, whose intensity is fixed, to the unknown pixels which will ultimately reach a steady state in which their intensity is determined by their surrounding known pixels. In this sense, Eq. (1) has a filling-in effect that can be exploited for image compression.

Different choices of the diffusivity function D lead to different kinds of diffusion. Linear diffusion [20] and nonlinear diffusion [24] were widely used for image smoothing and image enhancement. Edge structures in images can be enhanced by employing a diffusion tensor which allows diffusion in the direction perpendicular to the local gradient, while applying a nonlinear diffusivity function along the gradient direction. This idea has led to the development of anisotropic nonlinear edge-enhancing diffusion (EED) [34]. Among the six variants that were evaluated for image compression by Galić et al. [13], EED led to the most accurate reconstructions. Subsequently, this idea was applied to three-dimensional data compression [25], and combined with motion compensation in order to obtain a framework for video compression [2]. When combined with a suitable scheme for selecting and storing the preserved pixels, a few additional optimizations, and at sufficiently high compression rates, anisotropic diffusion has been shown to beat the quality even of JPEG2000 [27].

In this paper, we introduce a novel fourth-order PDE that generalizes second-order EED, and achieves even more accurate reconstructions. We build on prior works that proposed fourth-order analogs of the diffusion equation, and used them for image processing [10, 19, 21, 22, 26, 35]. In particular, we extend a work by Gorgi Zadeh et al. [15], who introduced the idea of steering anisotropic fourth-order diffusion with a fourth-order diffusion tensor. However, their method focuses on the curvature enhancement property of nonlinear fourth-order diffusion [10] in order to better localize ridge and valley structures. Deriving a suitable PDE for image inpainting requires a different definition of the diffusion tensor, more similar to the one in edge-enhancing diffusion [34]. Two anisotropic fourth-order PDEs for inpainting were previously introduced by Li et al. [21]. However, they only apply them to image restoration tasks in which small parts of an image are missing (such as in Fig. 8), not to the reconstruction from a small subset of pixels. Moreover, we demonstrate that the fourth-order diffusion tensor based framework is more general in the sense that it can be used to express anisotropic fourth-order diffusion as it was described by Hajiaboli [19] or by Li et al. [21], while providing additional flexibility.

2   Background and Related Work

We will now formalize the above-mentioned idea of diffusion-based inpainting (Sect. 2.1), and review two concepts that play a central role in our method: Anisotropic nonlinear diffusion (Sect. 2.2) and fourth-order diffusion (Sect. 2.3). Further details can be found in works by Galić et al. [13] and Weickert [34], respectively. Finally, we provide additional context with a brief discussion of alternative approaches to image compression (Sect. 2.4).

2.1   Diffusion-Based Inpainting

In order to apply Eq. (1) to image smoothing, we have to restrict it to the image domain Ω, and specify the behavior along its boundary $\partial\Omega$. It is common to assume that no heat is transferred through that boundary (homogeneous Neumann boundary condition). Moreover, the positive real line $(0, \infty)$ is typically taken as the time domain. The resulting PDE can be written as

$$
\begin{aligned}
\partial_t u &= \mathrm{div}(D \cdot \nabla u), \quad \Omega \times (0, \infty) , \\
\partial_n u &= 0, \quad \partial\Omega \times (0, \infty) ,
\end{aligned}
\tag{2}
$$

where n is the normal vector to the boundary $\partial\Omega$. The original image $f : \Omega \to \mathbb{R}$ is used to specify an initial condition $u = f$ at $t = 0$. For increasing diffusion time t, u will correspond to an increasingly smoothed version of the image.

In image inpainting, we know the pixel values on a subset $K \subset \Omega$ of the image, and aim to reconstruct plausible values in the unknown regions. A diffusion-based model for inpainting can be derived from the one for smoothing, by modeling the set of locations at which the pixel values are known with Dirichlet boundary conditions. In this case, $f : K \to \mathbb{R}$ will be used to model the known values. In inpainting-based image compression, K will consist of a small fraction of the pixels in the original image. With this, we obtain the following model for inpainting:

$$\partial_t u = \text{div}(D \cdot \nabla u), \quad \Omega \backslash K \times (0, \infty),$$
$$\partial_n u = 0, \quad \partial\Omega \times (0, \infty), \tag{3}$$
$$u = f, \quad K \times [0, \infty)$$

In this case, the diffusion process spreads out the information from the known pixels to their spatial neighborhood. For time $t \to \infty$, image smoothing and inpainting both converge to a steady state, i.e., $\lim_{t \to \infty} \partial_t u = 0$. However, the steady-state of smoothing is trivial (u approaches a constant image with average gray value), while the Dirichlet boundary conditions in the inpainting case ensure a non-trivial steady-state, which is taken as the final inpainting result: $u_{\text{inpainted}} = \lim_{t \to \infty} u$.

2.2   From Linear to Anisotropic Nonlinear Diffusion

So far, we assumed that the diffusion coefficient D is a scalar and constant, independent from the location within the image. This results in an inpainting model based on linear homogeneous diffusion [20]. With $D = 1$, it can be written as

$$\partial_t u = \Delta u, \quad \Omega \backslash K \times (0, \infty). \tag{4}$$

In this and all remaining equations in this section, the same boundary conditions are assumed as specified in Eq. (3). Despite its simplicity, it has been demonstrated that using this inpainting model for image compression can already beat the JPEG standard when applied to cartoon-like images, and selecting the retained pixels to be close to image edges [23].

When the diffusion coefficient is a scalar but depends on u, i.e., $D = g(u)$, then we call the model inpainting based on nonlinear isotropic diffusion [24]. A common variant is to make D depend on the local gradient magnitude, i.e.,

$$\partial_t u = \text{div}(g(||\nabla u_\sigma||^2)\nabla u), \quad \Omega \backslash K \times (0, \infty), \tag{5}$$

where g is a decreasing nonnegative diffusivity function, e.g., the Charbonnier diffusivity

$$g(s^2) = \frac{1}{\sqrt{1 + \frac{s^2}{\lambda^2}}}, \tag{6}$$

and λ is a contrast parameter separating low from high diffusion areas [8]. In order to localize edges better and to make the problem well-posed, the image is pre-smoothed with a Gaussian before taking its gradient, i.e., $g(||\nabla u_\sigma||^2)$ is used instead of $g(||\nabla u||^2)$ [6].

In the above-discussed models, the diffusion occurs only in the gradient direction. This can be changed by replacing the scalar diffusivity with a second-order diffusion tensor, i.e., a symmetric positive definite matrix. This is the basis of anisotropic nonlinear diffusion [34],

$$\partial_t u = \text{div}(\mathbf{D} \cdot \nabla u), \quad \Omega \backslash K \times (0, \infty) . \tag{7}$$

In edge-enhancing diffusion (EED), the diffusion tensor $\mathbf{D}$ is defined as

$$\mathbf{D} = g(||\nabla u_\sigma||^2) \cdot \mathbf{v}_1 \mathbf{v}_1^{\mathrm{T}} + 1 \cdot \mathbf{v}_2 \mathbf{v}_2^{\mathrm{T}}, \tag{8}$$

where $\mathbf{v}_1 = \frac{\nabla u_\sigma}{||\nabla u_\sigma||_2}$ and $\mathbf{v}_2 = \frac{\nabla u_\sigma^\perp}{||\nabla u_\sigma||_2}$. This means that diffusion across the edge ($\mathbf{v}_1$) is decreased depending on the gradient magnitude, while diffusion along the edge ($\mathbf{v}_2$) is allowed. Examples of EED based inpainting are included in our experimental results. In general, EED based inpainting results in better interpolated images than linear homogeneous or nonlinear isotropic PDEs. This makes it a current state-of-the-art choice for PDE-based image compression.

2.3 From Second to Fourth Order Diffusion

All models discussed above, as well as several others that have been proposed for inpainting [36], share a common property: They rely on second order PDEs. In image denoising, higher-order PDEs have a long history, going back to work by Scherzer [26]. You and Kaveh [35] propose fourth-order PDEs as a solution to the so-called staircasing problem that arises in edge-enhancing second-order PDEs, such as the filter proposed by Perona and Malik [24]: While the second-order Perona-Malik equation creates visually unpleasant step edges from continuous variations of intensity, corresponding fourth-order methods move these discontinuities into the gradients, where they are less noticeable to the human eye [16]. Subsequently, other fourth-order PDE-based models have been introduced, and have mostly been applied for denoising [18, 19, 22].

For a specific family of higher-order diffusion filters, Didas et al. [10] have shown that, in addition to preserving average gray value, they also preserve higher moments of the initial image. Moreover, depending on the diffusivity function, they can lead to adaptive forward and backward diffusion, and therefore to the enhancement of image features such as curvature. Gorgi Zadeh et al. [15] made use of this property in order to enhance ridges and valleys, by steering fourth-order diffusion with a fourth-order diffusion tensor. Our work adapts their method in order to achieve accurate inpainting and reconstruction from a small subset of pixels.

2.4 Alternative Approaches to Image Compression

The dominant lossy image compression techniques today are JPEG and JPEG2000. They are based on the discrete cosine transform (DCT) and wavelet transform, respectively. However, they are not sensitive to the geometry of an image, i.e. those standards are not tailored to their geometrical behavior [5]. Especially, the JPEG standard involves dividing the image into small square blocks. This can cause a degradation called "blocking effect" [30], and can result in unsatisfactory reconstructions especially at high compression rates.

It is an ongoing research trend to apply machine learning methods to image compression, such as convolutional and recurrent neural networks [3, 31, 32]. Learning based approaches tend to perform very well on the specific class of images on which they were trained, but require a huge amount of data. For example, Toderici et al. [33] used for training a dataset of 6 million 1280×720 images taken from the web.

3 Method

We will now introduce our novel PDE (Sect. 3.1), investigate its relationship to previously proposed anisotropic fourth-order diffusion (Sect. 3.2), and comment on our chosen discretization, as well as numerical stability (Sect. 3.3).

3.1 Anisotropic Edge-Enhancing Fourth Order PDE

Our fourth-order PDE builds on a model that was proposed by Gorgi Zadeh et al. [15] for ridge and valley enhancement. It can be stated concisely using Einstein notation, where summation is implied for indices appearing twice in the same expression:

$$\partial_t u = -\partial_{ji} \left[\mathcal{D}(\mathbf{H}_\rho(u_\sigma)) : \mathbf{H}(u) \right]_{ij} \qquad (9)$$

In this equation, $\mathbf{H}(u)$ denotes the Hessian matrix of image u. The "double dot product" $\mathbf{T} = \mathcal{D} : \mathbf{H}$ indicates that matrix $\mathbf{T}$ is obtained by applying a linear map $\mathcal{D}$ to $\mathbf{H}$, and the square bracket notation $[\mathbf{T}]_{ij}$ indicates taking the (i, j)th component:

$$[\mathbf{T}]_{ij} = \left[\mathcal{D}(\mathbf{H}_\rho(u_\sigma)) : \mathbf{H}(u) \right]_{ij} = \left[\mathcal{D}(\mathbf{H}_\rho(u_\sigma)) \right]_{ijkl} [\mathbf{H}(u)]_{kl} \qquad (10)$$

Since $\mathcal{D}$ maps matrices to matrices, it is a fourth-order tensor. Since its role is analogous to the second-order diffusion tensor in Eq. (7), it is referred to as a fourth-order diffusion tensor.

The diffusion tensor $\mathcal{D}$ in Eq. (9) is a function of the local normalized Hessian $\mathbf{H}_\rho(u_\sigma)$ which contains the information that is relevant to achieve curvature enhancement. For image inpainting, we propose to instead steer the fourth-order diffusion in analogy to edge-enhancing diffusion, i.e., as a function of the structure tensor $\mathbf{J}(u_\sigma)$, which is obtained from image u after Gaussian pre-smoothing with bandwidth σ. We construct our fourth order diffusion tensor $\mathcal{D}$ from its eigenvalues μ_i and eigentensors $\mathbf{E}_i$ via the spectral decomposition:

$$\mathcal{D}(\mathbf{J}(u_\sigma)) = \mu_1 \mathbf{E}_1 \otimes \mathbf{E}_1 + \mu_2 \mathbf{E}_2 \otimes \mathbf{E}_2 + \mu_3 \mathbf{E}_3 \otimes \mathbf{E}_3 + \mu_4 \mathbf{E}_4 \otimes \mathbf{E}_4 \qquad (11)$$

The eigenvalues and eigentensors are defined as

$$\begin{aligned}
\mu_1 &= g(\lambda_1), \ \mathbf{E}_1 = \mathbf{v}_1 \otimes \mathbf{v}_1 \ , \\
\mu_2 &= 1, \ \mathbf{E}_2 = \mathbf{v}_2 \otimes \mathbf{v}_2 \ , \\
\mu_3 &= \sqrt{g(\lambda_1)}, \ \mathbf{E}_3 = \frac{1}{\sqrt{2}}(\mathbf{v}_1 \otimes \mathbf{v}_2 + \mathbf{v}_2 \otimes \mathbf{v}_1) \ , \\
\mu_4 &= 0, \ \mathbf{E}_4 = \frac{1}{\sqrt{2}}(\mathbf{v}_1 \otimes \mathbf{v}_2 - \mathbf{v}_2 \otimes \mathbf{v}_1) \ ,
\end{aligned} \qquad (12)$$

where g is a nonnegative decreasing diffusivity function, λ_i and $\mathbf{v}_i$ are eigenvalues and eigenvectors of the structure tensor $\mathbf{J}(u_\sigma) = \nabla u_\sigma \nabla u_\sigma^\mathsf{T}$, i.e., $\lambda_1 = ||\nabla u_\sigma||_2^2$, $\mathbf{v}_1 = \frac{\nabla u_\sigma}{||\nabla u_\sigma||_2}$ and $\lambda_2 = 0$, $\mathbf{v}_2 = \frac{\nabla u_\sigma^\perp}{||\nabla u_\sigma||_2}$. The above-defined eigentensors are orthonormal with respect to the dot product $\mathbf{A} : \mathbf{B} = \text{trace}(\mathbf{B}^\mathsf{T}\mathbf{A})$ [15].

Combining this new definition of the fourth-order diffusion tensor with Dirichlet boundary conditions as in Eq. (3) results in our proposed model:

$$\begin{aligned}
\partial_t u &= - \partial_{ji}[\mathcal{D}(\mathbf{J}(u_\sigma)) : \mathbf{H}(u)]_{ij}, \quad \Omega \backslash K \times (0, \infty) \ , \\
u &= f, \quad K \times [0, \infty)
\end{aligned} \qquad (13)$$

As it is customary in PDE-based inpainting, we allow Eq. (13) to evolve until a steady state has been reached, i.e., the time derivative becomes negligible. In our numerical implementation, we use a Fast Semi-Iterative Scheme (FSI) [17] to greatly accelerate convergence to a large stopping time.

In the definition of our fourth-order diffusion tensor $\mathcal{D}$, the choice of μ_1 and μ_2 is analogous to anisotropic edge enhancing diffusion [34]. However, two additional terms occur in the fourth-order case, μ_3 and μ_4. As noted in [15], μ_4 is irrelevant, since the corresponding eigentensor $\mathbf{E}_4$ is anti-symmetric, and the dot product $\mathbf{E}_4 : \mathbf{H}$ with the Hessian of any sufficiently smooth image will be zero due to its symmetry. To better understand the role of μ_3, we observe that

$$\mathbf{E}_3 : \mathbf{H} = \frac{1}{\sqrt{2}} \left(\mathbf{v}_1^T \mathbf{H} \mathbf{v}_2 + \mathbf{v}_2^T \mathbf{H} \mathbf{v}_1 \right)$$

$$= \frac{1}{\sqrt{2}} \left(u_{\left(\frac{v_1+v_2}{\sqrt{2}}\right)\left(\frac{v_1+v_2}{\sqrt{2}}\right)} - u_{\left(\frac{v_1-v_2}{\sqrt{2}}\right)\left(\frac{v_1-v_2}{\sqrt{2}}\right)} \right), \tag{14}$$

which amounts to a mixed second derivative of u, in directions along and orthogonal to the regularized image gradient $\frac{\nabla u_\sigma}{||\nabla u_\sigma||_2}$ or, equivalently, to the difference of second derivatives in the directions that are exactly in between the two. This term vanishes if the Hessian is isotropic, or if the gradient is parallel to one of the Hessian eigenvectors. Therefore, the role of μ_3 can be seen as steering the amount of diffusion in cases of a Hessian anisotropy that goes along with a misalignment between gradient and Hessian eigenvectors.

Gorgi Zadeh et al. [15] simply set μ_3 to the arithmetic mean of μ_1 and μ_2. In our work, we empirically evaluated several alternative options for μ_3 by reconstructing the test image shown in Fig. 1a, which contains one rectangle, one circle, and two stars, from a randomly selected subset of 5% of its pixels. In this experiment, we compare EED based inpainting with our novel fourth-order edge enhancing diffusion (FOEED) with different settings of μ_3: Specifically, $\mu_3 = 1$ corresponds to the max-

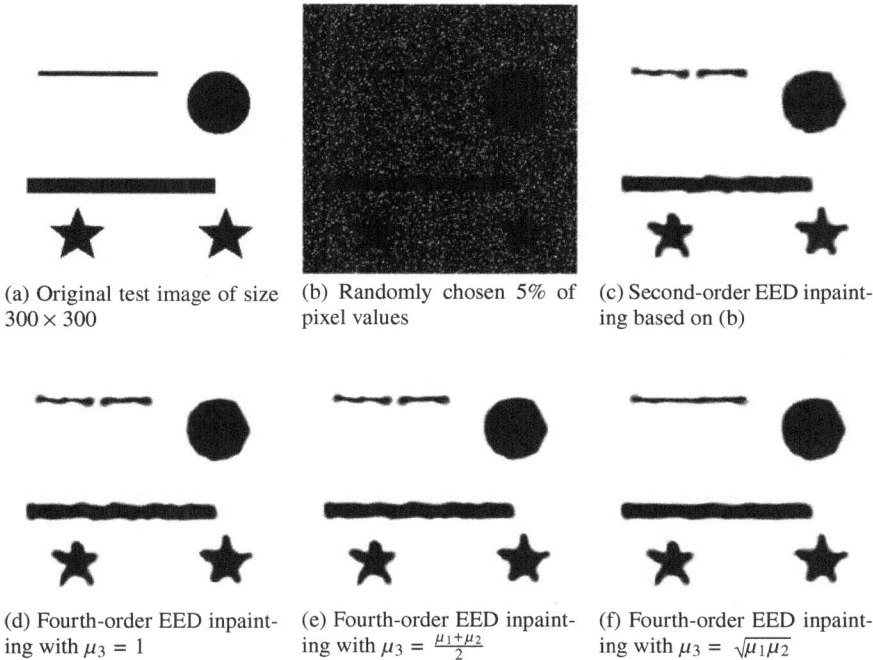

(a) Original test image of size 300×300

(b) Randomly chosen 5% of pixel values

(c) Second-order EED inpainting based on (b)

(d) Fourth-order EED inpainting with $\mu_3 = 1$

(e) Fourth-order EED inpainting with $\mu_3 = \frac{\mu_1+\mu_2}{2}$

(f) Fourth-order EED inpainting with $\mu_3 = \sqrt{\mu_1\mu_2}$

Fig. 1 Reconstruction of a synthetic test image (**a**) from 5% of its pixels (**b**) based on second-order diffusion (**c**) and fourth-order diffusion with different coefficients for the mixed term μ_3 (**d–f**). Visually, the reconstruction in (**f**) is most similar to the original image

Table 1 Numerical reconstruction errors on the test image (Fig. 1)

Errors	EED	FOEED ($\mu_3 = 1$)	FOEED $(\mu_3 = \frac{\mu_1+\mu_2}{2})$	FOEED $(\mu_3 = \sqrt{\mu_1\mu_2})$
MSE	647.183	660.588	634.321	533.987
AAE	5.043	4.581	4.505	4.140

imum of μ_1 and μ_2 (Fig. 1d), $\mu_3 = (1 + g(\lambda_1))/2$ corresponds to their arithmetic mean (Fig. 1e), and $\mu_3 = \sqrt{g(\lambda_1)}$ corresponds to their geometric mean (Fig. 1f). In all cases, we used the Charbonnier diffusivity (Eq. (6)), which is popular for image compression [13], the same contrast parameter ($\lambda = 0.1$) and smoothing parameter ($\sigma = 1$). The only difference is time step size, where second-order EED permitted a stable step size of 0.25, while a smaller step size of 0.05 was chosen for FOEED. A more detailed theoretical and empirical analysis of stability will be given in Sect. 3.3.

A numerical comparison of the results is given in Table 1. For evaluation, we used the well-known mean squared error (MSE) and average absolute error (AAE) between original and reconstructed images. For two-dimensional gray-valued images u and v with the same dimensions $m \times n$, the MSE and AAE are defined as

$$\mathrm{MSE}(u, v) = \frac{1}{mn} \sum_{i,j} (u_{i,j} - v_{i,j})^2 \,,$$

$$\mathrm{AAE}(u, v) = \frac{1}{mn} \sum_{i,j} |u_{i,j} - v_{i,j}| \,. \tag{15}$$

According to Table 1, the most accurate results are achieved by setting μ_3 to the geometric mean of μ_1 and μ_2. Fourth-order EED with this setting produces higher accuracy than second-order EED. Visually, Fig. 1 supports this conclusion. Specifically, fourth-order EED with $\mu_3 = \sqrt{\mu_1\mu_2}$ is the only variant that correctly connects the thin bar at the top of the test image, and it leads to a straighter shape of the thicker bar below, which is more similar to its original rectangular shape. In all subsequent experiments, we set $\mu_3 = \sqrt{\mu_1\mu_2}$.

3.2 A Unifying Framework for Fourth-Order Diffusion

Several fourth-order diffusion PDEs have been used for image processing previously. We can better understand how they relate to our newly proposed PDE by observing that the fourth-order diffusion tensor $\mathcal{D}$ introduces a unifying framework for fourth-order diffusion filters. In particular, given its coefficients d_{ijkl}, we can expand Eq. (9) by using Einstein notation as

$$\partial_t u = -\partial_{ji}[d_{ijkl} u_{kl}] \tag{16}$$

Effectively, the fourth-order diffusion tensor allows us to separately set the diffusivities for all $2^4 = 16$ fourth-order derivatives of the two-dimensional image u. We will now demonstrate how several well-known fourth-order PDEs can be expressed in this framework, starting with the You-Kaveh PDE [35]

$$\partial_t u = -\Delta(g(|\Delta u|)\Delta u) ,\tag{17}$$

which can be rewritten as

$$\begin{aligned}
\partial_t u = &-\partial_{xx}[g(|\Delta u|)u_{xx} + 0 \cdot u_{xy} + 0 \cdot u_{yx} + g(|\Delta u|)u_{yy}] \\
&-\partial_{yx}[0 \cdot u_{xx} + 0 \cdot u_{xy} + 0 \cdot u_{yx} + 0 \cdot u_{yy}] \\
&-\partial_{xy}[0 \cdot u_{xx} + 0 \cdot u_{xy} + 0 \cdot u_{yx} + 0 \cdot u_{yy}] \\
&-\partial_{yy}[g(|\Delta u|)u_{xx} + 0 \cdot u_{xy} + 0 \cdot u_{yx} + g(|\Delta u|)u_{yy}] .
\end{aligned}\tag{18}$$

Here and in all subsequent examples, many terms have zero coefficients. For brevity, we will omit them from now on.

Hajiaboli's anisotropic fourth-order PDE [19] is

$$\partial_t u = -\Delta\left(g(\|\nabla u\|)^2 u_{NN} + g(\|\nabla u\|)u_{TT}\right) ,\tag{19}$$

where N and T are unit vectors parallel and orthogonal to the gradient, respectively. It can be rewritten as

$$\begin{aligned}
\partial_t u = &-\partial_{xx}\Bigg[\left(\frac{g(\|\nabla u\|)^2 u_x^2 + g(\|\nabla u\|)u_y^2}{u_x^2 + u_y^2}\right)u_{xx} + \left(\frac{g(\|\nabla u\|)^2 u_x u_y - g(\|\nabla u\|)u_x u_y}{u_x^2 + u_y^2}\right)u_{xy} \\
&+ \left(\frac{g(\|\nabla u\|)^2 u_x u_y - g(\|\nabla u\|)u_x u_y}{u_x^2 + u_y^2}\right)u_{yx} + \left(\frac{g(\|\nabla u\|)^2 u_y^2 + g(\|\nabla u\|)u_x^2}{u_x^2 + u_y^2}\right)u_{yy}\Bigg] \\
&-\partial_{yy}\Bigg[\left(\frac{g(\|\nabla u\|)^2 u_x^2 + g(\|\nabla u\|)u_y^2}{u_x^2 + u_y^2}\right)u_{xx} + \left(\frac{g(\|\nabla u\|)^2 u_x u_y - g(\|\nabla u\|)u_x u_y}{u_x^2 + u_y^2}\right)u_{xy} \\
&+ \left(\frac{g(\|\nabla u\|)^2 u_x u_y - g(\|\nabla u\|)u_x u_y}{u_x^2 + u_y^2}\right)u_{yx} + \left(\frac{g(\|\nabla u\|)^2 u_y^2 + g(\|\nabla u\|)u_x^2}{u_x^2 + u_y^2}\right)u_{yy}\Bigg]
\end{aligned}\tag{20}$$

From this method, Li et al. [21] derived two anisotropic fourth-order PDEs that, to our knowledge, are the only anisotropic fourth-order models that have been applied to inpainting previously. We will refer to them as Li 1

$$\partial_t u = -\Delta(g(\|\nabla u\|)u_{NN} + u_{TT})\tag{21}$$

and Li 2

$$\partial_t u = -\Delta(u_{TT}) .\tag{22}$$

Li 1 can be re-written as

$$
\begin{aligned}
\partial_t u = -\partial_{xx}&\left[\left(\frac{g(||\nabla u||)u_x^2 + u_y^2}{u_x^2 + u_y^2}\right)u_{xx} + \left(\frac{g(||\nabla u||)u_x u_y - u_x u_y}{u_x^2 + u_y^2}\right)u_{xy}\right.\\
&\left.+\left(\frac{g(||\nabla u||)u_x u_y - u_x u_y}{u_x^2 + u_y^2}\right)u_{yx} + \left(\frac{g(||\nabla u||)u_y^2 + u_x^2}{u_x^2 + u_y^2}\right)u_{yy}\right]\\
-\partial_{yy}&\left[\left(\frac{g(||\nabla u||)u_x^2 + u_y^2}{u_x^2 + u_y^2}\right)u_{xx} + \left(\frac{g(||\nabla u||)u_x u_y - u_x u_y}{u_x^2 + u_y^2}\right)u_{xy}\right.\\
&\left.+\left(\frac{g(||\nabla u||)u_x u_y - u_x u_y}{u_x^2 + u_y^2}\right)u_{yx} + \left(\frac{g(||\nabla u||)u_y^2 + u_x^2}{u_x^2 + u_y^2}\right)u_{yy}\right],
\end{aligned}
\tag{23}
$$

while Li 2 becomes

$$
\begin{aligned}
\partial_t u = -\partial_{xx}&\left[\left(\frac{u_y^2}{u_x^2 + u_y^2}\right)u_{xx} + \left(\frac{-u_x u_y}{u_x^2 + u_y^2}\right)u_{xy} + \left(\frac{-u_x u_y}{u_x^2 + u_y^2}\right)u_{yx} + \left(\frac{u_x^2}{u_x^2 + u_y^2}\right)u_{yy}\right]\\
-\partial_{yy}&\left[\left(\frac{u_y^2}{u_x^2 + u_y^2}\right)u_{xx} + \left(\frac{-u_x u_y}{u_x^2 + u_y^2}\right)u_{xy} + \left(\frac{-u_x u_y}{u_x^2 + u_y^2}\right)u_{yx} + \left(\frac{u_x^2}{u_x^2 + u_y^2}\right)u_{yy}\right].
\end{aligned}
\tag{24}
$$

We observe that Li 1 is based on a similar idea as our proposed PDE: It permits fourth-order diffusion along the edge, while applying a nonlinear diffusivity function across the edge. However, expressing Li et al.'s models in terms of fourth-order diffusion tensors $\mathcal{D}_1$ and $\mathcal{D}_2$ reveals that our approach is more general. In particular, we can observe that

$$
\begin{aligned}
\mathcal{D}_1 &: \mathbf{H} = g(||\nabla u||)u_{NN}\mathbf{I} + u_{TT}\mathbf{I},\\
\mathcal{D}_2 &: \mathbf{H} = u_{TT}\mathbf{I},
\end{aligned}
\tag{25}
$$

where $\mathbf{I}$ is the 2×2 identity matrix. In our model, $\mathcal{D} : \mathbf{H}$ can yield arbitrary anisotropic tensors. In this sense, our model more fully accounts for anisotropy compared to the ones by Hajiaboli and Li et al.

The fourth-order Eq. (16) involves inner second derivatives of the image, which then get scaled by diffusivities, before outer second derivatives are taken. We observe that, in both cases, our model accounts for mixed derivatives that are ignored by previous approaches to anisotropic fourth-order diffusion: In the outer derivatives, this can be seen from the fact that Eq. (9) involves mixed derivatives, while Hajiaboli and Li et al. only consider the Laplacian.

Similarly, our definition of a fourth-order diffusion tensor $\mathcal{D}$ accounts for mixed derivatives also in the inner derivatives. Following Eq. (14), we obtain

$$
\begin{aligned}
\mathcal{D} : \mathbf{H} &= \mu_1(\mathbf{E}_1 \otimes \mathbf{E}_1) : \mathbf{H} + \mu_2(\mathbf{E}_2 \otimes \mathbf{E}_2) : \mathbf{H} + \mu_3(\mathbf{E}_3 \otimes \mathbf{E}_3) : \mathbf{H}\\
&= \mu_1 u_{v_1 v_1}\mathbf{E}_1 + \mu_2 u_{v_2 v_2}\mathbf{E}_2 + \frac{\mu_3}{\sqrt{2}}\left(u_{\left(\frac{v_1+v_2}{\sqrt{2}}\right)\left(\frac{v_1+v_2}{\sqrt{2}}\right)} - u_{\left(\frac{v_1-v_2}{\sqrt{2}}\right)\left(\frac{v_1-v_2}{\sqrt{2}}\right)}\right)\mathbf{E}_3.
\end{aligned}
\tag{26}
$$

Comparing Eqs. (25) and (26) reveals differences in the considered directions: First, N and T are derived from the unregularized gradient, while the corresponding directions $\mathbf{v}_1$ and $\mathbf{v}_2$ in our model include a Gaussian pre-smoothing. A second difference is that our model involves an additional term, which is steered by μ_3, and accounts for the directions in between the regularized gradient and its orthogonal vectors, i.e., $\left(\frac{\mathbf{v}_1+\mathbf{v}_2}{\sqrt{2}}\right)$ and $\left(\frac{\mathbf{v}_1-\mathbf{v}_2}{\sqrt{2}}\right)$. As it was demonstrated in the previous section, this term can have a noticeable effect on the outcome. Overall, we conclude that our newly proposed model is more general than the previously published ones.

3.3 Discretization and Stability

When discretizing Eq. (13) with standard finite differences

$$
\begin{aligned}
u_{xx} &\approx \frac{(u_{i-1,j} - 2u_{i,j} + u_{i+1,j})}{(\Delta x)^2} \, , \\
u_{yy} &\approx \frac{(u_{i,j-1} - 2u_{i,j} + u_{i,j+1})}{(\Delta y)^2} \, , \\
u_{xy} &\approx \frac{(u_{i-1,j-1} + u_{i+1,j+1} + u_{i-1,j+1} + u_{i+1,j-1})}{4(\Delta x)(\Delta y)} \, , \\
u_{yx} &= u_{xy} \, ,
\end{aligned}
\tag{27}
$$

we can write it down in matrix-vector form as in [15],

$$
\mathbf{u}^{k+1} = \mathbf{u}^k (\mathbf{I} - \tau \, \mathbf{P}_k) \, ,
\tag{28}
$$

where u^k is an mn dimensional image vector at iteration k. m, n are image width and height respectively; Δx and Δy are the corresponding pixel edge lengths. $\mathbf{P}_k$ is a positive semi-definite matrix that, with step size τ, leads to the system matrix $(\mathbf{I} - \tau \, \mathbf{P}_k)$. The notation $\mathbf{P}_k$ indicates that it is iteration dependent, i.e., $\mathbf{P}_k = \mathbf{P}(\mathbf{u}_k)$.

Stability of fourth-order PDEs for image processing is typically formalized in an L_2 sense, i.e., a time step is chosen such that

$$
||\mathbf{u}^{k+1}||_2 \leq ||\mathbf{u}^k||_2 \, .
\tag{29}
$$

In an inpainting scenario, it depends on our initialization of the unknown pixels whether we can expect Eq. (29) to hold. Therefore, we rely on a stability analysis of the smoothing variant of our proposed PDE. This variant is obtained by removing the

Dirichlet boundary conditions and instead solving a standard initial value problem. In this case, the stability analysis presented by Gorgi Zadeh et al. [15] carries over. It ensures that time step sizes

$$\tau \leq \frac{2}{16(\Delta x)^2 + 16(\Delta y)^2 + 2(\Delta x \Delta y)} \tag{30}$$

are stable in the L_2 sense. For a spatial discretization $\Delta x = \Delta y = 1$, this yields $\tau \leq 1/17 \approx 0.0588$. In inpainting, we empirically obtained a useful steady state with a time step size $\tau \leq 0.066$, independent of the initialization. The fact that this slightly exceeds the theoretical step size reflects the fact that Eq. (30) results from deriving a sufficient, not a necessary condition for stability.

Stability of fourth-order schemes generally requires a quite small time step τ. This makes it computationally expensive to reach the steady state by evaluating Eq. (28). Hafner et al. [17] propose a remedy to this problem, the so-called Fast Semi-Iterative Scheme (FSI). It extrapolates the basic solver iteration with the previous iterate and serves as an accelerated explicit scheme. The acceleration of the explicit scheme (28) is given as

$$\mathbf{u}^{m,k+1} = \alpha_k \cdot (\mathbf{I} - \tau \mathbf{P}(\mathbf{u}^{m,k}))\mathbf{u}^{m,k} + (1 - \alpha_k) \cdot \mathbf{u}^{m,k-1} , \tag{31}$$

where $\mathbf{u}^{m,-1} := \mathbf{u}^{m,0}$ and $\alpha_k = \frac{4k+2}{2k+3}$ for $k = 0, \ldots, n-1$. Here m stands for outer cycle, i.e. m-th cycle with inner cycle of length n. And for passing to the next outer cycle, we set $\mathbf{u}^{m+1,0} := \mathbf{u}^{m,n}$. The stability analysis requires the matrix $\mathbf{P}$ to be symmetric. This is satisfied since the diffusion tensor $\mathcal{D}$ is symmetric, and symmetric central discretizations are used. In our implementation, we used $n = 40$, and stopped iterating after the first outer cycle for which $\|\mathbf{u}^m - \mathbf{u}^{m-1}\|_2 < 10^{-4}$.

4   Experimental Results

To establish the usefulness of our proposed new model, we applied it to the reconstruction of images from a sparse subset of pixels (Sect. 4.1). Moreover, we evaluate performance for a more classic inpainting task, scratch removal (Sect. 4.2). We also demonstrate how results depend on the chosen diffusivity function and contrast parameter (Sect. 4.3).

4.1 Reconstruction From a Sparse Set of Pixels

Improving image reconstruction from a sparse set of known pixels was the main motivation behind our work. Therefore, we applied it to two well-known natural images, *toucan* and *peppers,* as well as to a medical image, a slice of a T_1 weighted brain MR scan *(t1slice).* For *toucan,* we kept a random subset of only 2% of the pixels. Due to the lower resolution of the *peppers* and *t1slice* images, we kept 5% and 20%, respectively.

In all three cases, results from our approach (FOEED) were compared to results from second-order EED, as well as from the two anisotropic fourth-order PDEs proposed by Li et al. [21]. In all experiments, we used the Charbonnier diffusivity function, we set the contrast parameter to $\lambda = 0.1$, and the pre-smoothing parameter to $\sigma = 1$.

Results for *toucan* are shown in Fig. 2, for *peppers* in Fig. 3, and for *t1slice* in Fig. 4. A quantitative evaluation in terms of MSE and AAE is presented in Table 2. In terms of the numerical results, our proposed method produced a more accurate reconstruction than any of the competing approaches. Visually, there is a clear difference between second-order (EED) and fourth-order approaches (Li1, Li2, FOEED). Especially, we found that the shapes of edges were reconstructed more accurately. For example, we noticed this around the body and face in the *toucan* image (Fig. 2). Similarly, the white and grey matter boundaries were better separated in the *t1slice* (Fig. 4).

As we expected based on the theoretical analysis in Sect. 3.2, visual differences between the fourth-order methods are more subtle. However, in the *peppers* image (Fig. 3), the tall and thin and the small and thick peppers in the foreground are much more clearly separated in the FOEED result than in any of the others.

In addition to experimenting with grayscale versions of the *toucan* and *peppers* images, we also applied EED and our FOEED filter channel-wise to the original RGB color versions. Results for *toucan* can be found in Fig. 5, for *peppers* in Fig. 6. Table 3 again provides a quantitative comparison. Similar observations can be made as in the grayscale images: Again, FOEED leads to lower reconstruction errors than EED, it visually reconstructs edges more accurately, and separates the peppers more clearly.

Finally, we reconstructed images from a larger number of pixels, to obtain visually cleaner results. Qualitative and numerical results are presented in Fig. 7 and Table 4, respectively. FOEED still yields lower numerical errors than EED. Unsurprisingly, the differences become smaller and less visually prominent as the mask density increases. The table also reveals that FOEED requires more CPU time compared to standard EED. However, due to the use of FSI in both cases, the difference in running times until convergence is much lower than the difference in time step sizes.

Fig. 2 1st row left: original *toucan* image of size 512 × 512; right: randomly chosen 2% of pixel values; 2nd row left: EED based inpainted image; right: Li1 based inpainted image; 3rd row left: Li2 based inpainted image; right: FOEED based inpainted image

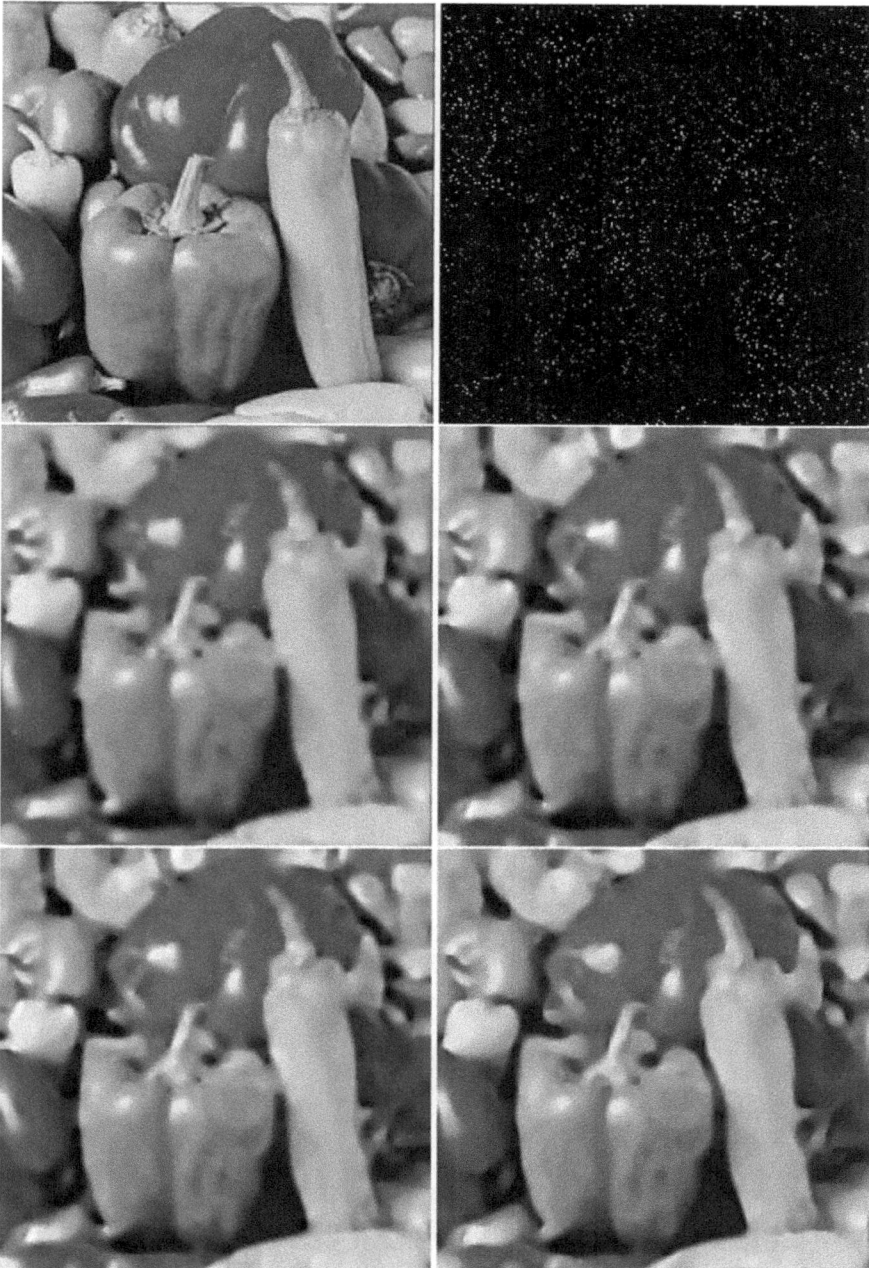

Fig. 3 1st row left: original *peppers* image of size 225 × 225; Right: randomly chosen 5% of pixel values; 2nd row left: EED based inpainted image; Right: Li1 based inpainted image; 3rd row left: Li2 based inpainted image; Right: FOEED based inpainted image

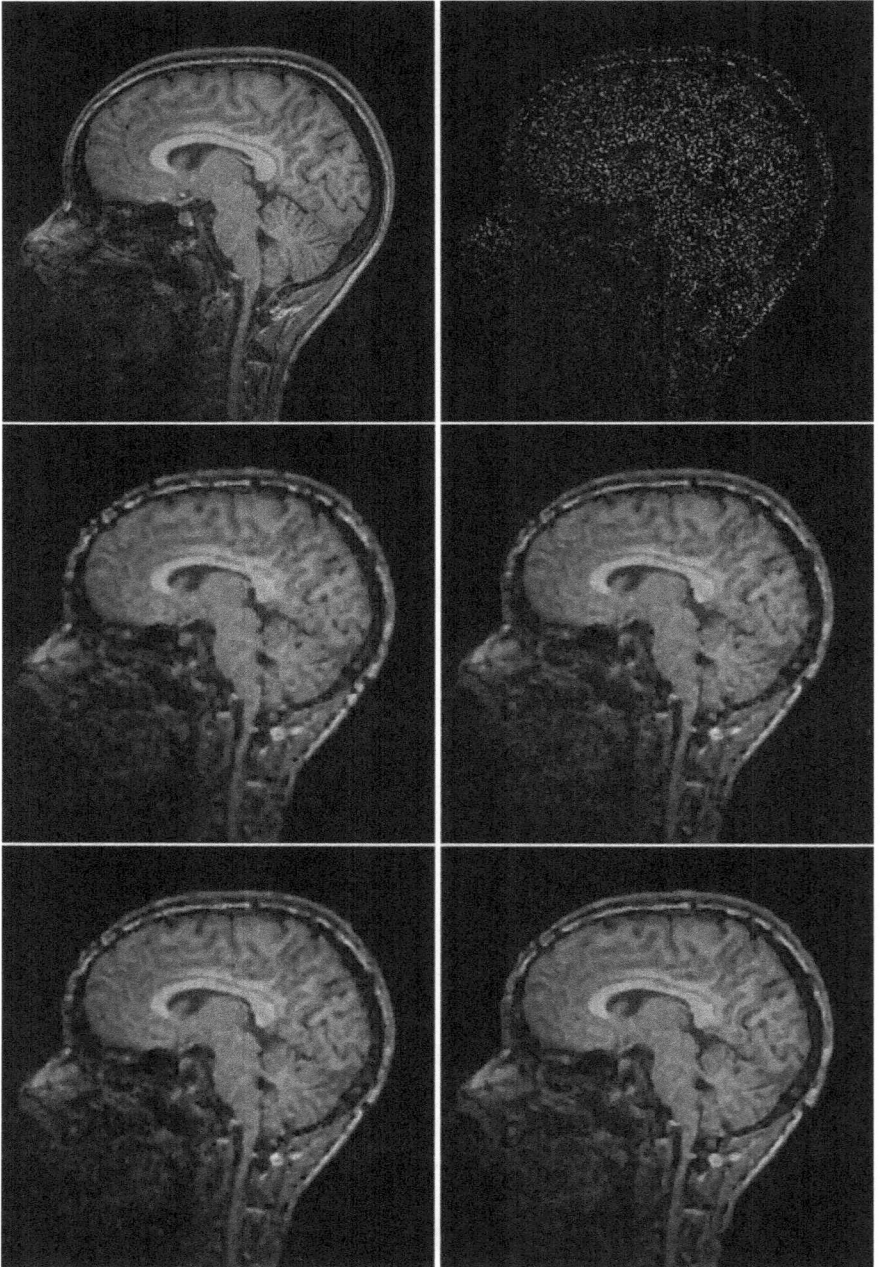

Fig. 4 1st row left: original *t1slice* image of size 256×256; Right: randomly chosen 20% of pixel values; 2nd row left: EED based inpainted image; Right: Li1 based inpainted image; 3rd row left: Li2 based inpainted image; Right: FOEED based inpainted image

Table 2 Numerical comparison of inpainting models for gray-valued images

Image	Errors	EED	FOEED	Li1	Li2
Toucan	MSE	105.37	**96.228**	100.994	102.665
	AAE	4.488	**4.164**	4.397	4.465
Peppers	MSE	467.261	**443.129**	455.633	459.606
	AAE	10.94	**10.523**	11.042	11.107
t1-slice	MSE	166.356	**150.002**	152.698	155.955
	AAE	5.895	**5.698**	5.789	5.853

Fig. 5 RGB *toucan* image, reconstructed from randomly chosen 2% of pixel values using EED (left) or FOEED (right)

Fig. 6 RGB *peppers* image, reconstructed from randomly chosen 5% of pixel values using EED (left) or FOEED (right)

Table 3 Numerical comparison of inpainting models for RGB images

Image	Errors	EED	FOEED
Toucan	MSE	119.062	**108.061**
	AAE	4.819	**4.594**
Peppers	MSE	478.799	**441.203**
	AAE	11.049	**10.543**

Table 4 Numerical comparison and computation times corresponding to Fig. 7

Image	Errors	EED	FOEED	CPU time
Toucan	MSE	18.029	**17.295**	53.060 (FOEED)
	AAE	1.696	**1.686**	21.259 (EED)
Peppers	MSE	113.5	**110.885**	20.999 (FOEED)
	AAE	4.565	**4.441**	19.79 (EED)
t1-slice	MSE	114.845	**107.323**	24.74 (FOEED)
	AAE	4.610	**4.553**	10.64 (EED)

4.2 Scratch Removal

Li et al. [21] proposed their anisotropic fourth-order PDE for more classical image inpainting tasks, such as scratch removal. We evaluated whether our more general filter can also provide a benefit in such a scenario by reconstructing a scratched version of the *peppers* image. Similar to Li et al., we first made the scratches rather thin, covering only 6% of all pixels. Results are shown in Fig. 8 and in Table 5. In this case, all methods work well: Numerical errors are small and similar between methods, and even though FOEED achieves the best numerical result, differences are difficult to discern visually.

Therefore, we created a more challenging version with thicker scratches, covering 18% of all pixels (Fig. 9). The corresponding numerical comparison is shown in Table 6. Here, FOEED achieves the most accurate reconstruction. Visually, we again observe that edges are reconstructed more accurately, and objects are more clearly separated, with fourth-order compared to second-order diffusion, and that steering it with a fourth-order diffusion tensor again provides small additional benefits over the previous methods.

Table 5 Numerical comparison for peppers with thinner scratches (Fig. 8)

Image	Errors	EED	FOEED	Li1	Li2
Peppers	MSE	9.520	**7.813**	8.161	8.132
	AAE	0.363	**0.326**	0.346	0.346

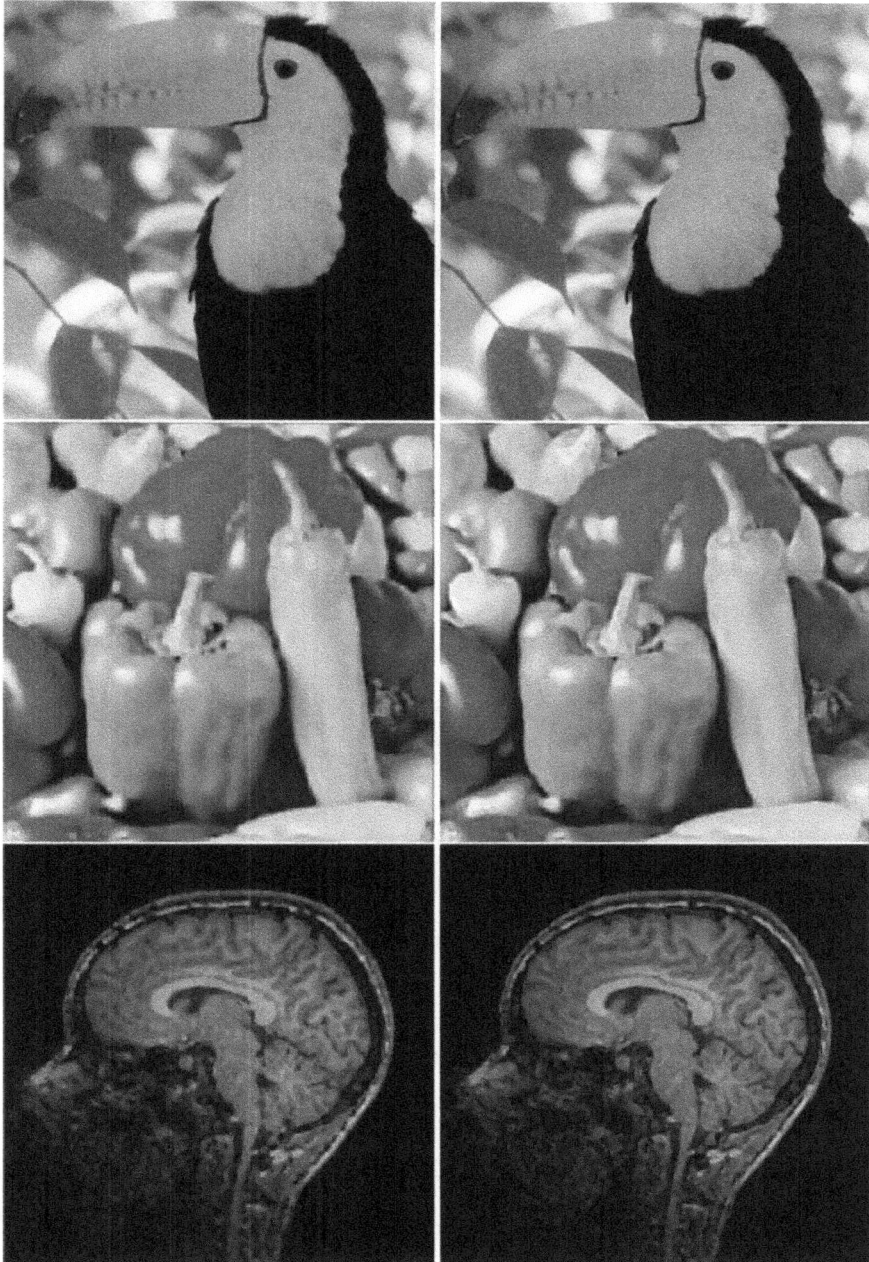

Fig. 7 Higher quality reconstructions from a larger subset of pixels. 1st row: *toucan* image, reconstructed with EED (left) or FOEED (right) from randomly chosen 14% of pixels; 2nd row left: same for 20% of pixels from *peppers*; 3rd row left: same for 30% of pixels from *t1slice*. As expected, increasing the fraction of known pixels reduces the differences in the results of the two schemes

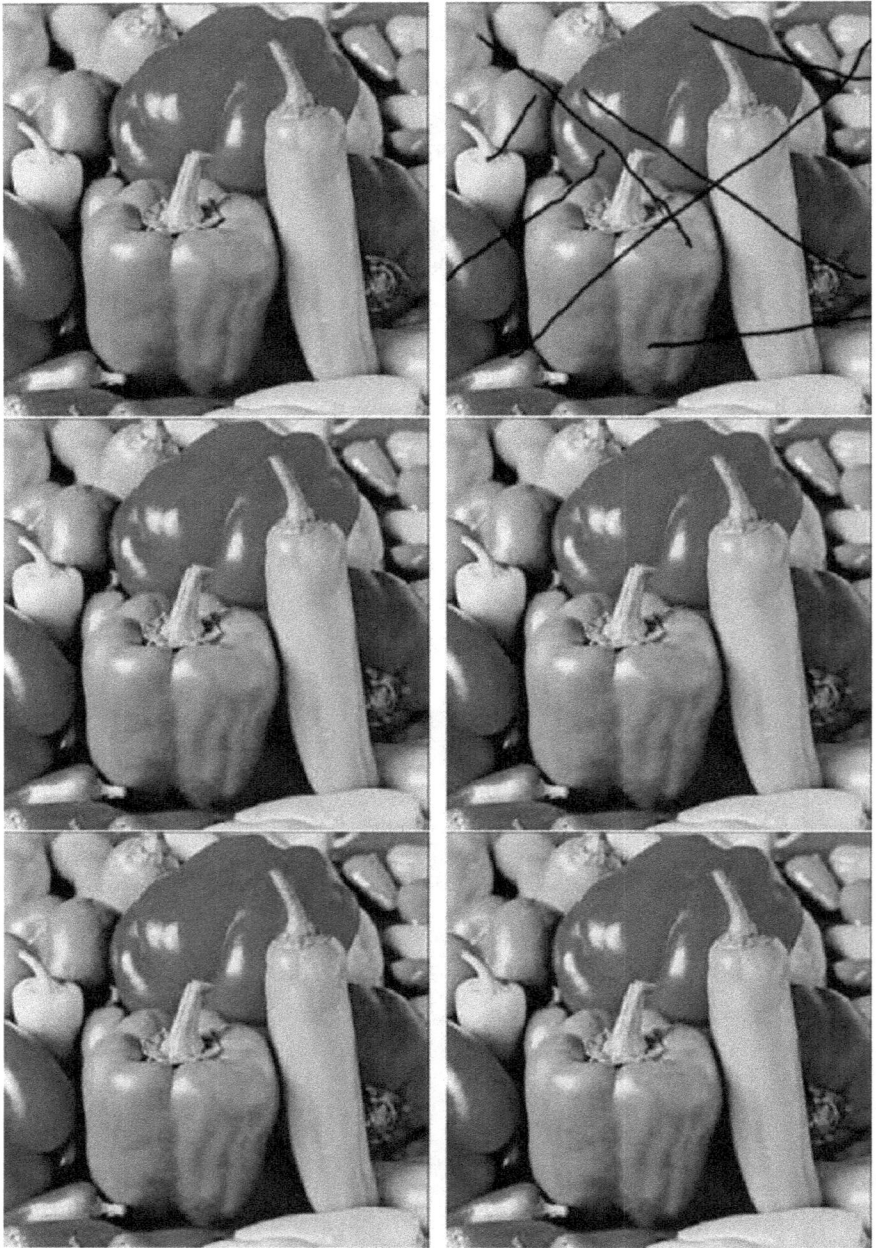

Fig. 8 1st row left: original *peppers* image of size 225×225; Right: corrupted image. 2nd row left: EED based inpainting; Right: Li1 based inpainting. 3rd row left: Li2 based inpainting; Right: FOEED based inpainting

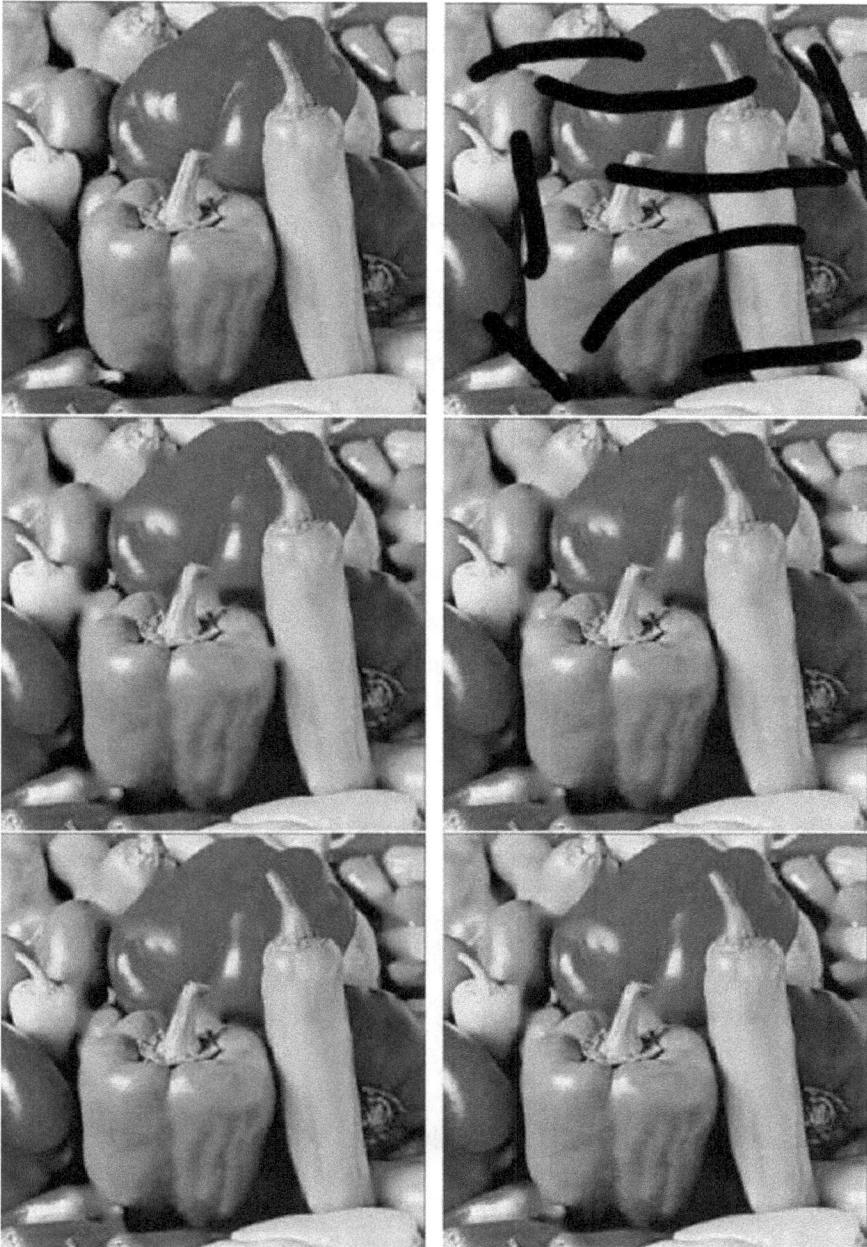

Fig. 9 1st row left: original *peppers* image of size 225 × 225; Right: corrupted image. 2nd row left: EED based inpainting; Right: Li1 based inpainting. 3rd row left: Li2 based inpainting; Right: FOEED based inpainting

Table 6 Numerical comparison for peppers with thicker scratches (Fig. 9)

Image	Errors	EED	FOEED	Li1	Li2
Peppers	MSE	104.744	**78.761**	101.670	101.592
	AAE	2.455	**2.099**	2.465	2.450

4.3　Effect of Diffusivity Function and Contrast Parameter

For image inpainting with second-order PDEs, the Charbonnier diffusivity was previously found to work better than other established diffusivity functions. To assess whether this is still true in the fourth-order case, we repeated the reconstruction of the *peppers* image as shown in Fig. 3 with different diffusivities. Table 7 summarizes the results. We conclude that the Charbonnier diffusivity still appears to be optimal.

Finally, in Fig. 10, we illustrate how the reconstructed image depends on the contrast parameter λ. As expected, increasing λ leads to an increased blurring of edges. In the limit, the diffusivity function takes on values close to 1 over a substantial part of the image, and our model starts to approximate homogeneous fourth-order diffusion.

Table 7 Numerical comparison of FOEED with different diffusivity functions

Image	Errors	Charbonnier [8] $\frac{1}{\sqrt{1+(\frac{s}{\lambda})^2}}$	Aubert [7] $\frac{(\frac{s}{\lambda})^2}{(s^2+\lambda^2)^2}$	Perona-Malik [24] $\frac{1}{1+(\frac{s}{\lambda})^2}$	Perona-Malik2 [24] $e^{-(\frac{s}{\lambda})^2}$	Geman-Reynolds [14] $\frac{2\lambda^2}{(s^2+\lambda^2)^2}$
Peppers	MSE	**443.129**	458.961	478.411	491.153	491.186
	AAE	**10.523**	10.587	10.943	11.157	11.007

Fig. 10 From left to right: FOEED based inpainted image with $\lambda = 0.1$, $\lambda = 0.5$, $\lambda = 15.5$

5 Conclusions

We introduced a novel fourth-order PDE for edge enhancing diffusion (FOEED), steered by a fourth-order diffusion tensor. We implemented it using a fast semi-iterative scheme, and demonstrated that it achieved improved accuracy in several inpainting tasks, including reconstructing images from a small fraction of pixels, or removing scratches.

Our main motivation for using fourth-order diffusion in this context is the increased smoothness of results compared to second-order PDEs [35], which we expected to result in visually more pleasant reconstructions. The model in our current work is still based on a single edge direction at each pixel, extracted via a traditional second-order structure tensor. It is left as a separate research goal for future work to combine this with approaches for the estimation of complex structures such as crossings or bifurcations [1, 29], and with their improved reconstruction, e.g., by operating on the space of positions and orientations [4, 9, 11].

Finally, our current work only considered reconstructions from a random subset of pixels. A practical image compression codec that uses our novel PDE should investigate how it interacts with more sophisticated approaches for selecting and coding inpainting masks [27].

Acknowledgements This research was supported by the German Academic Exchange Service (DAAD).

References

1. Aach, T., Mota, C., Stuke, I., Mühlich, M., Barth, E.: Analysis of superimposed oriented patterns. IEEE Trans. Image Process. **15**(12), 3690–3700 (2006)
2. Andris, S., Peter, P., Weickert, J.: A proof-of-concept framework for PDE-based video compression. In: Picture Coding Symposium (PCS), pp. 1–5. IEEE (2016)
3. Ballé, J., Laparra, V., Simoncelli, E.P.: End-to-end optimized image compression. In: International Conference on Learning Representations (ICLR) (2017)
4. Boscain, U., Chertovskih, R.A., Gauthier, J.P., Remizov, A.: Hypoelliptic diffusion and human vision: a semidiscrete new twist. SIAM J. Imaging Sci. **7**(2), 669–695 (2014)
5. Cagnazzo, M., Poggi, G., Verdoliva, L.: Region-based transform coding of multispectral images. IEEE Trans. Image Process. **16**(12), 2916–2926 (2007)
6. Catté, F., Lions, P.L., Morel, J.M., Coll, T.: Image selective smoothing and edge detection by nonlinear diffusion. SIAM J. Numer. Anal. **29**(1), 182–193 (1992)
7. Charbonnier, P., Blanc-Féraud, L., Aubert, G., Barlaud, M.: Two deterministic half-quadratic regularization algorithms for computed imaging. In: International Conference on Image Processing, pp. 168–172 (1994)
8. Charbonnier, P., Blanc-Féraud, L., Aubert, G., Barlaud, M.: Deterministic edge-preserving regularization in computed imaging. IEEE Trans. Image Process. **6**(2), 298–311 (1997)
9. Citti, G., Sarti, A.: A cortical based model of perceptual completion in the roto-translation space. J. Math. Imaging Vis. **24**(3), 307–326 (2006)
10. Didas, S., Weickert, J., Burgeth, B.: Properties of higher order nonlinear diffusion filtering. J. Math. Imaging Vis. **35**(3), 208–226 (2009)
11. Franken, E., Duits, R.: Crossing-preserving coherence-enhancing diffusion on invertible orientation scores. Int. J. Comput. Vis. **85**(3), 253 (2009)

12. Galić, I., Weickert, J., Welk, M., Bruhn, A., Belyaev, A., Seidel, H.P.: Towards PDE-based image compression. In: International Workshop on Variational, Geometric, and Level Set Methods in Computer Vision, pp. 37–48. Springer, Berlin (2005)
13. Galić, I., Weickert, J., Welk, M., Bruhn, A., Belyaev, A., Seidel, H.P.: Image compression with anisotropic diffusion. J. Math. Imaging Vis. **31**(2–3), 255–269 (2008)
14. Geman, D., Reynolds, G.: Constrained restoration and the recovery of discontinuities. IEEE Trans. Pattern Anal. Mach. Intell. **14**(3), 367–383 (1992)
15. Gorgi Zadeh, S., Didas, S., Wintergerst, M.W.M., Schultz, T.: Multi-scale anisotropic fourth-order diffusion improves ridge and valley localization. J. Math. Imaging Vis. **59**(2), 257–269 (2017)
16. Greer, J.B., Bertozzi, A.L.: h^1 solutions of a class of fourth order nonlinear equations for image processing. Discret. Continuous Dyn. Syst. **10**, 349–366 (2003)
17. Hafner, D., Ochs, P., Weickert, J., Reißel, M., Grewenig, S.: FSI schemes: Fast semi-iterative solvers for PDEs and optimisation methods. In: German Conference on Pattern Recognition (GCPR), pp. 91–102 (2016)
18. Hajiaboli, M.R.: A self-governing hybrid model for noise removal. In: Pacific-Rim Symposium on Image and Video Technology, pp. 295–305. Springer, Berlin (2009)
19. Hajiaboli, M.R.: An anisotropic fourth-order diffusion filter for image noise removal. Int. J. Comput. Vis. **92**(2), 177–191 (2011)
20. Iijima, T.: Basic theory on the normalization of pattern (in case of typical one-dimensional pattern). Bull. Electro-Tech. Lab. **26**, 368–388 (1962)
21. Li, P., Li, S.J., Yao, Z.A., Zhang, Z.J.: Two anisotropic fourth-order partial differential equations for image inpainting. IET Image Process. **7**(3), 260–269 (2013)
22. Lysaker, M., Lundervold, A., Tai, X.C.: Noise removal using fourth-order partial differential equation with applications to medical magnetic resonance images in space and time. IEEE Trans. Image Process. **12**(12), 1579–1590 (2003)
23. Mainberger, M., Weickert, J.: Edge-based image compression with homogeneous diffusion. In: International Conference on Computer Analysis of Images and Patterns, pp. 476–483 (2009)
24. Perona, P., Malik, J.: Scale-space and edge detection using anisotropic diffusion. IEEE Trans. Pattern Anal. Mach. Intell. **12**(7), 629–639 (1990)
25. Peter, P.: Three-dimensional data compression with anisotropic diffusion. In: German Conference on Pattern Recognition, pp. 231–236. Springer, Berlin (2013)
26. Scherzer, O.: Denoising with higher order derivatives of bounded variation and an application to parameter estimation. Computing **60**(1), 1–27 (1998)
27. Schmaltz, C., Peter, P., Mainberger, M., Ebel, F., Weickert, J., Bruhn, A.: Understanding, optimising, and extending data compression with anisotropic diffusion. Int. J. Comput. Vis. **108**(3), 222–240 (2014)
28. Schmaltz, C., Weickert, J., Bruhn, A.: Beating the quality of JPEG 2000 with anisotropic diffusion. In: Joint Pattern Recognition Symposium, pp. 452–461. Springer, Berlin (2009)
29. Schultz, T., Weickert, J., Seidel, H.P.: A higher-order structure tensor. In: Laidlaw, D.H., Weickert, J. (eds.) Visualization and Processing of Tensor Fields—Advances and Perspectives, pp. 263–280. Springer, Berlin (2009)
30. Thayammal, S., Selvathi, D.: A review on segmentation based image compression techniques. J. Eng. Sci. Technol. Rev. **6**(3) (2013)
31. Theis, L., Shi, W., Cunningham, A., Huszár, F.: Lossy image compression with compressive autoencoders. In: International Conference on Learning Representations (ICLR) (2017)
32. Toderici, G., O'Malley, S.M., Hwang, S.J., Vincent, D., Minnen, D., Baluja, S., Covell, M., Sukthankar, R.: Variable rate image compression with recurrent neural networks. In: Bengio, Y., LeCun, Y. (eds.) International Conference on Learning Representations (ICLR) (2016)
33. Toderici, G., Vincent, D., Johnston, N., Jin Hwang, S., Minnen, D., Shor, J., Covell, M.: Full resolution image compression with recurrent neural networks. In: IEEE Conference on Computer Vision and Pattern Recognition (CVPR), pp. 5306–5314 (2017)

34. Weickert, J.: Anisotropic Diffusion in Image Processing. Teubner Stuttgart (1998)
35. You, Y.L., Kaveh, M.: Fourth-order partial differential equations for noise removal. IEEE Trans. Image Process. **9**(10), 1723–1730 (2000)
36. Zhang, F., Chen, Y., Xiao, Z., Geng, L., Wu, J., Feng, T., Liu, P., Tan, Y., Wang, J.: Partial differential equation inpainting method based on image characteristics. In: International Conference on Image and Graphics, pp. 11–19. Springer, Berlin (2015)

Modeling Anisotropy

8

Magnetic Resonance Assessment of Effective Confinement Anisotropy with Orientationally-Averaged Single and Double Diffusion Encoding

Cem Yolcu, Magnus Herberthson, Carl-Fredrik Westin, and Evren Özarslan

Abstract Porous or biological materials comprise a multitude of micro-domains containing water. Diffusion-weighted magnetic resonance measurements are sensitive to the anisotropy of the thermal motion of such water. This anisotropy can be due to the domain shape, as well as the (lack of) dispersion in their orientations. Averaging over measurements that span all orientations is a trick to suppress the latter, thereby untangling it from the influence of the domains' anisotropy on the signal. Here, we consider domains whose anisotropy is modeled as being the result of a Hookean (spring) force, which has the advantage of having a Gaussian diffusion propagator while still confining the spatial range for the diffusing particles. In fact, this confinement model is the effective model of restricted diffusion when diffusion is encoded via gradients of long durations, making the model relevant to a broad range of studies aiming to characterize porous media with microscopic subdomains. In this study, analytical expressions for the powder-averaged signal under this assumption are given for so-called single and double diffusion encoding schemes, which sensitize the MR signal to the diffusive displacement of particles in, respectively, one or two consecutive time intervals. The signal for one-dimensional diffusion is shown to exhibit power-law dependence on the gradient strength while its coefficient bears signatures of restricted diffusion.

C. Yolcu (✉) · E. Özarslan
Department of Biomedical Engineering, Linköping University, Linköping, Sweden
e-mail: cem.yolcu@liu.se

E. Özarslan
e-mail: evren.ozarslan@liu.se

M. Herberthson
Department of Mathematics, Linköping University, Linköping, Sweden
e-mail: magnus.herberthson@liu.se

C.-F. Westin
Department of Radiology, Brigham & Women's Hospital, Harvard Medical School, Boston, Massachusetts, USA
e-mail: westin@bwh.harvard.edu

E. Özarslan
Center for Medical Image Science and Visualization, Linköping University, Linköping, Sweden

1   Introduction

Magnetic resonance has proved to be an extremely effective tool to peer into materials and tissues noninvasively. It manipulates the magnetic orientation of molecules pervading the material into a natural precession which emits radio waves. By imposing a spatially-varying magnetic field (hence frequency of precession), the emitted radio frequency signal is made to encode in its spectrum the coordinates of the molecules that emit it. This way, the signal can be spectrally decomposed to trace back what proportion of it originates from where, that is, from which 'voxel.' Commonly achieved voxel sizes are in the neighborhood of a millimeter.

Another use of a spatially-varying precession rate involves sensitizing the signal to the motion of the molecules; diffusion in particular. When the molecules trace out random (Brownian) paths where different locations they visit impart different precessional angles on them, their precessions lose coherence, attenuating the sum of their emitted radio waves. This attenuation of MR signal, specifically its response to the direction of the gradient in precession rate, can reveal or quantify how mobile the molecules are along different directions. The difference of mobility can arise from pore boundaries, impurities, cell membranes, etc. Hence the diffusion-attenuated signal encodes the influence of structures that can be significantly smaller than voxel dimensions. Tracing out axon bundles in human brain white matter is for instance a widely employed application of this principle. This modality of magnetic resonance imaging, which we refer to as diffusion MR, is the subject of this contribution.

While the anisotropy (i.e., variance under a rotation transformation) of a single pore or a cell may be easily visualized, one would be mistaken to make a one-to-one connection with that and the anisotropy of the signal (i.e., the response of the signal to orientations of the specimen or the apparatus). For instance, the signal of a voxel consisting of an unaligned mixture of cylindrical aqueous compartments will be less sensitive to rotations than one consisting of an aligned bundle of cylinders. The anisotropy of the individual compartments is common in the two examples, but the aligned case has more ensemble anisotropy.

In some sense, then, ensemble anisotropy confounds compartment anisotropy at the signal level, and eliminating it pronounces features at the subvoxel level. One way to achieve this is to take an average of the signal over all orientations. If the material allows it, it can be ground into a powder to that effect; hence the term *powder* averaging. However this is generally impossible in bio-medical applications. Then, repeated applications of a measurement protocol in different orientations is the avenue to follow.

In this contribution, we are concerned with two particular diffusion MR schemes, single and double diffusion encoding (SDE and DDE), orientationally averaged in the aforementioned fashion to eliminate ensemble anisotropy. The single encoding scheme [44] is the bread and butter of most applications, employing a magnetic field gradient that remains on for a specified duration in one direction, and then the opposite direction after a specified delay (Fig. 1). The signal then encodes the probability of Brownian displacement between the application of the two pulses.

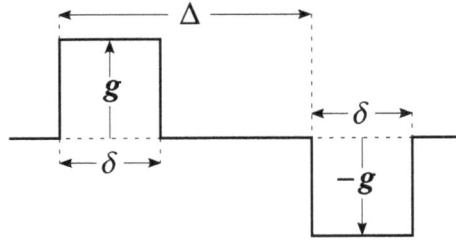

Fig. 1 The pulse sequence for single diffusion encoding, with time running along the horizontal direction

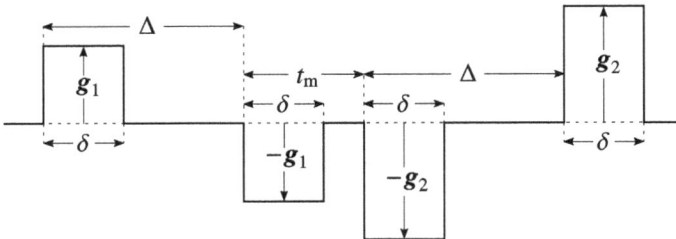

Fig. 2 The pulse sequence for double diffusion encoding

SDE has been the encoding scheme employed since some of the earliest studies of the powder averaged signal [5]. However, very recent theoretical [13] as well as experimental [1] studies have considered the orientationally-averaged signal for more general gradient waveforms.

The double diffusion encoding (DDE) scheme employs two single-encoding blocks in succession [9], in different directions in general, before the signal is read (Fig. 2). This method has been studied extensively in recent years mostly because it allows the anisotropy of the microdomains to be quantified [8], which has important implications for medical imaging. The reader is referred to the reviews [4, 11, 27, 40, 42] for in-depth presentation of the method. In a nutshell, the DDE technique encodes into the signal the joint probability of two Brownian displacements taking place between the two pulses of each block. For freely diffusing molecules, the two displacements are uncorrelated [36]. However, when restrictions, and arguably inhomogeneities and forces, are present, this is no longer the case [24], hence imparting signatures of compartment size onto the signal. As mentioned above, DDE employing gradient blocks in different directions is sensitive to anisotropy of subdomains inside the voxel [6, 8], but not independently of ensemble anisotropy [29].

While we do not consider restricted diffusion in the strict sense of restriction by hard walls in this article, we do respect the finite range of motion of the molecules by the aid of a harmonic attractive force [7, 20, 25, 46, 49]. This is an *effective* model which can mimic restrictive walls with reasonably tractable mathematics, and is actually approached when the gradient pulses are long [34].

This chapter is organized as follows. We first present the signal arising from an effectively confined domain under double diffusion encoding. Afterwards we derive analytical expressions for its powder average and consider the "stick" geometry, wherein diffusion is fully confined in two dimensions. We then treat the case of single diffusion encoding as an extreme case of double encoding and give its corresponding powder average expressions. The special cases of "stick" and "pancake" geometry are compared to their counterparts for free diffusion. The article is concluded after discussions based on the results of the previous sections.

2 Double Diffusion Encoding at the Compartment Level

Here we derive the double-encoded diffusion MR signal arising from a compartment which is characterized by an effective spring force attracting the molecules toward the center. Such a force mimics the confining effect of walls, membranes, etc., with minimal mathematical burden.

The spring (Hookean) force influences the motion of the molecules via a quadratic potential[1]

$$V(x) = \tfrac{1}{2}x^{\mathsf{T}}\mathbf{C}x,\qquad(1)$$

defining the *confinement tensor* $\mathbf{C}$, which can be taken to have $\mathbf{C}^{\mathsf{T}} = \mathbf{C}$ without loss of generality. However, in order for the steady state molecule number density $p_{st}(x) \propto e^{-V(x)}$ to be normalizable, $\mathbf{C}$ must be positive (semi)definite.[2] Under this (or any) potential, the magnetization density $\rho(x, t)$ evolves according to

$$\partial_t \rho(x, t) = (\mathbf{D}\nabla) \cdot e^{-V(x)}\nabla e^{V(x)}\rho(x, t) - i g(t) \cdot x \rho(x, t).\qquad(2)$$

Here, we have assumed that diffusion is governed by a spatially-uniform, possibly anisotropic diffusivity tensor $\mathbf{D}$ and that diffusion encoding is achieved by a magnetic field gradient waveform $g(t)$. Note that we absorb the gyromagnetic ratio γ into $g(t)$ so that it has dimensions of time^{-1} length^{-1}.[3]

The signal $S_c = \int d^3x\, \rho(x, t)$ arising from a single such confined compartment under a general encoding waveform $g(t)$ can be found thanks to the Brownian paths having a Gaussian probability measure under the potential (2) [49]:

$$S_c = \exp\left(-\tfrac{1}{2}q_c^{\mathsf{T}}(0)\mathbf{D}\mathbf{\Omega}^{-1}q_c(0) - \int_0^{t_e} d\tau\, q_c^{\mathsf{T}}(\tau)\mathbf{D}q_c(\tau)\right),\qquad(3)$$

---

[1] In units of the thermal energy scale $k_B T$, where k_B is the Boltzmann constant and T is the absolute temperature.

[2] Vanishing confinement (i.e., free diffusion) has an unnormalizable steady state, but it can be handled.

[3] Alternatively, $g(t)$ can be called a precession rate gradient waveform.

with the generalized encoding wave vector

$$q_c(t) = \int_t^{t_e} d\tau \, e^{-\Omega(\tau-t)} g(\tau). \tag{4}$$

Here, t_e is the duration of the encoding protocol, and

$$\Omega = DC \tag{5}$$

is a matrix of equilibration rates. According to an equivalence between the signals of diffusion under a Hookean confinement and under restricting walls [34], a restriction of linear size L should have a confinement value of around $C \approx 11/L^2$. For instance, the confinement value within walls 5 µm apart would be about $C \approx 0.44$ µm^{-2}. With water diffusing at $D = 3$ µm^2/ms, this would imply an equilibration rate of $\Omega \approx 1.3$ ms^{-1}. In other words, it would take random walkers a few milliseconds to spread out roughly to their eventual distribution.

As depicted in Fig. 2, double-diffusion encoding is achieved by a gradient waveform consisting of two pairs of bipolar rectangular pulses, each with a given duration δ and separation Δ, and magnitudes g_1 and $-g_2$; the minus sign is customary. The time between the leading edges of the second pulse of the first pair and the first pulse of the second pair, t_m, is called the mixing time. The calculation of the signal (3) therefore entails very simple integrals, but in a cumbersome piecewise fashion. Upon significant simplification one finds

$$S_c(g_1, g_2) = \exp\left(-g_1^\mathsf{T} T_\circ g_1 - g_2^\mathsf{T} T_\circ g_2 - 2g_1^\mathsf{T} T_\times g_2\right). \tag{6}$$

One may refer to the tensors

$$T_\circ = D\Omega^{-3} \left[(1 - e^{-\Omega\Delta})(1 - e^{-\Omega\delta})^2 e^{\Omega\delta} - (1 - e^{-2\Omega\delta})e^{\Omega\delta} + 2\Omega\delta\right], \tag{7a}$$

$$2T_\times = D\Omega^{-3} e^{-\Omega(t_m-\delta)} (1 - e^{-\Omega\delta})^2 (1 - e^{-\Omega\Delta})^2, \tag{7b}$$

respectively, as the *self-coupling* and *cross-coupling* tensors between encoding blocks.[4] For the orientational average below, we consider $|g_1| = |g_2| = g$, with a fixed angle ψ between them.

As the free diffusion limit is approached, one can see that the cross-coupling of encoding blocks vanishes, as $T_\times \sim (D/2)\Omega\delta^2\Delta^2(1 + \Omega\delta - \Omega t_m)$, due to displacements in separate time intervals being uncorrelated in pure Brownian motion [36]. The dependence on the mixing time t_m does not enter until second order in confinement, in line with approximate calculations done for a spherical wall [29].

---

[4] $T_\circ$ is the same tensor that appears in the single-encoding signal $S_c = e^{-g^\mathsf{T} T_\circ g}$ [49]. In the free diffusion ($\Omega \to 0$), it is easily shown that $T_\circ \to D\delta^2(\Delta - \delta/3)$, which recovers the free diffusion signal [43, 44]. .

3 Double Diffusion Encoding: Powder Average

The orientational (powder) average of the compartment signal (6) is performed as follows: The vectors g_1 and g_2 form a plane with unit normal vector

$$\hat{n} = \frac{g_1 \times g_2}{|g_1 \times g_2|} . \tag{8}$$

One constructs the average by integrating over all orientations of (the plane normal to) $\hat{n}$, and for each of these integrate over all orientations of the pair $\{g_1, g_2\}$ within the plane, with their relative angle ψ fixed. The procedure hence described can be written as[5]

$$\bar{S} = \int \frac{d\hat{n}}{4\pi} \int \frac{d\beta}{2\pi} S_c(g_{\hat{n}}(\beta), g_{\hat{n}}(\beta + \psi)) . \tag{9}$$

Here, $g_{\hat{n}}(\beta)$ is a vector of magnitude g in the plane normal to $\hat{n}$, whose orientation is parameterized by the (in-plane) azimuthal angle β, according to which we can identify $g_1 \to g_{\hat{n}}(\beta)$ and $g_2 \to g_{\hat{n}}(\beta + \psi)$.

We first take the in-plane integral

$$\tilde{S}(\hat{n}) \stackrel{\text{def}}{=} \int \frac{d\beta}{2\pi} S_c(g_{\hat{n}}(\beta), g_{\hat{n}}(\beta + \psi)) \stackrel{\text{def}}{=} \int \frac{d\beta}{2\pi} e^{-\sigma_{\hat{n}}(\beta)} , \tag{10}$$

which serves as the definitions for the intermediate quantities $\tilde{S}(\hat{n})$ and $\sigma_{\hat{n}}(\beta)$. According to Eq. (6), the latter is given by

$$\sigma_{\hat{n}}(\beta) = g_{\hat{n}}^{\mathsf{T}}(\beta) T_\circ g_{\hat{n}}(\beta) + g_{\hat{n}}^{\mathsf{T}}(\beta + \psi) T_\circ g_{\hat{n}}(\beta + \psi) + 2 g_{\hat{n}}^{\mathsf{T}}(\beta) T_\times g_{\hat{n}}(\beta + \psi) . \tag{11}$$

To anchor the angle coordinate β, we define the in-plane cartesian coordinates u, v such that $\hat{u} \parallel g_{\hat{n}}(0)$ and $\hat{v} = \hat{n} \times \hat{u}$, yielding

$$g_{\hat{n}}(\beta) = \hat{u} g \cos\beta + \hat{v} g \sin\beta . \tag{12}$$

After an exercise in trigonometric simplification, and denoting $\hat{u}^{\mathsf{T}} T_\circ \hat{v} = T_{uv}^\circ$ etc., one finds

$$\sigma_{\hat{n}}(\beta) = \varsigma_{\hat{n}} + \rho_{\hat{n}} \cos 2\beta - \varrho_{\hat{n}} \sin 2\beta , \tag{13}$$

---

[5]We suppress the obvious limits of these integrals.

with

$$\frac{\varsigma_{\hat{n}}}{g^2} = \left(T_{uu}^{\circ} + T_{vv}^{\circ}\right) + \left(T_{uu}^{\times} + T_{vv}^{\times}\right)\cos\psi,\tag{14a}$$

$$\frac{\rho_{\hat{n}}}{g^2} = \left(T_{uu}^{\circ} - T_{vv}^{\circ}\right)\frac{1 + \cos 2\psi}{2} + T_{uv}^{\circ}\sin 2\psi + \left(T_{uu}^{\times} - T_{vv}^{\times}\right)\cos\psi + 2T_{uv}^{\times}\sin\psi,\tag{14b}$$

$$\frac{\varrho_{\hat{n}}}{g^2} = \left(T_{uu}^{\circ} - T_{vv}^{\circ}\right)\frac{\sin 2\psi}{2} - T_{uv}^{\circ}\left(1 + \cos 2\psi\right) + \left(T_{uu}^{\times} - T_{vv}^{\times}\right)\sin\psi - 2T_{uv}^{\times}\cos\psi.\tag{14c}$$

These depend on $\hat{n}$ through $\hat{u}$ and $\hat{v}$, of course. Hence via Eq. (10),

$$\tilde{S}(\hat{n}) = e^{-\varsigma_{\hat{n}}} \int \frac{d\beta}{2\pi} e^{-\rho_{\hat{n}}\cos 2\beta + \varrho_{\hat{n}}\sin 2\beta}$$

$$= e^{-\varsigma_{\hat{n}}} I_0\left(\sqrt{\rho_{\hat{n}}^2 + \varrho_{\hat{n}}^2}\right),\tag{15}$$

where an integral representation of the modified Bessel function of order 0 was recognized [10]. The argument of the square root can be found after a semi-tedious calculation as

$$\rho_{\hat{n}}^2 + \varrho_{\hat{n}}^2 = g^4\left[\left(T_{uu}^{\circ} - T_{vv}^{\circ}\right)\cos\psi + \left(T_{uu}^{\times} - T_{vv}^{\times}\right)\right]^2 + 4g^4\left(T_{uv}^{\circ}\cos\psi + T_{uv}^{\times}\right)^2,\tag{16}$$

and we rewrite the powder-averaged double-encoding signal (9) via Eqs. (10) and (15) as

$$\bar{S} = \int \frac{d\hat{n}}{4\pi} e^{-\varsigma_{\hat{n}}} I_0\left(\sqrt{\rho_{\hat{n}}^2 + \varrho_{\hat{n}}^2}\right).\tag{17}$$

For the actual evaluation of the integral, explicit expressions in terms of the polar and azimuthal angles (θ, φ) of $\hat{n}$ need to be substituted,[6] which are steps we omit here. Eyeballing how entries of $\mathbf{T}_{\circ}$ and $\mathbf{T}_{\times}$ appear in Eq. (17) via Eqs. (14a) and (16), one notes that it is useful to define the intermediate tensors

$$\mathbf{M} = g^2\left(\mathbf{T}_{\circ}\cos\psi + \mathbf{T}_{\times}\right)\tag{18a}$$

$$\tilde{\mathbf{M}} = g^2\left(\mathbf{T}_{\circ} + \mathbf{T}_{\times}\cos\psi\right).\tag{18b}$$

Referring to their eigenvalues, in any preferred order (but the same for both), as m_i and $\tilde{m}_i$, and using the following shorthand

---

[6]$\hat{n} = (\sin\theta\cos\varphi, \sin\theta\sin\varphi, \cos\theta)$, $\hat{u} = (\cos\theta\cos\varphi, \cos\theta\sin\varphi, -\sin\theta)$,   and $\hat{v} = (-\sin\varphi, \cos\varphi, 0)$.

$$m_{ij}^- = m_i - m_j \,, \tag{19a}$$

$$\tilde{m}_{ij}^- = \tilde{m}_i - \tilde{m}_j \,, \tag{19b}$$

$$\tilde{m}_{ij}^+ = \tilde{m}_i + \tilde{m}_j \,, \tag{19c}$$

the powder-averaged signal (17) attains the form

$$\bar{S} = e^{-\tilde{m}_{12}^+} \int \frac{d\cos\theta \, d\varphi}{4\pi} e^{-\left(\tilde{m}_{31}^- + \tilde{m}_{12}^- \sin^2\varphi\right)\sin^2\theta} \tag{20}$$

$$\times \ I_0\left(\sqrt{\left(m_{12}^-\right)^2 + 2m_{12}^- \left[m_{31}^- + \left(m_{13}^- + m_{23}^-\right)\sin^2\varphi\right]\sin^2\theta + \left(m_{13}^- + m_{21}^- \sin^2\varphi\right)^2 \sin^4\theta}\right),$$

upon substantial algebraic manipulation.

Reference [13] found series expansions for this form in a different context. We can use its results. Namely,

$$\bar{S} = e^{-\tilde{m}_{12}^+} \sum_{k=0}^{\infty} \sum_{m=0}^{k} q_{mk} Y_{mk} \,, \quad \text{where} \tag{21a}$$

$$Y_{mk} = \sum_{n=0}^{\infty} \sum_{l=0}^{n} \frac{\sqrt{\pi}(-1)^n}{l!(n-l)!} \frac{\left(\tilde{m}_{31}^-\right)^{n-l}\left(\tilde{m}_{12}^-\right)^{l}}{2^{2(m+l)+1}} \binom{2m+2l}{m+l} \frac{(n+k)!}{\left(n+k+\frac{1}{2}\right)!} \,, \quad \text{and} \tag{21b}$$

$$q_{mk} = \frac{\left(m_{31}^-\right)^{k-m}}{(k-m)!} \sum_{j=0}^{k/2} \frac{\left(m_{12}^-\right)^{j} I_{k-j}\left(m_{12}^-\right)}{2^j j! \left(m_{13}^- + m_{23}^-\right)^{2j-m}} \, {}_2\tilde{F}_1\left(m-k,-2j; m+1-2j; \frac{m_{13}^- + m_{23}^-}{m_{12}^-}\right) \,, \tag{21c}$$

with the following three alternatives for Eq. (21b):

$$Y_{mk}^{(1)} = \frac{\sqrt{\pi}}{2^{2m+1}} \binom{2m}{m} \sum_{n=0}^{\infty} \frac{\left(\tilde{m}_{13}^-\right)^n (k+n)!}{n!\left(k+n+\frac{1}{2}\right)!} \, {}_2F_1\left(m+\tfrac{1}{2}, -n; m+1; \frac{\tilde{m}_{21}^-}{\tilde{m}_{31}^-}\right), \tag{22a}$$

$$Y_{mk}^{(2)} = \frac{\sqrt{\pi}}{2^{2m+1}} \sum_{n=0}^{\infty} \frac{\left(\tilde{m}_{21}^-\right)^n (k+n)!\binom{2n+2m}{n+m}}{2^{2n} n!\left(k+n+\frac{1}{2}\right)!} \, {}_1F_1\left(k+n+1; k+n+\tfrac{3}{2}; \tilde{m}_{13}^-\right), \tag{22b}$$

$$Y_{mk}^{(3)} = \frac{\sqrt{\pi}}{2^{2m+1}} \binom{2m}{m} \sum_{n=0}^{\infty} \frac{\left(\tilde{m}_{13}^-\right)^n (k+n)!}{n!\left(k+n+\frac{1}{2}\right)!} \, {}_2F_2\left(m+\tfrac{1}{2}, k+n+1; m+1, k+n+\tfrac{3}{2}; \tilde{m}_{21}^-\right). \tag{22c}$$

Here, $_iF_j(\ldots)$ are the confluent hypergeometric functions [3], tilde denoting regularization. Note that these expressions apply to the most general case, in which the confinement tensor (and thus the tensors that are functions of it) have three *distinct* eigenvalues. Special cases associated with coinciding eigenvalues are discussed in the appropriate occasion later.

Which alternative among Eq. (22) yields better convergence is a matter of how the eigenvalues $\tilde{m}_i$ are chosen to be ordered, by way of the sizes and signs of $\tilde{m}_{ij}$. As a general guideline it would be wise to order the eigenvalues so as to avoid a sequence that alternates in sign, and with a large expansion parameter. Take Eq. (22b) for instance. Given that $_1F_1(k+l+1; k+l+\frac{3}{2}; \tilde{m}_{13}^-) > 0$ and increasing for all $\tilde{m}_{13}^-$, it would be beneficial to make $\tilde{m}_{13}^-$ negative (and large if possible), while keeping $\tilde{m}_{21}^-$ positive (and small if possible). For a given set of eigenvalues $\tilde{m}_i$, ordering them in the fashion $\tilde{m}_1 < \tilde{m}_2 < \tilde{m}_3$ would be along this guideline, whereas the ordering $\tilde{m}_2 < \tilde{m}_3 < \tilde{m}_1$ would result in a sequence with larger terms and alternating sign.

Note however that while $\mathbf{T}_\circ$ is a monotonic (decreasing) function of $\mathbf{\Omega}$, $\mathbf{T}_\times$ is not; see Eq. (7). Through Eq. (18) this means that their mixtures $\mathbf{M}$ and $\tilde{\mathbf{M}}$ are not necessarily monotonic in the confinement $\mathbf{\Omega}$. Hence what ordering of the confinement eigenvalues Ω_i achieves what ordering in the eigenvalues m_i and $\tilde{m}_i$ is a question which has an answer only on a case-by-case basis. Furthermore, the ordering of Ω_i that yields a desirable ordering of $\tilde{m}_i$ for a particular one of Eq. (22) may not produce an ordering of m_i as desirable for the convergence of q_{mk} in Eq. (21c).

3.1 Axisymmetric Confinement

We refer to the condition when two of the eigenvalues Ω_i of the confinement tensor coincide as *axisymmetric* confinement. Under this condition, the series expansions above undergo simplifications.

The most drastic simplification occurs when $\Omega_1 = \Omega_2$. That is, given that two eigenvalues coincide, assigning first and second place to them is the wisest choice as far as the evaluation of the series expansions (21) is concerned.

First, the coefficient q_{mk} simplifies as[7]

---

[7]For this, it needs to be noted [13, supplementary information] that the way q_{mk} arises in the calculation—before ever arriving at Eq. (21c)—is that it is the coefficient in the (double) series expansion of the Bessel function in Eq. (20):

$$I_0(\ldots) = \sum_{k=0}^{\infty} \sum_{m=0}^{k} q_{mk} \sin^{2m}\varphi \, \sin^{2k}\theta.$$

Upon all coefficients in the argument except for m_{13}^- vanishing due to axisymmetry, one has

$$I_0\left(m_{13}^- \sin^2\theta\right) = \sum_{k=0}^{\infty} \frac{(m_{13}^-)^{2k}}{(k!)^2 2^{2k}} \sin^{2k}\theta,$$

which, comparing to the previous (double) expansion, implies Eq. (23).

$$q_{mk} = \begin{cases} \frac{\delta_{m0}(m_{13}^{-})^k}{(\frac{k}{2})!(\frac{k}{2})!2^k}, & \text{even } k \\ 0, & \text{odd } k \end{cases}. \tag{23}$$

The coefficient Y_{mk}, on the other hand, loses the interior summation in Eq. (21b) due to m_{12}^{-} vanishing. The remaining summation can be identified according to the definition of hypergeometric functions [3] as

$$Y_{mk} = \frac{\sqrt{\pi}}{2^{m+1}} \binom{2m}{m} \frac{k!}{(k + \frac{1}{2})!} \, {}_1F_1\big(k + 1; k + \tfrac{3}{2}; \tilde{m}_{13}^{-}\big), \tag{24}$$

which yields via Eq. (23) and Eq. (21a)

$$\bar{S}_{\text{axy}} = \sqrt{\pi} e^{-\tilde{m}_{12}^{+}} \sum_{n=0}^{\infty} \binom{2n}{n} \frac{(m_{13}^{-})^{2n}}{2^{2n+1}} \frac{{}_1F_1\big(2n + 1; 2n + \tfrac{3}{2}; \tilde{m}_{13}^{-}\big)}{(2n + \frac{1}{2})!}. \tag{25}$$

This expansion is not so sensitive to the ordering of eigenvalues number 1 and 3, as the expansion parameter (m_{13}^{-}) is squared, and the hypergeometric function ${}_1F_1(2n+1; 2n+3/2; \tilde{m}_{13}^{-}) > 0$ for all arguments.

We depict in Fig. 3 the evaluation of the powder averaged signal (25) for axisymmetric confinement for representative values of encoding parameters. In Fig. 3a, the confinement is anisotropic (prolate), whereas in Fig. 3b, it is (nearly) isotropic. The bell-shaped dependence on the relative angle ψ between the gradient directions is seen in both cases, which is a sign that diffusion is not free [29]. This dependence is due mainly to the exponential prefactor in Eq. (25) that has nothing to do with the difference between the confinement eigenvalues (anisotropy). When the mixing time is increased, the bell-shaped modulation, indicating confinement regardless of anisotropy , stops overwhelming the relatively smaller influence of the rest of the expression (25): see Fig. 3a where the confinement is anisotropic. In an isotropic confinement, on the other hand, angular modulation simply disappears when the mixing time is increased (Fig. 3b), illustrating that the angular modulation that survives the increase in mixing time is due only to compartmental anisotropy, and not due to the fact of confinement (or to ensemble anisotropy, which was already eliminated by powder averaging).

3.2   Insights from Two Dimensions

The signal expressions for double encoding are a bit unwieldy to get a conceptual handle on. However, some insight can be gleaned from considering the orientational averaging in two dimensions.

Obviously, in the spirit of Eq. (9), the 2D orientationally averaged signal can be written as

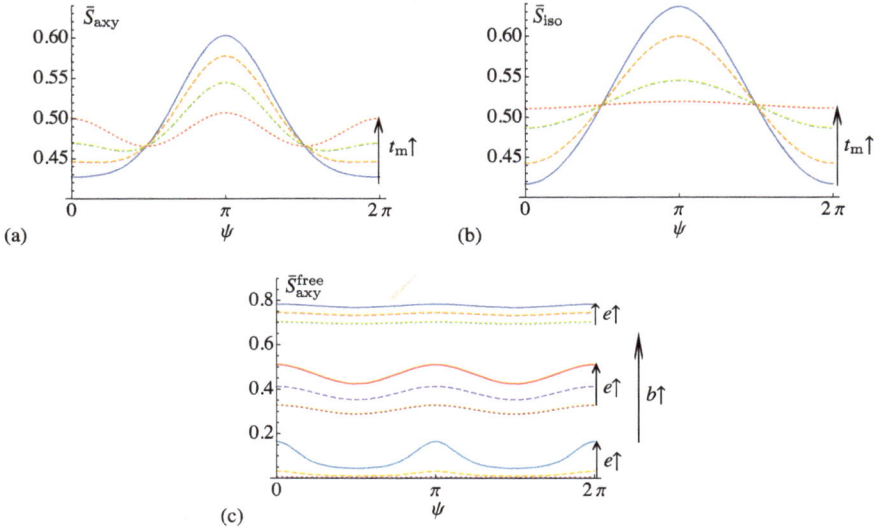

(a)

(b)

(c)

Fig. 3 Dependence of powder averaged signal (25) on the angle ψ between double-encoding gradient directions. **a** Prolate confinement $\boldsymbol{\Omega}$, with eigenvalues $\{1, 1, 0.1\}$ in arbitrary units. $\Delta = 10$ and $\delta = 1$, in units of Ω_1^{-1}. The mixing time t_m was set to the following multiples $\{1, 2, 6, 30\}$ of δ. The gradient strength is set such that $D_0 g^2 / \Omega_1^3 = 0.16$, with D_0 being the unit of diffusivity. **b** Isotropic confinement $\boldsymbol{\Omega}$, with eigenvalues $\{1, 1, 1\}$ in arbitrary units. $\Delta = 3$ and $\delta = 2/3$, in units of Ω_1^{-1}. The mixing time t_m was set to the following multiples $\{1, 3/2, 3, 6\}$ of δ. The gradient strength is set such that $D_0 g^2 / \Omega_1^3 = 1$. **c** Free diffusion with varying anisotropy (prolate) for comparison, with diffusivity $\mathbf{D}$ eigenvalues $\{D_0, eD_0, eD_0\}$. Across the three groups of plots, diffusion weighting $b = g^2 \delta^2 (\Delta - \delta/3)$ takes the values $\{0.4, 1.4, 10\}$ in units of D_0^{-1}, whereas each group of plots have the eccentricity parameter set to $\{0.01, 0.1, 0.2\}$

$$\bar{S}_{2D} = \int \frac{d\beta}{2\pi} \, S_c(\boldsymbol{g}(\beta), \boldsymbol{g}(\beta + \psi)) \,. \tag{26}$$

That is, there is no normal vector to integrate over; everything takes place in an (x, y) plane. The same steps Eq. (10) through Eq. (15) apply, and one has

$$\bar{S}_{2D} = e^{-g^2 \mathrm{Tr}\left(\mathbf{T}_\circ + \mathbf{T}_\times \cos \psi\right)} I_0\big(g^2 (T_1^\circ - T_2^\circ) \cos \psi + g^2 (T_1^\times - T_2^\times)\big) \,. \tag{27}$$

The dependence on the relative angle ψ of the gradient directions occurs both in the exponential attenuation factor and in the Bessel function. The angular dependence in the Bessel function has to do with anisotropy $(T_1^\circ - T_2^\circ)$ which the exponential factor is insensitive to due to the trace. In the exponent, on the other hand, the angular dependence is controlled by $\mathrm{Tr}\mathbf{T}_\times$, whose presence is due to confinement (since $\mathbf{T}_\times \to 0$ in the free diffusion limit). For large mixing times $(t_m \Omega_i \gg 1)$ the latter drops out and the angular modulation due to anisotropy is liberated. However, for smaller mixing time, it turns out that the angular dependence of the exponential factor dominates, which is due to confinement, suppressing the signature of anisotropy.

An interesting double-encoding scheme is its "symmetrized" version [35], which fixes $\psi = \pi/2$, but varies the magnitudes $g_1 = g \cos \alpha$ and $g_2 = g \sin \alpha$ as a function of some parameter α. In that scenario, one can easily calculate that

$$\bar{S}_{2D}^{sym} = e^{-\frac{g^2}{2}\operatorname{Tr}(T_o)} I_0\left(\frac{g^2}{2\sqrt{2}}\sqrt{\left(\Delta T_o^2 + \Delta T_x^2\right) + \left(\Delta T_o^2 - \Delta T_x^2\right)\cos 4\alpha}\right), \quad (28)$$

showing that the modulation due to confinement in the exponential factor drops out, and one of pure anisotropy remains (with ΔT_o denoting $T_1^o - T_2^o$ etc.). The result is a 'cleaner' version of the signal modulation wherein the confinement anisotropy is the only source of angular modulation characterized by the $\cos 4\alpha$ dependence. It should be remembered though that the confinement model is purely Gaussian, as such does not account for compartmental kurtosis. When truly restricted diffusion is considered, the compartment anisotropy and compartmental kurtosis both yield the same type of angular dependence compromising the interpretation of such angular dependence except when the compartments are isotropic [35].

3.3   One-Dimensional Diffusion Under High Gradient: g^{-2} Scaling

A special case of interest is when the compartment has an extremely elongated shape, resembling a "stick". Such extremely anisotropic shapes combined with an asymptotically large gradient strength ($g \to \infty$) tend to exhibit power-law dependence on the gradient strength g or the wave vector $q = \delta g$. In what follows, we confirm that the orientationally averaged signal (25) is no exception. Physical interpretation of the limits or extremes involved is remarked on at the end.

The two transverse confinement eigenvalues $\Omega_1 = \Omega_2 = \Omega_\perp$ approach infinity while $\Omega_3 = \Omega_\parallel$ is finite. Accordingly, $m_1 = m_2 = m_\perp = 0$ and $m_3 = m_\parallel$ is finite (similarly for $\tilde{m}_i$). Combined with a gradient $g \to \infty$, noting that $m_\parallel \sim g^2$ and $\tilde{m}_\parallel \sim g^2$ due to the definition (18), the orientationally averaged signal (25) assumes the asymptotic form

$$\bar{S}_{stick} \sim \left(2\tilde{m}_\parallel\sqrt{1 - \left(\frac{m_\parallel}{\tilde{m}_\parallel}\right)^2}\right)^{-1}; \quad (29)$$

see footnote 8 for details.[8] When we substitute from Eq. (18), the power-law dependence ($\sim g^{-2}$) becomes obvious:

$$\bar{S}_{\text{stick}} \sim \left(2g^2 \left(T_{\parallel}^{\circ} + T_{\parallel}^{\times} \cos \psi \right) \sqrt{1 - \left(\frac{T_{\parallel}^{\circ} \cos \psi + T_{\parallel}^{\times}}{T_{\parallel}^{\circ} + T_{\parallel}^{\times} \cos \psi} \right)^2} \right)^{-1}. \tag{30}$$

This result should be compatible with the findings of Ref. [13] in the free diffusion limit ($\Omega \to 0$). It is easy to see that Eq. (7) implies $T_{\parallel}^{\times} \to 0$ and $T_{\parallel}^{\circ} \to D_{\parallel} \delta^2 (\Delta - \delta/3)$ as $\Omega \to 0$, yielding

$$\bar{S}_{\text{stick}}^{\text{free}} \sim \left(2g^2 \delta^2 D_{\parallel} \left(\Delta - \tfrac{\delta}{3} \right) \sin \psi \right)^{-1}, \tag{31}$$

which is in fact the powder averaged signal of a rank-1 diffusivity tensor $\left(\begin{smallmatrix} D_{\parallel} & 0 & 0 \\ 0 & 0 & 0 \\ 0 & 0 & 0 \end{smallmatrix} \right)$ under the rank-2 measurement matrix $g^2 \delta^2 \left(\Delta - \tfrac{\delta}{3} \right) \left(\begin{smallmatrix} 1+\cos \psi & 0 & 0 \\ 0 & 1-\cos \psi & 0 \\ 0 & 0 & 0 \end{smallmatrix} \right)$ of DDE, as predicted by Ref. [13]. This scaling ($\sim g^{-2}$) of the signal of stick-compartments under rank-2 measurements (planar encoding) has been recently confirmed in vivo [1]. Note that the expression is invalid for $\psi = 0$ and $\psi = \pi$ where the two encoding directions coincide, and the corresponding measurement matrix is rank-1.

Lastly, we address the physical interpretation of the limits of *large* confinement (Ω) and gradient (g) values. Note that the final expressions such as Eqs. (21)–(22) and (25) come out in terms of the eigenvalues of the matrices $\mathbf{M}$ and $\tilde{\mathbf{M}}$, which are in turn defined in Eqs. (18) and (7). The sense in which a confinement eigenvalue Ω and gradient strength g is large follows in particular from Eq. (7): In the same order as the limits have been carried out, infinite (transverse) confinement ($\Omega_{\perp} \to \infty$) is to be interpreted physically as $g^2 D_{\perp} \Omega_{\perp}^{-3} \ll 1$, while infinite gradient ($g \to \infty$)

---

[8] At the extreme of $m_1 = m_2 = m_{\perp} = 0$, the signal (25) becomes

$$\bar{S}_{\text{stick}} \sim \sqrt{\pi} \sum_{n=0}^{\infty} \binom{2n}{n} \frac{m_{\parallel}^{2n}}{2^{2n+1}} \frac{{}_1 F_1 \left(2n + 1; 2n + \tfrac{3}{2}; -\tilde{m}_{\parallel} \right)}{\left(2n + \tfrac{1}{2} \right)!}.$$

The hypergeometric function can be rewritten by a so-called Kummer transformation as

$${}_1 F_1 \left(2n + 1; 2n + \tfrac{3}{2}; -\tilde{m}_{\parallel} \right) = e^{-\tilde{m}_{\parallel}} {}_1 F_1 \left(\tfrac{1}{2}; 2n + \tfrac{3}{2}; \tilde{m}_{\parallel} \right) \sim \left(2n + \tfrac{1}{2} \right)! / \sqrt{\pi} \, \tilde{m}_{\parallel}^{2n+1},$$

where the second step uses the asymptotic form ($\tilde{m}_{\parallel} \to \infty$) of the Hypergeometric function [10]. Hence, the signal attains the form

$$\bar{S}_{\text{stick}} \sim \sum_{n=0}^{\infty} \binom{2n}{n} \frac{m_{\parallel}^{2n}}{2^{2n+1} \tilde{m}_{\parallel}^{2n+1}},$$

converging to Eq. (29) for $|m_{\parallel}/\tilde{m}_{\parallel}| < 1$, which can be verified by carefully considering the functional dependences arising from the explicit form (18) of the matrices $\mathbf{M}$ and $\tilde{\mathbf{M}}$ (with the exception of the encoding angles $\psi = 0$ and $\psi = \pi$ where the two encoding directions coincide).

is to be taken to mean $g^2 D_\parallel \Omega_\parallel^{-3} \gg 1$. One may write $D_\parallel^{-1}\Omega_\parallel^3 \ll g^2 \ll D_\perp^{-1}\Omega_\perp^3$ in short. It should be noted that this order-of-magnitude treatment ignores potential exceptions or extremes which may arise from very particular combinations of the protocol parameters δ, Δ, t_{m}, and ψ.

4 Single Diffusion Encoding

The powder-averaged single-encoding signal can be obtained by way of "hacking" its double-encoded counterpart. One notes that $S_c = \mathrm{e}^{-g^{\mathsf{T}}\mathbf{T}_\circ g}$ is the single-encoding signal at the compartment level [49]. Then the features of the form (6) to get rid of are (i) the presence of a second vector g_2 unequal to the first, (ii) the presence of cross-coupling $(\mathbf{T}_\times)$, and (iii) the double occurrence of self-coupling. The first is hacked away by setting $\psi = 0$. For the second, one simply sets $\mathbf{T}_\times = 0$.[9] The result is $S_c = \mathrm{e}^{-2g^{\mathsf{T}}\mathbf{T}_\circ g}$, suffering from the third problem above, which is fixed by replacing $g \to g/\sqrt{2}$. All this yields via Eq. (18)

$$\mathbf{M} = \tfrac{1}{2}g^2 \mathbf{T}_\circ = \tilde{\mathbf{M}}, \tag{32}$$

which is the substitution that converts the powder-average expressions for double-encoding into those of single-encoding.

With this condition applied, the $\tilde{m}_{ij}^-$'s appearing in Eq. (21) turn into m_{ij}^-'s. However this alone does not produce drastic simplifications such as reducing summations. Rather than using Eq. (21), in fact, it is better to note that other results in Ref. [13] are quite suitable in the case of single-encoding. Since the compartment signal has the form $S_c = \mathrm{e}^{-\mathrm{Tr}(\mathbf{T}_\circ g g^{\mathsf{T}})}$, with the matrix $g g^{\mathsf{T}}$ axisymmetric by dint of being rank-1, the results of Ref. [13] for axisymmetric diffusion or measurement tensor can be applied. Written in terms of the parameters of the present discussion, the relevant formulas of Ref. [13] indicate the following alternative expressions

$$\bar{S}^{(1)} = \frac{\sqrt{\pi}}{2}\mathrm{e}^{-g^2 T_3^\circ} \sum_{n=0}^{\infty} \frac{g^{2n}\left(T_3^\circ - T_1^\circ\right)^n}{\left(n+\frac{1}{2}\right)!} \, {}_2F_1\left(\tfrac{1}{2}, -n; 1; \tfrac{T_1^\circ - T_2^\circ}{T_1^\circ - T_3^\circ}\right), \tag{33a}$$

$$\bar{S}^{(2)} = \mathrm{e}^{-g^2 T_3^\circ} \sum_{n=0}^{\infty} \frac{g^{2n}\left(T_1^\circ - T_2^\circ\right)^n}{n!(2n+1)!} \, {}_1F_1\left(n+1; n+\tfrac{3}{2}; g^2(T_3^\circ - T_1^\circ)\right), \tag{33b}$$

$$\bar{S}^{(3)} = \frac{\sqrt{\pi}}{2}\mathrm{e}^{-g^2 T_3^\circ} \sum_{n=0}^{\infty} \frac{g^{2n}\left(T_3^\circ - T_1^\circ\right)^n}{\left(n+\frac{1}{2}\right)!} \, {}_2F_2\left(\tfrac{1}{2}, n+1; 1, n+\tfrac{3}{2}; g^2(T_1^\circ - T_2^\circ)\right), \tag{33c}$$

---

[9]Physically, this can be imagined as the limit $t_{\mathrm{m}} \to \infty$. However, physics is not necessary. We simply have a set of expressions, e.g. Eq. (21), containing $\mathbf{T}_\circ$ and $\mathbf{T}_\times$ via Eq. (18) via Eq. (19), and we want to remove instances of $\mathbf{T}_\times$.

$$\bar{S}^{(4)} = \frac{\sqrt{\pi}}{2} e^{-g^2 T_3^\circ} \sum_{n=0}^{\infty} \frac{\binom{2n}{n} g^{4n} \left(T_1^\circ - T_2^\circ\right)^{2n}}{16^n \left(2n+\frac{1}{2}\right)!} \, {}_1F_1\!\left(2n+1;\, 2n+\tfrac{3}{2};\, \tfrac{g^2}{2}\left(2T_3^\circ - T_2^\circ - T_1^\circ\right)\right),$$

(33d)

for fully anisotropic $\boldsymbol{\Omega}$ (hence $\mathbf{T}_\circ$); the axisymmetric case is taken up below. These summations are subject to the same guidelines that followed Eq. (22) except that the caveats about conflicting eigenvalue ordering do not apply here. The single-encoding expressions here only involve the self-coupling tensor $\mathbf{T}_\circ$ which is a monotonically decreasing function of $\boldsymbol{\Omega}$; see Eq. (7). Therefore the ordering of the eigenvalues T_i° and Ω_i are certain to be in exactly the opposite sense of each other.

4.1   Axisymmetry and the Power-Laws for Confined diffusion

The first alternative in Eq. (33) needs special care when $T_1^\circ = T_3^\circ$ and the argument of the hypergeometric function $_2F_1(\ldots)$ diverges. Invoking a property of the hypergeometric function,[10] and renaming $T_1^\circ = T_\perp^\circ$ and $T_2^\circ = T_\parallel^\circ$, one finds[11]

$$\bar{S}_{\mathrm{axy}} = \frac{\sqrt{\pi}}{2} e^{-g^2 T_\perp^\circ(\delta, \Delta)} \frac{\mathrm{erf}\!\left(g\sqrt{T_\parallel^\circ(\delta, \Delta) - T_\perp^\circ(\delta, \Delta)}\right)}{g\sqrt{T_\parallel^\circ(\delta, \Delta) - T_\perp^\circ(\delta, \Delta)}}.$$

(34)

Here, we have also explicitly denoted the dependence on the encoding protocol's timing parameters as per Eq. (7). Note that in the free diffusion limit ($\boldsymbol{\Omega} \to 0$, see footnote 4), the known single-encoding powder-average signal [2, 18, 48]

10

$$\lim_{x \to 0} x^n {}_2F_1\!\left(m+\tfrac{1}{2}, -n; m+1; \tfrac{y}{x}\right) = (-y)^n \frac{\left(m+n-\tfrac{1}{2}\right)!\, m!}{\left(m-\tfrac{1}{2}\right)!\,(m+n)!}.$$

[11] Alternatively to using the results of Ref. [13] here, one may go back to the integral expression (20). First, we note that axisymmetry, with the choice $\Omega_1 = \Omega_2$, makes the integrand independent of φ, yielding

$$\bar{S} = \tfrac{1}{2} e^{-\tilde{m}_{12}^+} \int \mathrm{d}\cos\theta\, e^{\tilde{m}_{13}^- \sin^2\theta} I_0\!\left(m_{13}^- \sin^2\theta\right).$$

We are not aware of a closed-form evaluation of this integral. However, its special case (32) relevant here, making the arguments of the exponential and Bessel function match, has the result [12, 6.625–4]

$$\bar{S} = \frac{\sqrt{\pi}}{2} e^{-2m_1} \frac{\mathrm{erf}\!\left(\sqrt{2m_{31}^-}\right)}{\sqrt{2m_{31}^-}} = \frac{\sqrt{\pi}}{2} e^{-g^2 T_1^\circ} \frac{\mathrm{erf}\!\left(g\sqrt{T_3^\circ - T_1^\circ}\right)}{g\sqrt{T_3^\circ - T_1^\circ}}.$$

$$\bar{S}_{\text{axy}} = \frac{\sqrt{\pi}}{2} e^{-q^2(\Delta - \frac{\delta}{3})D_\perp} \frac{\text{erf}\left(q\sqrt{\left(\Delta - \frac{\delta}{3}\right)\left(D_\parallel - D_\perp\right)}\right)}{q\sqrt{\left(\Delta - \frac{\delta}{3}\right)\left(D_\parallel - D_\perp\right)}}, \tag{35}$$

with $q = g\delta$, is recovered.

Two special limiting cases follow.

4.1.1 Stick: $\bar{S} \propto g^{-1}$ Scaling

When particles have negligible latitude to move in the transverse direction, the orientationally-averaged signal (34) assumes a special form. In terms of the confinement model, this corresponds to $\Omega_\perp \to \infty$, which implies $T_\perp^\circ \to 0$ via Eq. (7).[12] The signal (34) then becomes

$$\bar{S}_{\text{stick}} = \frac{\sqrt{\pi}}{2} \frac{\text{erf}\left(g\sqrt{T_\parallel^\circ(\delta, \Delta)}\right)}{g\sqrt{T_\parallel^\circ(\delta, \Delta)}} \overset{g \to \infty}{\sim} \frac{\sqrt{\pi}}{2g\sqrt{T_\parallel^\circ(\delta, \Delta)}} \overset{\Omega_\parallel \to 0}{\sim} \frac{\sqrt{\pi}}{2q\sqrt{\left(\Delta - \frac{\delta}{3}\right)D_\parallel}}. \tag{36}$$

The large gradient regime has been important in identifying stick-like compartments via the q^{-1} scaling, which has been observed in white-matter areas of the brain and has been interpreted with the assumption of a free one-dimensional diffusion [23], which is adequate for channels of straight long channels of infinitesimal diameter. A notable exception is Ref. [33], which has incorporated the effects of finite size and curvature to offer an explanation for any deviation from the q^{-1} scaling, most apparent within gray matter. Ref. [33] also pointed out that such a scaling is not the true asymptotic behavior of the signal; the latter is rather dictated by the Debye-Porod law yielding q^{-4} scaling for narrow pulses [38], while an even steeper attenuation is predicted when the pulses are wide.

The above expression suggests that a similar decay is expected for the gradient magnitude rather than the q-value. The crucial difference is in the dependence on the timing parameters $\{\delta, \Delta\}$ of the SDE sequence, see Fig. 1. It would thus be interesting, e.g. in white-matter, to investigate whether the dependence on the timing parameters is more like $1/\sqrt{\Delta - \delta/3}$ (free difusion along the fiber) or $1/\sqrt{T_\parallel^\circ(\delta, \Delta)}$ (confined diffusion along the fiber), which would inform about the diffusion process along the axons.

---

[12]The limits $\Omega_\perp \to \infty$ and $g \to \infty$ are to be interpreted in the same sense as Sect. 3.3.

4.1.2   Pancake: $\bar{S} \propto g^{-2}$ Scaling

For completeness, we consider the opposite case where particles are able to spread in a plane, but not along the normal. Here, $\Omega_\| \to \infty$, implying $T_\|^\circ \to 0$ via Eq. (7).[13] One then has[14]

$$\bar{S}_{\text{pancake}} = \frac{\sqrt{\pi}}{2} e^{-g^2 T_\perp^\circ(\delta, \Delta)} \frac{\operatorname{erfi}\left(g\sqrt{T_\perp^\circ(\delta, \Delta)}\right)}{g\sqrt{T_\perp^\circ(\delta, \Delta)}} \overset{g \to \infty}{\sim} \frac{1}{2g^2 T_\perp^\circ(\delta, \Delta)} \overset{\Omega_\perp \to 0}{\sim} \frac{1}{2q^2\left(\Delta - \frac{\delta}{3}\right)D_\perp}. \tag{37}$$

Thus, the orientationally-averaged signal attenuates at a faster rate than in the case of sticks. Similarly though, the dependence on the timing parameters are different in free and confined diffusion scenarios.

5   Discussion

We have provided, for the first time, explicit expressions for the orientationally-averaged SDE and DDE MR signal intensity for structures represented by confinement tensors [49]. The latter is the effective model of restricted diffusion when pulses are long enough for the diffusing particles to traverse distances larger than the pore size [34]. As such, our findings are relevant for a broad range of porous materials featuring isolated, small pores.

The counterpart of these results for compartments of free anisotropic diffusion were given in Ref. [13] for arbitrary encoding waveforms. Reference [13] showed *inter alia* that the absence of axisymmetry could be discerned from the decay of the orientationally-averaged signal by employing the most general formulas. Though not rigorously studied here, we expect such arguments to be valid in the case of confined geometries as well by employing Eqs. (21)–(22) and (33).

For the time being, considering arbitrary waveforms for confined compartments seems extremely challenging. However, taking confinement into account is important, since the free diffusion model lacks features (such as the dependence on relative angle in double encoding) that a realistic signal will bear, see Fig. 3. Such bell-shaped angular modulation was related to the radius of gyration of the pores [24] and has since been used to estimate the apparent size of pores [4, 17, 29, 30, 41] via DDE measurements. The underlying reason was thought to be a restriction effect [11] that not only leads to an anisotropy of the diffusion process at a length scale smaller than the pore size [26, 29, 31, 32], but also makes the apparent diffusion coefficient depend on the diffusion time [15]. Thus, this effect is absent when free diffusion is thought to take place within individual pores, see Fig. 3c. Our results indicate that

---

[13] The limits $\Omega_\| \to \infty$ and $g \to \infty$ are to be interpreted in a sense analogous to Sect. 3.3, with $\perp$ and $\|$ interchanged.

[14] The imaginary error function has the properties $\operatorname{i}\operatorname{erfi}(z) = \operatorname{erf}(\operatorname{i}z)$, and $\operatorname{erfi}(z) \sim e^{z^2}/z\sqrt{\pi}$.

the confinement tensor framework is capable of capturing such angular modulation, which makes it a suitable representation of diffusion within microdomains [22, 34, 49]. This is important in applications like q-space trajectory imaging [47] and diffusion tensor distribution imaging [45], which aim to characterize the structure of subdomains using general gradient waveforms.

Another angular modulation that is apparent in Fig. 3 is w-shaped, which was pointed out by Mitra [24] for randomly distributed sticks. Later, it was proved that the dominant contribution to such angular modulation had the functional form $\cos 2\psi$ for fully restricted structures and cylinders of finite diameter [28]. Such modulation that manifests itself at twice the "angular frequency" [19, 24, 28] is present even when diffusion within the subdomains is envisioned to be free as long as it is anisotropic, see Fig. 3c. Thus, such modulation is truly indicative of the anisotropy of the subdomains, be it free (Fig. 3c), confined (Fig. 3a, b), or truly restricted [28, 32]. For more information on such anisotropy, the reader is referred to the review on this topic by Ianuş et al. [14] in this book series.

From a mathematical point of view, the expressions in Ref. [13] provided the Laplace transform of a tensor distribution, which includes rotated copies of a given diffusion tensor wherein all orientations are equally likely, thus extending the modeling approach that employs parametric diffusion tensor distributions [16, 21, 37, 39] to a new type of tensor distribution. Evaluating the signal, in a similar fashion, for confinement rather than diffusion tensors can be regarded as the evaluation of a transform whose kernel is the compartmental signal given in (6) instead of the kernel of the matrix Laplace transform $e^{-\mathbf{BD}}$.

6 Conclusion

We have given analytical expressions for the orientationally averaged diffusion MR signal originating from confined anisotropic compartments for two relatively simple encoding schemes. A number of observations related to signal modulation and power-law tails were made for such confined pores. These findings complement and extend the exact expressions for locally free diffusion provided in Ref. [13] to confined diffusion albeit for SDE and DDE measurements.

Acknowledgments We acknowledge the following sources for funding: Swedish Foundation for Strategic Research AM13-0090, the Swedish Research Council 2016-04482, Linköping University Center for Industrial Information Technology (CENIIT), VINNOVA/ITEA3 17021 IMPACT, and National Institutes of Health P41EB015902 and R01MH074794.

References

1. Afzali, M., Aja-Fernández, S., Jones, D.K.: Direction-averaged diffusion-weighted MRI signal using different axisymmetric B-tensor encoding schemes. Magn. Reson. Med. (in press) (2020). https://doi.org/10.1002/mrm.28191, https://onlinelibrary.wiley.com/doi/abs/10.1002/mrm.28191
2. Anderson, A.W.: Measurement of fiber orientation distributions using high angular resolution diffusion imaging. Magn. Reson. Med. **54**(5), 1194–1206 (2005). https://doi.org/10.1002/mrm.20667

3. Arfken, G.B., Weber, H.J.: Mathematical Methods for Physicists. Academic Press, San Diego (2001)
4. Callaghan, P.T.: Translational Dynamics and Magnetic Resonance: Principles of Pulsed Gradient Spin Echo NMR. Oxford University Press, New York (2011)
5. Callaghan, P.T., Jolley, K.W., Lelievre, J.: Diffusion of water in the endosperm tissue of wheat grains as studied by pulsed field gradient nuclear magnetic resonance. Biophys. J. **28**, 133–142 (1979)
6. Callaghan, P.T., Komlosh, M.E.: Locally anisotropic motion in a macroscopically isotropic system: displacement correlations measured using double pulsed gradient spin-echo NMR. Magn. Reson. Chem. **40**, S15–S19 (2002)
7. Callaghan, P.T., Pinder, D.N.: Dynamics of entangled polystyrene solutions studied by pulsed field gradient nuclear magnetic resonance. Macromolecules **13**, 1085–1092 (1980)
8. Cheng, Y., Cory, D.G.: Multiple scattering by NMR. J. Am. Chem. Soc. **121**, 7935–7936 (1999)
9. Cory, D.G., Garroway, A.N., Miller, J.B.: Applications of spin transport as a probe of local geometry. Polym. Preprints **31**, 149 (1990)
10. *NIST Digital Library of Mathematical Functions*. http://dlmf.nist.gov/, Release 1.0.24 of 2019-09-15. http://dlmf.nist.gov/. Olver, F.W.J., Olde Daalhuis, A.B., Lozier, D.W., Schneider, B.I., Boisvert, R.F., Clark, C.W., Miller, B.R., Saunders, B.V., Cohl, H.S., McClain, M.A. (eds.)
11. Finsterbusch, J.: Multiple-wave-vector diffusion-weighted NMR. Ann. Rep. NMR Spectrosc. **72**, 225–299 (2011)
12. Gradshteyn, I.S., Ryzhik, I.M.: Table of Integrals, Series, and Products, 6th edn. Academic Press, London (2000)
13. Herberthson, M., Yolcu, C., Knutsson, H., Westin, C.F., Özarslan, E.: Orientationally-averaged diffusion-attenuated magnetic resonance signal for locally-anisotropic diffusion. Sci. Rep. **9**(1), 4899 (2019). https://doi.org/10.1038/s41598-019-41317-8
14. Ianuş, A., Shemesh, N., Alexander, D.C., Drobnjak, I.: Measuring microscopic anisotropy with diffusion magnetic resonance: from material science to biomedical imaging. In: Schultz, T., Özarslan, E., Hotz, I. (eds.) Modeling, Analysis, and Visualization of Anisotropy, Mathematics and Visualization, pp. 229–255. Springer International Publishing (2017)
15. Jespersen, S.N.: Equivalence of double and single wave vector diffusion contrast at low diffusion weighting. NMR Biomed. **25**(6), 813–818 (2012). https://doi.org/10.1002/nbm.1808
16. Jian, B., Vemuri, B.C., Özarslan, E., Carney, P.R., Mareci, T.H.: A novel tensor distribution model for the diffusion-weighted MR signal. NeuroImage **37**(1), 164–176 (2007). https://doi.org/10.1016/j.neuroimage.2007.03.074
17. Komlosh, M.E., Özarslan, E., Lizak, M.J., Horkayne-Szakaly, I., Freidlin, R.Z., Horkay, F., Basser, P.J.: Mapping average axon diameters in porcine spinal cord white matter and rat corpus callosum using d-PFG MRI. NeuroImage **78**, 210–6 (2013). https://doi.org/10.1016/j.neuroimage.2013.03.074
18. Kroenke, C.D., Ackerman, J.J.H., Yablonskiy, D.A.: On the nature of the NAA diffusion attenuated MR signal in the central nervous system. Magn. Reson. Med. **52**(5), 1052–9 (2004). https://doi.org/10.1002/mrm.20260
19. Lawrenz, M., Koch, M.A., Finsterbusch, J.: A tensor model and measures of microscopic anisotropy for double-wave-vector diffusion-weighting experiments with long mixing times. J. Magn. Reson. **202**(1), 43–56 (2010). https://doi.org/10.1016/j.jmr.2009.09.015
20. Le Doussal, P., Sen, P.N.: Decay of nuclear magnetization by diffusion in a parabolic magnetic field: An exactly solvable model. Phys. Rev. B **46**(6), 3465–3485 (1992)
21. Leow, A.D., Zhu, S., Zhan, L., McMahon, K., de Zubicaray, G.I., Meredith, M., Wright, M.J., Toga, A.W., Thompson, P.M.: The tensor distribution function. Magn. Reson. Med. **61**(1), 205–14 (2009). https://doi.org/10.1002/mrm.21852
22. Liu, C., Özarslan, E.: Multimodal integration of diffusion MRI for better characterization of tissue biology. NMR Biomed. **32**(4), e3939 (2019). https://doi.org/10.1002/nbm.3939
23. McKinnon, E.T., Jensen, J.H., Glenn, G.R., Helpern, J.A.: Dependence on b-value of the direction-averaged diffusion-weighted imaging signal in brain. Magn. Reson. Imaging **36**, 121–127 (2017). https://doi.org/10.1016/j.mri.2016.10.026

24. Mitra, P.P.: Multiple wave-vector extensions of the NMR pulsed-field-gradient spin-echo diffusion measurement. Phys. Rev. B **51**(21), 15074–15078 (1995)
25. Mitra, P.P., Halperin, B.I.: Effects of finite gradient-pulse widths in pulsed-field-gradient diffusion measurements. J. Magn. Reson. A **113**, 94–101 (1995)
26. Moutal, N., Maximov, I.I., Grebenkov, D.S.: Probing surface-to-volume ratio of an anisotropic medium by diffusion NMR with general gradient encoding. IEEE Trans. Med. Imaging **38**(11), 2507–2522 (2019). https://doi.org/10.1109/TMI.2019.2902957
27. Novikov, D.S., Fieremans, E., Jespersen, S.N., Kiselev, V.G.: Quantifying brain microstructure with diffusion MRI: Theory and parameter estimation. NMR Biomed. **32**(4), e3998 (2019). https://doi.org/10.1002/nbm.3998
28. Özarslan, E.: Compartment shape anisotropy (CSA) revealed by double pulsed field gradient MR. J. Magn. Reson. **199**(1), 56–67 (2009). https://doi.org/10.1016/j.jmr.2009.04.002
29. Özarslan, E., Basser, P.J.: Microscopic anisotropy revealed by NMR double pulsed field gradient experiments with arbitrary timing parameters. J. Chem. Phys. **128**(15), 154511 (2008). https://doi.org/10.1063/1.2905765
30. Özarslan, E., Komlosh, M., Lizak, M., Horkay, F., Basser, P.: Double pulsed field gradient (double-PFG) MR imaging (MRI) as a means to measure the size of plant cells. Magn. Reson. Chem. **49**, S79–S84 (2011). https://doi.org/10.1002/mrc.2797
31. Özarslan, E., Nevo, U., Basser, P.J.: Anisotropy induced by macroscopic boundaries: Surface-normal mapping using diffusion-weighted imaging. Biophys. J. **94**(7), 2809–2818 (2008). https://doi.org/10.1529/biophysj.107.124081
32. Özarslan, E., Shemesh, N., Basser, P.J.: A general framework to quantify the effect of restricted diffusion on the NMR signal with applications to double pulsed field gradient NMR experiments. J. Chem. Phys. **130**(10), 104702 (2009). https://doi.org/10.1063/1.3082078
33. Özarslan, E., Yolcu, C., Herberthson, M., Knutsson, H., Westin, C.F.: Influence of the size and curvedness of neural projections on the orientationally averaged diffusion MR signal. Front. Phys. **6**, 17 (2018)
34. Özarslan, E., Yolcu, C., Herberthson, M., Westin, C.F., Knutsson, H.: Effective potential for magnetic resonance measurements of restricted diffusion. Front. Phys. **5**, 68 (2017)
35. Paulsen, J.L., Özarslan, E., Komlosh, M.E., Basser, P.J., Song, Y.Q.: Detecting compartmental non-Gaussian diffusion with symmetrized double-PFG MRI. NMR Biomed. **28**(11), 1550–6 (2015). https://doi.org/10.1002/nbm.3363
36. Risken, H.: The Fokker-Planck Equation, 2nd edn. Springer, Berlin (1989)
37. Scherrer, B., Schwartzman, A., Taquet, M., Sahin, M., Prabhu, S.P., Warfield, S.K.: Characterizing brain tissue by assessment of the distribution of anisotropic microstructural environments in diffusion-compartment imaging (DIAMOND). Magn. Reson. Med. **76**(3), 963–77 (2016). https://doi.org/10.1002/mrm.25912
38. Sen, P.N., Hürlimann, M.D., de Swiet, T.M.: Debye-Porod law of diffraction for diffusion in porous media. Phys. Rev. B **51**(1), 601–604 (1995)
39. Shakya, S., Batool, N., Özarslan, E., Knutsson, H.: Multi-fiber reconstruction using probabilistic mixture models for diffusion MRI examinations of the brain. In: Schultz, T., Özarslan, E., Hotz, I. (eds.) Modeling, Analysis, and Visualization of Anisotropy, pp. 283–308. Springer International Publishing, Cham (2017)
40. Shemesh, N., Jespersen, S.N., Alexander, D.C., Cohen, Y., Drobnjak, I., Dyrby, T.B., Finsterbusch, J., Koch, M.A., Kuder, T., Laun, F., Lawrenz, M., Lundell, H., Mitra, P.P., Nilsson, M., Özarslan, E., Topgaard, D., Westin, C.F.: Conventions and nomenclature for double diffusion encoding NMR and MRI. Magn. Reson. Med. **75**(1), 82–7 (2016). https://doi.org/10.1002/mrm.25901
41. Shemesh, N., Özarslan, E., Basser, P.J., Cohen, Y.: Measuring small compartmental dimensions with low-q angular double-PGSE NMR: the effect of experimental parameters on signal decay. J. Magn. Reson. **198**(1), 15–23 (2009). https://doi.org/10.1016/j.jmr.2009.01.004
42. Shemesh, N., Özarslan, E., Basser, P.J., Cohen, Y.: Accurate noninvasive measurement of cell size and compartment shape anisotropy in yeast cells using double-pulsed field gradient MR. NMR Biomed. **25**(2), 236–246 (2012). https://doi.org/10.1002/nbm.1737

43. Stejskal, E.O.: Use of spin echoes in a pulsed magnetic-field gradient to study anisotropic, restricted diffusion and flow. J. Chem. Phys. **43**(10), 3597–3603 (1965)
44. Stejskal, E.O., Tanner, J.E.: Spin diffusion measurements: Spin echoes in the presence of a time-dependent field gradient. J. Chem. Phys. **42**(1), 288–292 (1965)
45. Topgaard, D.: Diffusion tensor distribution imaging. NMR Biomed. **32**(5), e4066 (2019). https://doi.org/10.1002/nbm.4066
46. Uhlenbeck, G.E., Ornstein, L.S.: On the theory of the Brownian motion. Phys. Rev. **36**, 823–841 (1930)
47. Westin, C.F., Knutsson, H., Pasternak, O., Szczepankiewicz, F., Özarslan, E., van Westen, D., Mattisson, C., Bogren, M., O'Donnell, L.J., Kubicki, M., Topgaard, D., Nilsson, M.: Q-space trajectory imaging for multidimensional diffusion MRI of the human brain. NeuroImage **135**, 345–62 (2016). https://doi.org/10.1016/j.neuroimage.2016.02.039
48. Yablonskiy, D.A., Sukstanskii, A.L., Leawoods, J.C., Gierada, D.S., Bretthorst, G.L., Lefrak, S.S., Cooper, J.D., Conradi, M.S.: Quantitative in vivo assessment of lung microstructure at the alveolar level with hyperpolarized ^{3}He diffusion MRI. Proc. Natl. Acad. Sci. U. S. A. **99**(5), 3111–6 (2002). https://doi.org/10.1073/pnas.052594699
49. Yolcu, C., Memiç, M., Şimşek, K., Westin, C.F., Özarslan, E.: NMR signal for particles diffusing under potentials: From path integrals and numerical methods to a model of diffusion anisotropy. Phys. Rev. E **93**, 052602 (2016)

9

Single Encoding Diffusion MRI: A Probe to Brain Anisotropy

Maëliss Jallais and Demian Wassermann

Abstract This chapter covers anisotropy in the context of probing microstructure of the human brain using single encoded diffusion MRI. We will start by illustrating how diffusion MRI is a perfectly adapted technique to measure anisotropy in the human brain using water motion, followed by a biological presentation of human brain. The non-invasive imaging technique based on water motions known as diffusion MRI will be further presented, along with the difficulties that come with it. Within this context, we will first review and discuss methods based on signal representation that enable us to get an insight into microstructure anisotropy. We will then outline methods based on modeling, which are state-of-the-art methods to get parameter estimations of the human brain tissue.

1 Accessing Brain Anisotropy Using Diffusion MRI

1.1 Introduction

Diffusion-weighted MR imaging is a non-invasive tool used to probe tissue microstructure. During a typical acquisition of tens of milliseconds in brain imaging, water molecules can displace up to tens of micrometers. Diffusion is therefore sensitive to a wide range of microstructural and physiological parameters in the tissue. The diffusing molecules get restricted by the boundaries of the underlying microstructure of tissues. Diffusion anisotropy corresponds to the hindrance of those molecules, otherwise free diffusing (i.e. isotropically). Changes in anisotropy have been related to brain diseases such as ischemia, multiple sclerosis, trauma, or brain tumors [2, 58]. Diffusion anisotropy is therefore considered as a potential biological marker for changes in tissue microstructure. A loss of anisotropy can also be the sign

M. Jallais (✉) · D. Wassermann
Université Paris-Saclay, Inria, CEA, 91120 Palaiseau, France
e-mail: maeliss.jallais@inria.fr

D. Wassermann
e-mail: demian.wassermann@inria.fr

of an increasing isotropy, as neurons grow in size. Panagiotaki et al. [57] notably use this change to study the evolution of tumor cell size in response to a drug. An understanding of the origin of anisotropy change and a combined study of both anisotropy and isotropy can therefore lead to great discoveries on a tissue microstructure and its evolution.

An ultimate goal of a Magnetic Resonance (MR) diffusion theory is then to relate the microstructural and physiological parameters quantitatively to the diffusion-weighted signal. This task appears to be complicated as deducing those parameters constitutes a complex inverse problem requiring careful modeling of the diffusion signal over a wide range of diffusion time and diffusion weightings (diffusion weightings will be further explained in Sect. 2.2).

Regarding its cellular composition, brain can be decomposed in two main parts: white matter and grey matter. The former designates regions that contain mainly long-range myelinated axons, which cross the brain connecting different parts of grey matter, and relatively few cell bodies. A method to study those connections is called tractography and has been well explored during the past few years [28]. Grey matter contains mainly cell bodies, connected by neurites, and relatively few myelinated axons. Anisotropy exists in both white matter and grey matter and is due to the presence of cells with long cylindrical processes (axons in white matter and neurites in grey matter). Its presence, or its absence, will provide us with key information about the tissue structure at the cellular level.

Fick et al. dedicated a review on existing diffusion anisotropy metrics [14], which includes Fractional Anisotropy (FA) [3], Generalized Fractional Anisotropy (GFA) [60], Propagator Anisotropy (PA) [50], Orientation Dispersion Index (ODI) [67], and microscopic Fractional Anisotropy (µFA) [33]. We present here a complementary approach, which considers anisotropy as a probe for accessing microstructure, either through signal representations or tissue modeling.

1.2 Anisotropy as Reflected by Water Motion

Particles suspended in a fluid are constantly undergoing small random movements, which is known as Brownian motion [42]. The physical process of a steadily spread of a substance is called diffusion. Diffusion can therefore be considered a macroscopic manifestation of Brownian motion on the microscopic level. When no barrier impedes diffusion preferentially in one direction over another, molecular displacements are equal in all directions. This is known as isotropic diffusion. However, in brain, molecule movements are hindered by cell membranes. Diffusion is then not equal along all directions anymore and has become anisotropic. The distance traveled by a water molecule depends on its interactions. Certain geometric characteristics of the underlying structure at the microscopic level can therefore be inferred from the molecule movements [32]. The further a molecule travels during the time of an acquisition, the greater signal attenuation we get. The objective is to use this attenuation to deduce the structure of the medium where the water molecules are trapped in.

Depending on the diffusion time, i.e. the amount of time between two gradient pulses (see Sect. 2.1), the information we get about the structure will be different. If the diffusion time was extremely short, only the local intrinsic diffusivity of the fluid, i.e. the rate at which particles can spread, would be measured. The hindrance effects would only become apparent at longer times. The degree of anisotropy hence also depends on the diffusion time.

1.3 Structural Brain Anisotropy

Brain tissue is very anisotropic due to the cylindrical shapes of axons and processes. Water molecules within those fibers will on average move further along them than across them due to their small diameter. Typical axon diameter in humans is of the order of 1–10 μm [5]. Process diameters in grey matter lie between 0.1 and 15 μm.

The strong anisotropy in white matter due to the axons encouraged its wide study over the past decades. The more complex tissue structure and weaker anisotropy in grey matter make its study harder. The presence of isotropy in grey matter is partly due to the numerous somas whose shapes resemble spheres (see Sect. 4.1). Soma diameters range between 20 and 120 μm. White matter models need to be adapted to account for the presence of somas in order to be applied to grey matter [38, 53]. Myelin also appears to modulate the degree of diffusion anisotropy between axons and processes (and so between white matter and grey matter), but has a smaller role in anisotropy than membrane [6].

Note that anisotropy is not only a property of neural fibers. Anisotropy has also been observed in liquid crystals, muscles and other tissues, even in fruits and vegetables [5]. The degree of fractional anisotropy (see Sect. 3.1) is however higher in healthy neural fibers than in other tissues such as skeletal muscle, kidney, and myocardium [11, 18].

1.4 Measuring Anisotropy Using Diffusion MRI

Using Diffusion Magnetic Resonance Imaging (dMRI) as a non-invasive probe in human brain, we aim at getting information about its structure. The acquired diffusion signal is a sum of the diffusion signals coming from each compartment weighted by their relative volume fractions [56], and is therefore modulated by the geometry of the tissue microstructure. Relevant information to infer from it are soma diameters, soma and process densities, and diffusivities. Two complementary approaches have emerged for extracting these information about the tissue microstructure from the diffusion signal: signal representation and tissue modelling (denomination from Novikov et al. [48]).

Signal representations aim at quantifying parameters and are model-independent mathematical expressions. Their parameters do not carry any particular physical meaning. Representations can be used to store, compress or compare measurements. There is an infinite way to represent a continuous function. One chooses a representation according to the need of a particular neuroimaging study [44]. Although signal representations are suited for all kind of tissues, they lack specificity and provide only an indirect characterization of the microstructure.

Biophysical tissue models rely on a schematic geometry of the underlying tissue. They are pictures representing a physical reality relying on assumptions meant to simplify the complexity of a biological tissue. A good model only keeps relevant features which characterize the tissue and discards irrelevant degrees of freedom. The designed analytical expression is then fit to the diffusion data in order to estimate these relevant features of the microstructure. This advantage of providing greater specificity and interpretation of biologically-relevant parameters appears to be the weakness of the method. Indeed the initial geometric assumption must be chosen as to accurately capture all of the features of the tissue that effectively impact the diffusion signal in a given acquisition range [48], but we also must be able to mathematically solve this inverse problem. Model validations are important because a wrong model could lead to wrong interpretations of a physical phenomenon.

Techniques from these two approaches, signal representations and tissue modelling, will be reviewed respectively in Sects. 3 and 4.

2 Diffusion MRI: Introduction to a Non-Invasive Imaging Technique

Nuclear Magnetic Resonance Imaging allows to non-invasively study the brain in-vivo, and in particular brain anisotropy, induced by its microstructure.

2.1 Diffusion MRI Acquisition Sequence

Consider an MRI acquisition sequence. After slice selection, all the nuclei on this plane are precessing at the same frequency. To obtain a diffusion MR image, two gradient pulses are added to the acquisition sequence. The first applied pulse is going to make the particles go off phase. We then apply a second gradient with the same strength in the opposite direction, during the same amount of time. If molecules stayed still between those two gradients, they would have all come back to their original phase, the two gradients cancelling each other. However, after turning the first gradient on, molecules are moving randomly (Brownian motion). After a certain evolution time, if molecules are not at the same location, the second gradient causes destructive interference, which results in a loss of signal. The further a molecule travels from its initial position during the time between the two diffusion gradients along the gradient direction, the greater signal attenuation we get. The ratio between the signal obtained with diffusion gradients and the one without them quantifies the amount of ongoing diffusion. The objective is then to deduce the structure of the medium where the water molecules are trapped in from those signal losses.

2.2 Mathematical Foundations

Stejskal and Tanner invented in 1965 the Pulsed Gradient Spin Echo (PGSE) sequence [59] to measure diffusion in a specific direction. In this sequence, two opposite diffusion gradients are applied during a time δ, separated by a time interval Δ. The diffusion-weighting is globally encoded by the b-value [37], and reflects the strength and timing of the gradients used to generate the diffusion weighted images. This factor is computed as follow:

$$b = \gamma^2 g^2 \delta^2 (\Delta - \delta/3), \tag{1}$$

where γ (MHzT^{-1}) is the nuclear gyromagnetic ratio of the water proton ^1H and g is the strength of the diffusion gradient. In the following sections, $\mathbf{g}$ encodes the direction of the applied diffusion weighting in addition to its strength ($g = ||\mathbf{g}||$), and $\hat{\mathbf{g}}$ is the corresponding unit vector ($\hat{\mathbf{g}} = \mathbf{g}/||\mathbf{g}||$).

The quantity $E(b) = S(b)/S_0$ expresses, for each voxel, the attenuation of the diffusion-weighted signal along the selected gradient direction, S_0 being the image acquired without diffusion gradients. In the absence of restrictions (free diffusion), the signal attenuation can be expressed as:

$$E(b) = e^{-bD}, \tag{2}$$

with D the diffusion coefficient.

If δ is assumed to be infinitely narrow, i.e. the diffusion during that time is negligible, the signal attenuation can be related to the ensemble average propagator (EAP) $P(\mathbf{r}, \tau)$ via a Fourier relationship under the q-space formalism [9, 59]:

$$E(\mathbf{q}, \tau) = \frac{S(\mathbf{q}, \tau)}{S_0} = \int_{\mathbb{R}^3} P(\mathbf{r}, \tau) e^{-2\pi i \mathbf{q} \cdot \mathbf{r}} d\mathbf{r}, \tag{3}$$

where $\mathbf{q}$ is the wave vector and τ the diffusion time, which, for the PGSE sequence, are expressed as

$$\mathbf{q} = \gamma \delta g / 2\pi \text{ and } \tau = \Delta - \delta/3. \tag{4}$$

The diffusion time τ expresses the time interval during which spins are allowed to diffuse before measurement. By increasing the spatial frequency $q = ||\mathbf{q}||$ it is possible to achieve a higher spatial resolution of $P(\mathbf{r}, \tau)$ in the displacement space described by $\mathbf{r}$.

2.3   Acquisition Strategies

Experimental parameters, and especially q and τ, influence the diffusion signal attenuation along different gradient directions, and therefore the estimation of diffusion anisotropy. Ideally, many gradient directions, q-values and diffusion times would be required to completely characterize diffusion anisotropy in a tissue. In practice, the sampling strategy depends on the application and on the chosen signal representation. This way, only one shell of gradient directions and a single b-value are usually used in DTI (see Sect. 3.1). Also using only one shell at a higher b-value and more directions, are the High Angular Resolution Diffusion Imaging (HARDI) schemes, which aim at increasing the angular resolution of the diffusion signal with the intent of resolving crossing tissue configurations [61]. Different diffusion-weightings signal acquisitions are also needed for some signal representations. In that case, multi-shell acquisitions are set up using different q-shells with fixed diffusion time. Each shell then represents a collection of samples in the three-dimensional space with the same q-value. An optimal spatial coverage is important to measure the diffusion signal as efficiently as possible. Expansions have been proposed such that all the acquired samples lie on different non-collinear directions [10]. This multi-shell design can be extended to τ-shells, called $q\tau$ acquisitions [13] in order to exploit different values for both q and τ. In that case, a complete q-shell scheme is acquired for each desired diffusion time. Ning et al [44] reviews and compares 16 reconstruction algorithms (single and multi-shells) to help determine an appropriate acquisition protocol (number of b-values) and the analysis method to use for a particular neuroimaging study.

2.4   Difficulties

A main drawback to take into consideration is inherent to the dMRI acquisition process. Due to the acquisition device limitations and the mesoscopic size of neurons, one voxel, at the macroscopic scale, includes thousands of somas and processes. This means that the acquired signal is an average of the signal coming from all those cells. Several issues have then to be considered.

First, the acquired signal in a voxel will be an average of the signal of all the diffusing molecules within this voxel, which could correspond to not less than 3000 axons in white matter. Features that will then be computed from it, such as anisotropy, will be an average of all the components in the tissue. One needs to note that every tissue is made of several compartments and that the signals from each of these compartments where water molecules are present are averaged. Investigations using diffusion-weighted spectroscopy, an imaging technique with increased cellular specificity, are also led to try and target specific compartment(s) [38, 54]. This average problem leads to a second issue: a small change in anisotropy (or other features) can actually reflect greater pathological differences. It means that there needs to be a big change in the voxel to be able to detect it in the acquired diffusion signal. Third issue is that anisotropic cellular elements might be considered as isotropic due to the tree pattern of processes within grey matter [26] or to crossing fibers in white matter. At least, as expected from an acquisition, the signal is noisy. Low concentration of water molecules in some tissue (and thus long scan times) can lead to a poor signal-to-noise ratio.

In addition to those issues, we must recall that the spacing between axons, axon diameter, myelin thickness, etc are all also variables, even within the same tract, which adds to the complexity of the problem. The barriers to diffusion have also not a simple nor regular geometry. The correspondence between the biological features of the tissue and the non-invasive diffusion measure is therefore not straightforward.

3 Quantifying Anisotropy via Signal Representation

Signal representation is an indirect method that aims at describing the diffusion signal with no assumptions about the underlying structure. It can therefore be applied to healthy or diseased tissues. Several methods are described, with an emphasis made on the cumulant expansion, which is the most widespread signal representation.

3.1 Cumulant Expansion

Common signal representations are based on the cumulant expansion [39, 65], which corresponds to a development of the logarithm of the signal in polynomials up to a given order in b:

$$\ln\left(\frac{S(b)}{S_0}\right) = -bD + \frac{1}{6}(bD)^2 K + \dots \tag{5}$$

where D is the diffusion coefficient and K the kurtosis. This formula can also be written in the tensor form:

$$\ln\frac{S(b)}{S_0} = -bC^{(2)}_{i_1 i_2}g_{i_1}g_{i_2} + b^2 C^{(4)}_{i_1 \dots i_4}g_{i_1}\dots g_{i_4} - \dots \tag{6}$$

where $C^{(l)}$ are the cumulant tensors, and $\mathbf{g}$ is the direction of the applied diffusion weighting (see Sect. 2.2). Note that Einstein's convention of summation over repeated indices is used here.

An expansion in moments, which corresponds to a Taylor expansion of the signal, is also possible. While expansions in moments and in cumulants are mathematically equivalent, for a similar order truncation at some fixed (low) order, the cumulant series provides a more accurate estimation of the dMRI signal than a moment expansion. Moment expansion is more optimal for analytical treatments because contributions from different tissue compartments add up. A combinatorial relation exists between the two expansions [40, 65]. Computing the cumulant tensors and converting them into moments is promoted to be the most numerically stable methodology to adopt [49].

One of the most popular MRI techniques in brain research as well as in clinical practice is Diffusion Tensor Imaging (DTI) [3], based on the cumulant expansion up to the first order in b. This technique is valid for low diffusion weighting ($b \ll (DK)^{-1}$). Note that this technique does not assume that the medium is homogeneous with unrestricted diffusion ($K = 0$), which appears to be not true for most biological tissues, but that it follows a Gaussian law when $b \ll (DK)^{-1}$.

Using tensor decomposition, three eigenvalues (λ_1, λ_2 and λ_3) reflecting axial and radial diffusivity of molecules within fibers and in the extra-cellular space are computed (see Fig. 1). The difference between those two diffusivities enable to define variables such as Mean Diffusivity (MD) (the average of all the eigenvalues) or Fractional Anisotropy (FA), defined as follow:

$$FA = \frac{\text{std}(\lambda)}{\text{rms}(\lambda)} = \sqrt{\frac{1}{2}} \frac{\sqrt{(\lambda_1 - \lambda_2)^2 + (\lambda_2 - \lambda_3)^2 + (\lambda_3 - \lambda_1)^2}}{\sqrt{\lambda_1^2 + \lambda_2^2 + \lambda_3^2}} \tag{7}$$

Those two measures are complementary, as they bring different information to the comprehension of a tissue (Fig. 2). Hofstetter et al. [19] used MD to hypothesize the presence of bigger cells in the brain after a learning session. Beaulieu et al. also investigated anisotropy in the human brain grey matter using DTI [4].

Diffusion Kurtosis Imaging (DKI) goes beyond DTI and its first order expansion by also estimating the kurtosis of the diffusion probability distribution function [25].

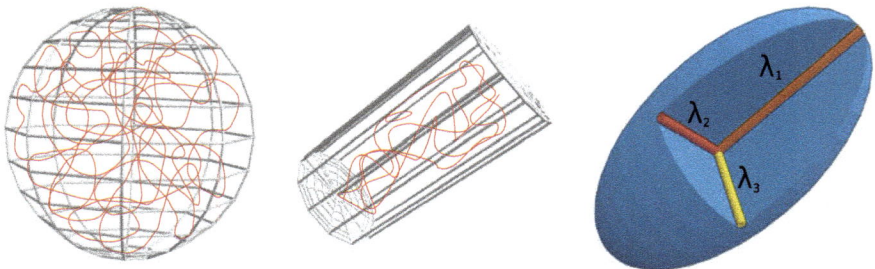

Fig. 1 Isotropic diffusion in somas can be modeled by a sphere (left). Anisotropic diffusion in neurites can be represented by an ellipsoid reflecting axial (λ_1) and radial (λ_2 and λ_3) diffusion. This image has been inspired by the book chapter written by Christian Beaulieu [4]

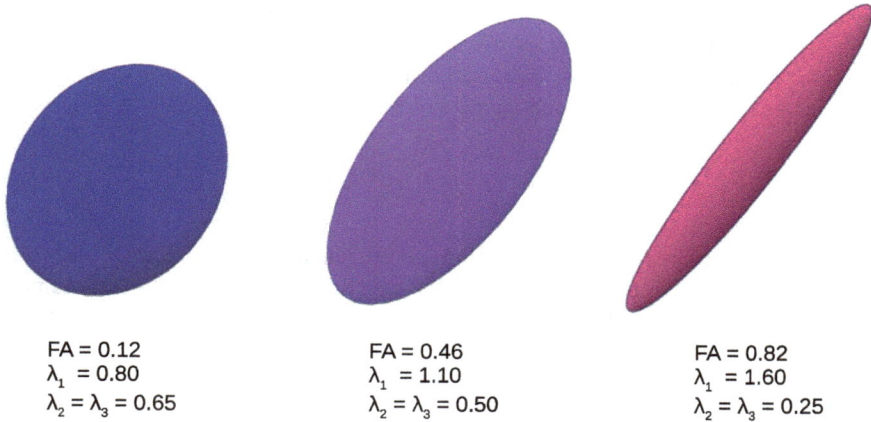

FA = 0.12	FA = 0.46	FA = 0.82
λ_1 = 0.80	λ_1 = 1.10	λ_1 = 1.60
$\lambda_2 = \lambda_3$ = 0.65	$\lambda_2 = \lambda_3$ = 0.50	$\lambda_2 = \lambda_3$ = 0.25

Fig. 2 Mean diffusivity (MD) and fractional anisotropy (FA) are two complementary measures. Here are three examples of ellipses ranging from isotropic to anisotropic that have the same mean diffusivity (0.7×10^{-3} mm^{2}s^{-1}). This image has been inspired by the book chapter written by Christian Beaulieu [4]

The kurtosis quantifies the non-Gaussianity of a distribution. The information that it provides is complementary to DTI metrics. Fitting the kurtosis tensor significantly improves the accuracy of the diffusion tensor estimation [64]. In a same way, extending the series to the sixth order cumulant (in b^3) also increases the accuracy of the kurtosis estimation, albeit with a penalty on precision.

In order to estimate the six independent components of the diffusion tensor, the minimal required data is one b = 0 (unweighted) image and six non-collinear directions on a single diffusion weighting, or "shell". The additional estimation of the 15 independent components of the kurtosis tensor requires a minimal acquisition of one b = 0 image and one or two nonzero shells with 15 non-collinear gradient directions, so that a total of 22 diffusion-weighted images are acquired [64]. The choice of the shell b-values is a trade-off between accuracy and precision. The b-values should be as low as possible to respect the validity of the cumulant expansion, but slightly higher values enable to limit the impact of noise [12]. Jelescu et al. [21] suggest a typical value around b = 1 ms µm^{-2} for DTI and 2 ms µm^{-2} for DKI in vivo. For further details on the optimization of acquisition parameters for precise measurement of diffusion in anisotropic systems, we invite the reader to have a look at the work of Jones et al. [31].

3.2 Other Representations

Yablonskiy et al. [66] hypothesize that the acquired diffusion signal is a sum of signals originating from many spin packets, present in different cell types, at different positions. Each spin packet having then different trajectories and facing different hindrances, they make the assumption that they all have a different apparent diffusion coefficient (ADC). Hence, they introduced a distribution of diffusion coefficients $\rho(D)$, and expressed the diffusion signal as following:

$$\frac{S}{S_0} = \int_0^\infty \rho(D)e^{-bD}dD \tag{8}$$

Theoretically, the distribution of diffusion coefficients can be estimated using the inverse Laplace transform. In practice, some functional form needs to be assumed for $\rho(D)$ due to the mathematical ill-conditioning of the inverse Laplace transform. In addition, a very strong diffusion weighted regime is needed for the estimated distribution to accurately reflect the tissue distribution of diffusion coefficients [34, 47].

Jian et al. [29] propose a statistical method to infer connectivity patterns based on the characterization of the water molecule diffusion by a continuous distribution of diffusion tensors. They described the MR signal attenuation as the Laplace transform of this probability distribution defined on the manifold of symmetric positive-definite tensors. Combined with a spherical deconvolution approach, displacement probability functions and distinct fiber orientations can be estimated in each voxel in a HARDI dataset.

The multi-shell Mean Apparent Propagator (MAP)-MRI method, as proposed by Özarslan et al. [50], expands the signal using harmonic oscillator basis functions. It indeed represents the diffusion-weighted signal by an anisotropic Gaussian modulated by a series of Hermite polynomials. This method allows the estimation of three-dimensional EAP, where both restricted (non-Gaussian) diffusion and crossing axons can be represented. However, according to [44], this method fails to estimate crossing angles correctly. The strength of the method resides on its capacity to accurately estimate diffusion properties such as return-to-origin probability, and mean-squared displacement. The propagator anisotropy (PA) metric was derived from this method, which is a measure of dissimilarity between the reconstructed EAP and its closest isotropic approximation EAP.

Hanyga et al. [17] proposed a new space-fractional diffusion model based on an anomalous anisotropic diffusion equation that preserves posivity. This method seems well-suited for applications to DTI [42].

Other representations exist, but have not been included in this chapter.

3.3 Limitations

The validity and therefore the usefulness of the cumulant expansion depends on its convergence towards the acquisition signal, characterized by the convergence radius b_c [35]. If $b < b_c$, then the series can be approximated using a couple of low terms in Eq. 6, higher order terms being flooded by the noise, i.e. small contributions to the signal can not be decoupled from the noise in experimental data. The number of parameters to estimate is then reduced, but a good accuracy does not assure its validity. Otherwise, if $b > b_c$, the series in Eq. 6 diverges which means that the model function cannot be reduced to a polynomial. A good quality fitting gives then more credit to the underlying model.

Hutchinson et al. [20] compare the DTI, DKI, MAP-MRI and NODDI (see Sect. 4.4.1) methods in different experimental conditions to study the influence of noise and sampling (among others) on parameter estimations. All methods proved to be influenced by the acquisition parameters such as the b-values, the resolution, the SNR and the diffusion time. The need of DKI to fit a higher order tensor explains its high sensitivity to noise.

Regional issues are also to be noted, related to crossing fibers, which can be detected as isotropic zones [1]. Indeed, several diffusion directions are possible in that case. The angular resolution needs to be high enough and the model designed to take this particular case into account.

4 Biophysical Modeling to Measure Anisotropy

This second approach is based on a biophysical model designed for a particular tissue geometry. This model is fit to the diffusion signal acquired, which allows the estimation of the relevant parameters of the microstructure. While it can provide a greater specificity of biological parameters, the design of the model remains difficult, as it needs to accurately capture all the features that effectively and substantially impact the diffusion signal in a given acquisition range (the coarse graining problem, see [46]).

Another big challenge of this approach comes from the number of unknowns to estimate after the definition of all effective parameters. To estimate them all we would need a lot of different b-values. This is unfeasible in clinical applications, because first the gradients used in clinical MRIs are not strong enough, and secondly it would require a patient to stay in the MRI device for a very long time. Some methods rely on constraint to bypass this problem, as presented in Sect. 4.4.

4.1 Multi-compartmental Model

Tissue in the brain can generally be decomposed into four compartments. The first one corresponds to the somas, which are the central part of the neurons, mainly present in grey matter. Glial cells are also comprised in this compartment, as done by Palombo et al. [54]. However, their possible high exchange rate with the extracellular space (ECS) is still a matter of discussion and this argument would argue in favor of their better modeling in the ECS compartment [15]. Somas can be modeled as spheres of different diameters. Neurites, the second compartment, connect those neurons together, either in short distances in grey matter (they are called dendrites), or long distance connections in white matter (axons). The diffusivity across the processes is considered zero due to the restriction implied by the fixed small diameter. Processes can therefore be modeled by cylinders with zero-radius ("sticks") (see part Sect. 4.2 below). The orientation of a collection of processes within a voxel is characterized by an orientation distribution function (ODF) [60]. The third compartment corresponds to extra-cellular space (ECS) and is modeled as Gaussian anisotropic. The last one is the cerebrospinal fluid (CSF), which could contribute if a voxel contains part of a ventricle, and corresponds to free diffusing molecules. It is hence modeled as free diffusion.

The acquired water signal originates from these four compartments and are weighted according to their relative signal fraction f:

$$S(b) = f_{\text{somas}} \cdot S_{\text{somas}}(b) + f_{\text{neurites}} \cdot S_{\text{neurites}}(b) + f_{\text{ECS}} \cdot S_{\text{ECS}}(b) + f_{\text{CSF}} \cdot S_{\text{CSF}}(b), \tag{9}$$

with $f_{\text{somas}} + f_{\text{neurites}} + f_{\text{ECS}} + f_{\text{CSF}} = 1$. Remark that f_{somas}, f_{neurites}, f_{ECS} and f_{CSF} are not the relative volume fractions due to the T2 differences between the compartments. In the following models presented, a combination of those compartments is used to model particular tissues and keep only the relevant compartments. Note that a common assumption is made that the exchanges between the compartments can be neglected at the time scales of clinical dMRI, at least in white matter. The estimation of exchange rate in vivo is challenging and more investigations are needed to validate this hypothesis in white and gray matter.

4.2 Neurites as Sticks

Neurites have been modeled by zero-radius impermeable cylinders, characterized by their longitudinal diffusivity, the transverse diffusivity being considered zero. These neurites are called "sticks" and correspond to the most anisotropic Gaussian compartment possible [7, 26, 36].

The intra-neurite response function, i.e. the diffusion signal from water inside a stick of diffusivity D_a pointing in the unit direction $\hat{\mathbf{n}}$, is defined as:

$$G_{\hat{n}}(\hat{\mathbf{g}}, b) = e^{-bD_a(\hat{\mathbf{g}} \cdot \hat{\mathbf{n}})^2}, \tag{10}$$

with $\hat{\mathbf{g}}$ being the unit gradient direction of the measurement. It is determined by $\cos\theta \equiv \hat{\mathbf{g}} \cdot \hat{\mathbf{n}}$, where θ is the angle between $\hat{\mathbf{n}}$ and $\hat{\mathbf{g}}$.

The signal, after being isotropically averaged over multiple gradient directions $\hat{\mathbf{g}}$, is the following [8, 24, 30]:

$$\bar{S} \simeq \beta \cdot b^{-1/2}, \tag{11}$$

with $\beta = \sqrt{\frac{\pi}{4}} \cdot f_{\text{neurites}}/(D_a^{\|})^{1/2}$.

In brain tissue, at sufficiently large b-values, the extra-axonal space signal is exponentially suppressed, its diffusivity being non-zero in any direction. The only remaining signal in white matter comes from the axons ($S_{neurites}$ in Eq. 9), and follows the power law from Eq. 11 [41, 62]. This equation captures the very anisotropy of white matter.

Veraart et al. [63] recently proved that the radius of the axons can be estimated for very high b-values, where the transverse diffusivity is not considered null anymore. The direction-averaged DWI signal then follows the following law:

$$\bar{S} \simeq \beta e^{-bD_a^{\perp}} \cdot b^{-1/2} \tag{12}$$

Such law however does not hold in gray matter, which indicates that white matter and grey matter require different models in order to accurately capture their microstructure. Several hypothesis have been elaborated to explain the different behaviour of gray matter DWI signal. McKinnon et al. [41] and Veraart et al. [63] suggest that an increased permeability in cell membranes of neurites in gray matter might be the cause of an increased exponent, while Palombo et al. [52] advocate the abundance of cell body in gray matter. Özarslan et al. [51] suggest that curvy projections, along with longer pulse duration, lead to a disappearance of the $b^{-1/2}$ decay.

4.3 Standard Model of Diffusion in Neural Tissue

The measured diffusion signal in brain white matter is a sum of anisotropic compartments. It can be modeled as a convolution between a response kernel $\mathcal{K}$ from a perfectly aligned fascicle pointing in the direction $\hat{\mathbf{n}}$ and the fiber orientation distribution (ODF) $\mathcal{P}(\hat{\mathbf{n}})$ normalized to $\int d\hat{\mathbf{n}}\mathcal{P}(\hat{\mathbf{n}}) \equiv 1$.

$$S_{\hat{\mathbf{g}}}(b) = \int_{|\hat{\mathbf{n}}|=1} \mathcal{P}(\hat{\mathbf{n}})\mathcal{K}(b, \hat{\mathbf{g}} \cdot \hat{\mathbf{n}})d\hat{\mathbf{n}}, \tag{13}$$

$\hat{\mathbf{g}}$ being defined in Sect. 2.2.

In the case of white matter, the kernel can be written as:

$$\mathcal{K}(b, \xi) = S_0 \left[f e^{-bD_a \xi^2} + (1 - f - f_{CSF}) e^{-bD_e^{\perp} - b\left(D_e^{\parallel} - D_e^{\perp}\right)\xi^2} + f_{CSF} e^{-bD_{CSF}} \right], \tag{14}$$

with $\xi = \hat{\mathbf{g}} \cdot \hat{\mathbf{n}}$. Those exponential contributions correspond to the intra-axonal space modeled by a stick compartment (Eq. 10), the extra-axonal space modeled by an axially symmetric Gaussian compartment with transverse and longitudinal diffusivities $D_e^{\perp}$ and $D_e^{\parallel}$, and the cerebrospinal fluid (CSF) compartment. All those compartments are represented in Fig. 3. This decomposition has been widely used in white matter

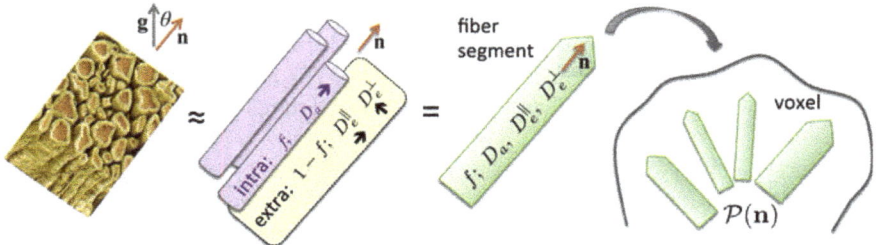

Fig. 3 Standard Model of diffusion in neuronal tissue. Two-compartment model (intra- and extra-neurite spaces) described by 4 independent parameters: f, D_a, $D_e^{\parallel}$ and $D_e^{\perp}$ and a fiber orientation distribution $\mathcal{P}(\hat{n})$. This figure is reproduced from Novikov et al. [49]

by the community. As a consequence, Novikov et al. suggested to call it the Standard Model (SM) [46]. For the sake of reference, we will also refer to it as the SM in this chapter.

4.4 Standard Model Parameter Estimation Using Constraints

In the previous sections we presented the SM of diffusion in neural tissue as a sum of anisotropic Gaussian compartments, as defined by Novikov et al. [46]. We will now introduce some methods based on the SM that rely on constraints to overcome the challenge of estimating many biological parameters of interest.

4.4.1 Neurite Orientation Dispersion and Density Imaging

In order to reduce the number of parameters that need to be estimated, Zhang et al. [67] proposed to impose restrictions on the intrinsic diffusivities. They introduced a method called Neurite Orientation Dispersion and Density Imaging (NODDI), which relies on a three-compartment SM (intra-axonal space, extra-axonal space and CSF), described by seven parameters: volume fractions f_{intra} and f_{iso}, diffusivities $D_a^{\parallel}$, $D_e^{\parallel}$, $D_e^{\perp}$ and D_{iso}, and the orientation dispersion modeled by a Watson distribution of concentration parameter κ. By fixing the diffusivities to the following values:

$$D_a^{\parallel} = D_e^{\parallel} = 1.7\,\mu\text{m}^2/\text{ms} \tag{15}$$

$$D_e^{\perp} = (1 - f_{\text{intra}}) \cdot D_e^{\parallel} \tag{16}$$

$$D_{iso} = 3\,\mu m^2/\text{ms} \tag{17}$$

only the two volume fractions and the orientation dispersion need to be estimated.

Although this method allows to estimate the parameters, the validity of those constraints need to be questioned. To begin with, if we admit that the equalities are correct, they imply that a small deviation from the fixed values, as occurs in cerebral ischemia, will induce a non-negligible bias in the other parameters estimation, leading to false interpretations. However, studies using Diffusion weighted spectroscopy MR which can quantify the diffusion of specific metabolites (e.g. [55]), suggest, through the study of metabolites specifically found on different sections of the neurons and extra cellular tissue, that such diffusivity is not constant across the whole brain. Whether and how these findings can be used to shed light on water diffusion in the brain, is an open question.

Another drawback of this method is that it leads to indetermination, which means that NODDI returns one possible result among a multiplicity of mathematical solutions by fixing $D_a^{\parallel} = D_e^{\parallel}$ [23, 49]. If we consider the case where all the parameter constraints are released and the CSF compartment neglected (called NODDIDA [22], which stands for NODDI with diffusivity assessment), two distinct solutions to the estimation problem exist: $D_a^{\parallel} > D_e^{\parallel}$ and $D_a^{\parallel} < D_e^{\parallel}$ (see Figs. 8 and 9 in Jelescu et al. [23]). Both solutions lie within biologically plausible ranges, and determining which solution is biologically correct is an active field of research, although most studies are suggesting $D_a^{\parallel} > D_e^{\parallel}$. At least, the tortuosity approximation that relates $D_e^{\perp}$ and $D_e^{\parallel}$ has been invalidated in the case of tight packings of axons [45].

4.4.2 White Matter Tract Integrity Metrics

Another approach to estimate the relevant features of interest in a tissue proposes to relate the scalar parameters to the DKI components. Called White Matter Tract Integrity (WMTI) [16], it is a two-compartment SM that relies on the assumption that sticks are highly aligned within a voxel.

The tissue is described as a sum of two Gaussian compartments (intra- and extra-axonal space, Eq. 9 with $f_{\text{somas}} = 0$ and $f_{\text{CSF}} = 0$), where axons are modeled as sticks embedded in a Gaussian anisotropic extra-axonal medium. Each compartment is characterized by a tensor (D_a and D_e) derived from the kurtosis tensors D and K. In any direction j:

$$D_j = f_{\text{intra}} D_{\text{a},j} + (1 - f_{\text{intra}}) D_{\text{e},j}, \tag{18}$$

$$K_j = 3 f_{\text{intra}} \cdot (1 - f_{\text{intra}}) \frac{(D_{\text{e},j} - D_{\text{a},j})^2}{D_j^2} \tag{19}$$

We retrieve the two possible mathematical solutions mentioned before, as demonstrated by the square in Eq. 19. The solution chosen in this method is $D_a^{\parallel} < D_e^{\parallel}$, which leads to:

$$f_{\text{intra}} = \frac{K_{\max}}{K_{\max} + 3}, \tag{20}$$

$$D_{\text{e},j} = D_j \left[1 + \sqrt{\frac{K_j \cdot f_{\text{intra}}}{3(1 - f_{\text{intra}})}} \right], \tag{21}$$

$$D_{a,j} = D_j \left[1 - \sqrt{\frac{K_j (1 - f_{intra})}{3 f_{intra}}} \right].$$ (22)

Although WMTI enables to capture the changes of diffusivities, it has two main limitations. First, this approach is limited to regions of highly aligned single fiber bundles, which are only present in some white matter regions. Jespersen et al. suggested a method that alleviates this assumption by assuming a Watson distribution of the axons (like in NODDI) [27]. Second, as it relies on the DKI decomposition, this method is only restricted to the low b-value regime, which could lead to some bias.

4.5 Lemonade

As explained in Sect. 4, estimating both compartment diffusivities and orientation dispersion of neurites simultaneously is problematic and tends to be biased. Some methods suggest fixing some parameters such as NODDI or to limit its application to coherent fibers only as WMTI to work around these problems. Releasing these constraints necessitates to estimate a larger number of parameters.

A very recent method in white matter estimates the scalar parameters of a two-compartment kernel separately from the ODF without any constraints. The method developed by Novikov et al. [49] is based on the modeling of the diffusion signal as a convolution of the ODF and the response kernel from a perfectly aligned fiber segment, as presented in Sect. 4.3. It can be decomposed into two steps. A first step solves an algebraic system of equations that relates the kernel parameters to the signal moments for low b-values. This part was called LEMONADE, which stands for Linearly Estimated Moments provide Orientations of Neurites And their Diffusivities Exactly. It requires at least 3 non-zero b-shells inferior to 2.5 ms μm^{-2} and returns estimates for f_{intra}, $D_a^{\parallel}$, $D_e^{\parallel}$, $D_e^{\perp}$ and $p_2 = \frac{3 \langle (\cos \psi)^2 \rangle - 1}{2}$, which gives an estimate of the orientation dispersion. In a second step, a rotationally invariant energy function of the system is minimized exploiting all available data and using the first estimates as initialization values.

This method emphasizes the existence of the two mathematical solutions as introduced before, and shows that, in principle, the degeneracy can be avoided using measurements up to the 3rd order of b-values. However, due to noise in the data, the solution selection remains difficult in practice and individual validation should be carried out.

The assumptions made in this approach are, as in the other methods previously presented, the existence of only two compartments, the uniformity of diffusivities across all axons in the voxel, and axial symmetry of the kernel. These assumptions are also the limitations of the model used. Validation in the case of pathological tissue also needs to be investigated.

We refer the reader to Jelescu and Budde's review on the accuracy and validation of biophysical parameters of different diffusion models in white matter, which includes the ones presented before [21].

5 Summary and Above

We have shown two main approaches to describe microstructure anisotropy using diffusion MRI: signal representation and biophysical modeling. While the former are general and make no assumptions about the underlying tissue, models are designed for a particular tissue and therefore provide greater specificity and interpretation of the estimated biological parameters. The difficulties in modeling reside in accurately capturing the features that effectively and substantially impact the diffusion signal in a given acquisition range, and being able to correctly fit the model (inverse problem).

Anisotropy provides great insight into a tissue structure, and its evolution can enlighten the progression of certain pathologies. The presence of isotropy must not be neglected either, as it can be a great marker of other microstructures, such as in grey matter where it denotes the presence of somas.

Although great progresses have been made during the last decade, some questions remain unresolved. To cite a few, we can wonder to which extent we can consider compartments as non-exchanging. Diffusion time, brain region and myelination of the tissue will most likely impact the answer of this question. Can we also come up with methods less sensitive to the signal-to-noise ratio or a way to disentangle thermal noise and artifacts from the signal of interest?

Acknowledgments This work was supported by the ERC-StG NeuroLang and the ANR/NSF NeuroRef grants.

References

1. Alexander, D.C., Seunarine, K.K.: Mathematics of Crossing Fibers. Diffusion MRI: Theory, Methods, and Application, pp. 451–464. Oxford (2010)
2. Assaf, Y., Pasternak, O.: Diffusion Tensor Imaging (DTI)-based white matter mapping in brain research: a review. J. Mol. Neurosci. **34**(1), 51–61 (2008). https://doi.org/10.1007/s12031-007-0029-0
3. Basser, P., Mattiello, J., Lebihan, D.: Estimation of the effective self-diffusion tensor from the nmr spin echo. J. Magn. Reson. Ser. B **103**(3), 247–254 (1994). https://doi.org/10.1006/jmrb.1994.1037, http://www.sciencedirect.com/science/article/pii/S1064186684710375
4. Beaulieu, C.: The basis of anisotropic water diffusion in the nervous system–a technical review. NMR Biomed. **15**(7–8), 435–455 (2002). https://doi.org/10.1002/nbm.782
5. Beaulieu, C.: Chapter 8—the biological basis of diffusion anisotropy. In: Johansen-Berg, H., Behrens, T.E. (eds.) Diffusion MRI (Second Edition), 2nd edn., pp. 155 – 183. Academic Press, San Diego (2014). https://doi.org/10.1016/B978-0-12-396460-1.00008-1. http://www.sciencedirect.com/science/article/pii/B9780123964601000081

6. Beaulieu, C., Allen, P.S.: Determinants of anisotropic water diffusion in nerves. Magn. Reson. Med. **31**(4), 394–400 (1994). https://doi.org/10.1002/mrm.1910310408

7. Behrens, T., Woolrich, M., Jenkinson, M., Johansen-Berg, H., Nunes, R., Clare, S., Matthews, P., Brady, J., Smith, S.: Characterization and propagation of uncertainty in diffusion-weighted MR imaging. Magn. Reson. Med. **50**(5), 1077–1088 (2003). https://doi.org/10.1002/mrm.10609

8. Callaghan, P., Jolley, K., Lelievre, J.: Diffusion of water in the endosperm tissue of wheat grains as studied by pulsed field gradient nuclear magnetic resonance. Biophys. J. **28**(1), 133–141 (1979). https://doi.org/10.1016/S0006-3495(79)85164-4

9. Callaghan, P.T., Coy, A., MacGowan, D., Packer, K.J., Zelaya, F.O.: Diffraction-like effects in NMR diffusion studies of fluids in porous solids. Nature **351**(6326), 467–469 (1991). https://doi.org/10.1038/351467a0

10. Caruyer, E., Lenglet, C., Sapiro, G., Deriche, R.: Design of multishell sampling schemes with uniform coverage in diffusion mri. Magn. Reson. Med. **69**(6), 1534–1540 (2013). https://doi.org/10.1002/mrm.24736

11. Cleveland, G., Chang, D., Hazlewood, C., Rorschach, H.: Nuclear magnetic resonance measurement of skeletal muscle: anisotrophy of the diffusion coefficient of the intracellular water. Biophys. J. **16**(9), 1043–1053 (1976). https://doi.org/10.1016/S0006-3495(76)85754-2

12. DK., J.: Precision and accuracy in diffusion tensor magnetic resonance imaging. Top Magn. Reson. Imaging **21**, 87–99 (2010). https://doi.org/10.1097/RMR.0b013e31821e56ac

13. Fick, R.H.J., Petiet, A., Santin, M., Philippe, A.C., Lehericy, S., Deriche, R., Wassermann, D.: Multi-spherical diffusion mri: Exploring diffusion time using signal sparsity. In: Fuster, A., Ghosh, A., Kaden, E., Rathi, Y., Reisert, M. (eds.) Computational Diffusion MRI, pp. 71–83. Springer International Publishing, Cham (2017)

14. Fick, R.H.J., Pizzolato, M., Wassermann, D., Deriche, R.: Diffusion MRI anisotropy: modeling, analysis and interpretation. In: Schultz, T., Özarslan, E., Hotz, I. (eds.) Modeling, Analysis, and Visualization of Anisotropy, pp. 203–228. Springer International Publishing, Cham (2017). https://doi.org/10.1007/978-3-319-61358-1. Series Title: Mathematics and Visualization

15. Fields, R., Woo, D., Basser, P.: Glial regulation of the neuronal connectome through local and long-distant communication. Neuron **86**(2), 374–386 (2015). https://doi.org/10.1016/j.neuron.2015.01.014, http://www.sciencedirect.com/science/article/pii/S0896627315000409

16. Fieremans, E., Jensen, J.H., Helpern, J.A.: White matter characterization with diffusional kurtosis imaging. NeuroImage **58**(1), 177–188 (2011). https://doi.org/10.1016/j.neuroimage.2011.06.006, http://www.sciencedirect.com/science/article/pii/S1053811911006148

17. Hanyga, A., Magin, R.L.: A new anisotropic fractional model of diffusion suitable for applications of diffusion tensor imaging in biological tissues. Proc. R. Soc. Math. Phys. Eng. Sci. **470**(2170), 20140319 (2014). https://doi.org/10.1098/rspa.2014.0319

18. Henkelman, R.M., Stanisz, G.J., Kim, J.K., Bronskill, M.J.: Anisotropy of nmr properties of tissues. Magn. Reson. Med. **32**(5), 592–601 (1994). https://doi.org/10.1002/mrm.1910320508

19. Hofstetter, S., Tavor, I., Tzur Moryosef, S., Assaf, Y.: Short-term learning induces white matter plasticity in the fornix. J. Neurosci. **33**(31), 12844–12850 (2013). https://doi.org/10.1523/JNEUROSCI.4520-12.2013, https://www.jneurosci.org/content/33/31/12844

20. Hutchinson, E.B., Avram, A.V., Irfanoglu, M.O., Koay, C.G., Barnett, A.S., Komlosh, M.E., Özarslan, E., Schwerin, S.C., Juliano, S.L., Pierpaoli, C.: Analysis of the effects of noise, dwi sampling, and value of assumed parameters in diffusion mri models. Magn. Reson. Med. **78**(5), 1767–1780 (2017). https://doi.org/10.1002/mrm.26575

21. Jelescu, I.O., Budde, M.D.: Design and validation of diffusion MRI models of white matter. Front. Phys. **5**, (2017). https://doi.org/10.3389/fphy.2017.00061

22. Jelescu, I.O., Veraart, J., Adisetiyo, V., Milla, S.S., Novikov, D.S., Fieremans, E.: One diffusion acquisition and different white matter models: how does microstructure change in human early development based on wmti and noddi? NeuroImage **107**, 242–256 (2015). https://doi.org/10.1016/j.neuroimage.2014.12.009, http://www.sciencedirect.com/science/article/pii/S1053811914010015

23. Jelescu, I.O., Veraart, J., Fieremans, E., Novikov, D.S.: Degeneracy in model parameter estimation for multi-compartment diffusion in neuronal tissue: Degeneracy in Model Parameter

Estimation of Diffusion in Neural Tissue. NMR Biomed. **29**(1), 33–47 (2016). https://doi.org/10.1002/nbm.3450

24. Jensen, J.H., Glenn, G.R., Helpern, J.A.: Fiber ball imaging. NeuroImage **124**, 824–833 (2016). https://doi.org/10.1016/j.neuroimage.2015.09.049, http://www.sciencedirect.com/science/article/pii/S105381191500871X

25. Jensen, J.H., Helpern, J.A., Ramani, A., Lu, H., Kaczynski, K.: Diffusional kurtosis imaging: the quantification of non-gaussian water diffusion by means of magnetic resonance imaging. Magn. Reson. Med. **53**(6), 1432–1440 (2005). https://doi.org/10.1002/mrm.20508

26. Jespersen, S.N., Kroenke, C.D., Østergaard, L., Ackerman, J.J., Yablonskiy, D.A.: Modeling dendrite density from magnetic resonance diffusion measurements. NeuroImage **34**(4), 1473–1486 (2007). https://doi.org/10.1016/j.neuroimage.2006.10.037, http://www.sciencedirect.com/science/article/pii/S1053811906010950

27. Jespersen, S.N., Olesen, J.L., Hansen, B., Shemesh, N.: Diffusion time dependence of microstructural parameters in fixed spinal cord. NeuroImage **182**, 329–342 (2018). https://doi.org/10.1016/j.neuroimage.2017.08.039, http://www.sciencedirect.com/science/article/pii/S1053811917306869. Microstructural Imaging

28. Jeurissen, B., Descoteaux, M., Mori, S., Leemans, A.: Diffusion mri fiber tractography of the brain. NMR Biomed. **32**(4), e3785 (2019). https://doi.org/10.1002/nbm.3785, https://onlinelibrary.wiley.com/doi/abs/10.1002/nbm.3785. E3785 NBM-17-0045.R2

29. Jian, B., Vemuri, B.C., Özarslan, E., Carney, P.R., Mareci, T.H.: A novel tensor distribution model for the diffusion-weighted mr signal. NeuroImage **37**(1), 164–176 (2007). https://doi.org/10.1016/j.neuroimage.2007.03.074, http://www.sciencedirect.com/science/article/pii/S105381190700273X

30. Joabsson, F., Nydén, M., Linse, P., Söderman, O.: Pulsed field gradient nmr studies of translational diffusion in cylindrical surfactant aggregates. J. Phys. Chem. B **101**(47), 9710–9716 (1997). https://doi.org/10.1021/jp971890g

31. Jones, D., Horsfield, M., Simmons, A.: Optimal strategies for measuring diffusion in anisotropic systems by magnetic resonance imaging. Magn. Reson. Med. **42**(3), 515–525 (1999). https://doi.org/10.1002/(SICI)1522-2594(199909)42:3<515::AID-MRM14>3.0.CO;2-Q

32. Kac, M.: Can one hear the shape of a drum? Am. Math. Mon. **73**(4P2), 1–23 (1966). https://doi.org/10.1080/00029890.1966.11970915

33. Kaden, E., Kruggel, F., Alexander, D.C.: Quantitative mapping of the per-axon diffusion coefficients in brain white matter. Magn. Reson. Med. **75**(4), 1752–1763 (2016). https://doi.org/10.1002/mrm.25734

34. Kiselev, V.G.: Fundamentals of diffusion mri physics. NMR Biomed. **30**(3), e3602 (2017). https://doi.org/10.1002/nbm.3602

35. Kiselev, V.G., Il'yasov, K.A.: Is the "biexponential diffusion" biexponential? Magn. Reson. Med. **57**(3), 464–469 (2007). https://doi.org/10.1002/mrm.21164

36. Kroenke, C.D., Ackerman, J.J., Yablonskiy, D.A.: On the nature of the naa diffusion attenuated mr signal in the central nervous system. Magn. Reson. Med. **52**(5), 1052–1059 (2004). https://doi.org/10.1002/mrm.20260

37. Le Bihan, D., Breton, E.: Imagerie de diffusion in-vivo par résonance magnétique nucléaire. Comptes-Rendus de l'Académie des Sciences **93**(5), 27–34 (1985). https://hal.archives-ouvertes.fr/hal-00350090

38. Ligneul, C., Palombo, M., Hernández-Garzón, E., Carrillo-de Sauvage, M.A., Flament, J., Hantraye, P., Brouillet, E., Bonvento, G., Escartin, C., Valette, J.: Diffusion-weighted magnetic resonance spectroscopy enables cell-specific monitoring of astrocyte reactivity in vivo. NeuroImage **191**, 457–469 (2019). https://doi.org/10.1016/j.neuroimage.2019.02.046, https://linkinghub.elsevier.com/retrieve/pii/S1053811919301429

39. Liu, C., Bammer, R., Acar, B., Moseley, M.E.: Characterizing non-gaussian diffusion by using generalized diffusion tensors. Magn. Reson. Med. **51**(5), 924–937 (2004). https://doi.org/10.1002/mrm.20071

40. Mayer, J.E., Montroll, E.: Molecular distribution. J. Chem. Phys. **9**(1), 2–16 (1941). https://doi.org/10.1063/1.1750822

41. McKinnon, E.T., Jensen, J.H., Glenn, G.R., Helpern, J.A.: Dependence on b-value of the direction-averaged diffusion-weighted imaging signal in brain. Magn. Reson. Imaging **36**, 121–127 (2017). https://doi.org/10.1016/j.mri.2016.10.026, https://linkinghub.elsevier.com/retrieve/pii/S0730725X16301965
42. Meerschaert, M.M., Magin, R.L., Ye, A.Q.: Anisotropic fractional diffusion tensor imaging. J. Vib. Control **22**(9), 2211–2221 (2016). https://doi.org/10.1177/1077546314568696. PMID: 27499605
43. M.R.S.E., R.B.F.H., V.P.L.S., R.A.: Xxvii. a brief account ofmicroscopicalobservations made in the months of june, july and august 1827, on theparticles contained in the pollen of plants; and on the general existence ofactive molecules in organic and inorganic bodies. Philoso. Mag. **4**(21), 161–173 (1828). https://doi.org/10.1080/14786442808674769
44. Ning, L., Laun, F., Gur, Y., DiBella, E.V., Deslauriers-Gauthier, S., Megherbi, T., Ghosh, A., Zucchelli, M., Menegaz, G., Fick, R., St-Jean, S., Paquette, M., Aranda, R., Descoteaux, M., Deriche, R., O'Donnell, L., Rathi, Y.: Sparse reconstruction challenge for diffusion mri: validation on a physical phantom to determine which acquisition scheme and analysis method to use? Med. Image Anal. **26**(1), 316–331 (2015). https://doi.org/10.1016/j.media.2015.10.012, http://www.sciencedirect.com/science/article/pii/S1361841515001541
45. Novikov, D.S., Fieremans, E.: Relating extracellular diffusivity to cell size distribution and packing density as applied to white matter. In: Proceedings of the 20th Annual Meeting of ISMRM, p. 1829 (2012)
46. Novikov, D.S., Fieremans, E., Jespersen, S.N., Kiselev, V.G.: Quantifying brain microstructure with diffusion MRI: theory and parameter estimation: brain microstructure with dMRI: theory and parameter estimation. NMR Biomed. e3998 (2018). https://doi.org/10.1002/nbm.3998, http://doi.wiley.com/10.1002/nbm.3998
47. Novikov, D.S., Kiselev, V.G.: Effective medium theory of a diffusion-weighted signal. NMR Biomed. **23**(7), 682–697 (2010). https://doi.org/10.1002/nbm.1584
48. Novikov, D.S., Kiselev, V.G., Jespersen, S.N.: On modeling. Magn. Reson. Med. **79**(6), 3172–3193 (2018). https://doi.org/10.1002/mrm.27101
49. Novikov, D.S., Veraart, J., Jelescu, I.O., Fieremans, E.: Rotationally-invariant mapping of scalar and orientational metrics of neuronal microstructure with diffusion MRI. NeuroImage **174**, 518–538 (2018). https://doi.org/10.1016/j.neuroimage.2018.03.006, https://linkinghub.elsevier.com/retrieve/pii/S1053811918301915
50. Özarslan, E., Koay, C.G., Shepherd, T.M., Komlosh, M.E., İrfanoğlu, M.O., Pierpaoli, C., Basser, P.J.: Mean apparent propagator (map) mri: a novel diffusion imaging method for mapping tissue microstructure. NeuroImage **78**, 16–32 (2013). https://doi.org/10.1016/j.neuroimage.2013.04.016, http://www.sciencedirect.com/science/article/pii/S1053811913003431
51. Özarslan, E., Yolcu, C., Herberthson, M., Knutsson, H., Westin, C.F.: Influence of the size and curvedness of neural projections on the orientationally averaged diffusion mr signal. Front. Phys. **6**, 17 (2018). https://doi.org/10.3389/fphy.2018.00017
52. Palombo, M., Alexander, D.C., Zhang, H.: A generative model of realistic brain cells with application to numerical simulation of the diffusion-weighted MR signal. NeuroImage **188**, 391–402 (2019). https://doi.org/10.1016/j.neuroimage.2018.12.025, https://linkinghub.elsevier.com/retrieve/pii/S1053811918321694
53. Palombo, M., Ianus, A., Guerreri, M., Nunes, D., Alexander, D.C., Shemesh, N., Zhang, H.: Sandi: A compartment-based model for non-invasive apparent soma and neurite imaging by diffusion mri. NeuroImage **215**, 116835 (2020). https://doi.org/10.1016/j.neuroimage.2020.116835, http://www.sciencedirect.com/science/article/pii/S1053811920303220
54. Palombo, M., Ligneul, C., Najac, C., Le Douce, J., Flament, J., Escartin, C., Hantraye, P., Brouillet, E., Bonvento, G., Valette, J.: New paradigm to assess brain cell morphology by diffusion-weighted mr spectroscopy in vivo. Proc. Natl. Acad. Sci. **113**(24), 6671–6676 (2016). https://doi.org/10.1073/pnas.1504327113, https://www.pnas.org/content/113/24/6671
55. Palombo, M., Ligneul, C., Valette, J.: Modeling diffusion of intracellular metabolites in the mouse brain up to very high diffusion-weighting: Diffusion in long fibers (almost) accounts

for non-monoexponential attenuation: modeling diffusion of brain metabolites in vivo up to very high diffusion weighting. Magn. Reson. Med. **77**(1), 343–350 (2017). https://doi.org/10. 1002/mrm.26548

56. Panagiotaki, E., Schneider, T., Siow, B., Hall, M.G., Lythgoe, M.F., Alexander, D.C.: Compartment models of the diffusion mr signal in brain white matter: a taxonomy and comparison. NeuroImage **59**(3), 2241–2254 (2012). https://doi.org/10.1016/j.neuroimage.2011.09.081, http://www.sciencedirect.com/science/article/pii/S1053811911011566

57. Panagiotaki, E., Walker-Samuel, S., Siow, B., Johnson, S.P., Rajkumar, V., Pedley, R.B., Lythgoe, M.F., Alexander, D.C.: Noninvasive quantification of solid tumor microstructure using verdict mri. Cancer Res. **74**(7), 1902–1912 (2014). https://doi.org/10.1158/0008-5472.CAN-13-2511, https://cancerres.aacrjournals.org/content/74/7/1902

58. Soares, J., Marques, P., Alves, V., Sousa, N.: A hitchhiker's guide to diffusion tensor imaging. Front. Neurosci. **7**, 31 (2013). https://doi.org/10.3389/fnins.2013.00031

59. Tanner, J.E., Stejskal, E.O.: Restricted self-diffusion of protons in colloidal systems by the pulsed-gradient, spin-echo method. J. Chem. Phys. **49**(4), 1768–1777 (1968). https://doi.org/10.1063/1.1670306

60. Tuch, D.S.: Q-ball imaging. Magn. Reson. Med. **52**(6), 1358–1372 (2004). https://doi.org/10.1002/mrm.20279

61. Tuch, D.S., Reese, T.G., Wiegell, M.R., Makris, N., Belliveau, J.W., Wedeen, V.J.: High angular resolution diffusion imaging reveals intravoxel white matter fiber heterogeneity. Magn. Reson. Med. **48**(4), 577–582 (2002). https://doi.org/10.1002/mrm.10268

62. Veraart, J., Fieremans, E., Novikov, D.S.: On the scaling behavior of water diffusion in human brain white matter. NeuroImage **185**, 379–387 (2019). https://doi.org/10.1016/j.neuroimage. 2018.09.075, https://linkinghub.elsevier.com/retrieve/pii/S1053811918319475

63. Veraart, J., Nunes, D., Rudrapatna, U., Fieremans, E., Jones, D.K., Novikov, D.S., Shemesh, N.: Noninvasive quantification of axon radii using diffusion mri. eLife **9**, e49855 (2020). https://doi.org/10.7554/eLife.49855

64. Veraart, J., Poot, D.H.J., Van Hecke, W., Blockx, I., Van der Linden, A., Verhoye, M., Sijbers, J.: More accurate estimation of diffusion tensor parameters using diffusion kurtosis imaging. Magn. Reson. Med. **65**(1), 138–145 (2011). https://doi.org/10.1002/mrm.22603

65. VG, K.: The cumulant expansion: an overarching mathematical framework for understanding diffusion nmr. In: DK, J. (ed.) Diffusion MRI: Theory, Methods, and Applications, pp. 152 – 68. Oxford University Press, Oxford (2010)

66. Yablonskiy, D.A., Bretthorst, G.L., Ackerman, J.J.: Statistical model for diffusion attenuated mr signal. Magn. Reson. Med. **50**(4), 664–669 (2003). https://doi.org/10.1002/mrm.10578

67. Zhang, H., Schneider, T., Wheeler-Kingshott, C.A., Alexander, D.C.: NODDI: Practical in vivo neurite orientation dispersion and density imaging of the human brain. NeuroImage **61**(4), 1000–1016 (2012). https://doi.org/10.1016/j.neuroimage.2012.03.072, https://linkinghub.elsevier.com/retrieve/pii/S1053811912003539

10

Riemann-DTI Geodesic Tractography Revisited

Luc Florack, Rick Sengers, Stephan Meesters, Lars Smolders, and Andrea Fuster

Abstract Clinical tractography is a challenging problem in diffusion tensor imaging (DTI) due to persistent validation issues. Geodesic tractography, based on a shortest path principle, is conceptually appealing, but has not produced convincing results so far. A major weakness is its rigidity with respect to candidate tracts it is capable of producing given a pair of endpoints, showing a tendency to produce false positives (such as shortcuts) and false negatives (e.g. if a shortcut supplants the correct solution). We propose a new geodesic paradigm that appears to overcome these problems, making a step towards semi-automatic clinical use. To this end we couple the DTI tensor field to a *family* of Riemannian metrics, governed by control parameters. In practice these parameters may allow for edits by an expert through manual selection among multiple tract suggestions, or for bringing in a priori knowledge. In this paper, however, we consider an automatic, evidence-driven procedure to determine optimal controls and corresponding tentative tracts, and illustrate the role of edits to remediate erroneous defaults.

L. Florack · R. Sengers (✉) · S. Meesters · L. Smolders · A. Fuster
Eindhoven University of Technology, Department of Mathematics & Computer Science,
NL-5600 Eindhoven, MB, The Netherlands
e-mail: H.J.C.E.Sengers@tue.nl

L. Florack
e-mail: L.J.M.Florack@tue.nl

S. Meesters
e-mail: S.P.L.Meesters@tue.nl

L. Smolders
e-mail: L.Smolders@student.tue.nl

A. Fuster
e-mail: A.Fuster@tue.nl

1 Introduction

Tractography aims at reconstructing bundles of nerve fibers in the brain (aka tracts) from diffusion weighted magnetic resonance imaging (DWI), the only neuroimaging technique enabling non-invasive in vivo imaging of the brain's fibrous structure. Unfortunately, persistent issues curb clinical progress [14, 53]. Decades after its inception [5, 7, 23, 30, 38, 47], lack of consensus and even skepsis as to its clinical feasibility prevail. This has sparked new incentives to specifically address current limitations in quantitative evaluation studies and international competitions [9, 12, 13, 37, 39, 51].

The geometric rationale for Diffusion Tensor Imaging (DTI), in its original form proposed by O'Donnell et al. [43] and, from a somewhat different perspective, by Lenglet et al. [31], and subsequently adapted by Fuster et al. [21], Hao et al. [24, 25], and several others, stipulates that one can 'geometrize away' local diffusivity patterns inside the brain. The idea is to incorporate anisotropic diffusivity of water in brain white matter, viewed as a porous medium with an orientational preference along axon bundles [57], into the intrinsic geometry of a suitably defined, curved manifold (akin to the geometrization of gravitational forces in general relativity theory). A similar approach is taken by Aumentado-Armstrong et al. [4], where a curved manifold is used to model conductivity of electrical signals in the heart muscles. In the case of DTI a Riemannian manifold presents itself, since its defining metric represents a positive definite quadratic form (or inner product) that can be formally mapped one-to-one onto the DTI tensor. Due to its modest performance, however, geodesic tractography has been largely abandoned in return for other approaches. A notable exception is the probabilistic approach by Hauberg et al. [26] and Schober et al. [55].

Our goal is to provide a versatile deterministic Riemann-DTI geometric paradigm for DTI geodesic tractography, revisiting original ideas (*loc. cit.*), which overcomes some of the main weaknesses, such as producing false positives (e.g. shortcuts) and false negatives (e.g. a shortcut is obscured by the correct solution). Since the mapping of DTI data to biologically meaningful tracts is generally ill-posed, we aim for a flexible metric equipped with control parameters. This admits adaptation to fiducial 'ground truth' tracts, in which data-extrinsic knowledge may be incorporated (e.g. via manual edits by an expert or via machine learning). The parameters control a locally smooth, spatially varying so-called '3-bein', or *triad*, detailed in Sect. 2 (cf. Savadjiev et al. [54] and Piuze et al. [48] for a similar idea in the context of myofiber geometry from DTI, based on Cartan's method of moving frames). In Sect. 3 we perform experiments to illustrate the theory, provide proof of principle, and present results in the context of a simple DTI-tractography phantom for the sake of illustration. There are no obstacles for application on more sophisticated simulated or real data.

2 Theory

Differential geometric approaches in DWI are not new [1, 3, 15, 20–22, 24, 25, 31–33, 43, 46, 48]. The premise underlying the Riemann-DTI paradigm is that tissue microstructure imparts non-random barriers to water diffusion [8, 57]. The fibrous nature of brain white matter, comprising bundles of elongated axons connecting nerve cells in surrounding grey matter regions, facilitates mobility of water molecules along fiber directions. Since DTI captures the main diffusion anisotropy, it is natural to stipulate a Riemannian metric proportional to the inverse of the diffusion tensor. In this way, a relatively large mean free path is tantamount to a relatively short Riemannian distance, so that the problem of tractography can be related to a geodesic ('shortest path') problem. Candidate tracts can then be obtained by direct integration of the geodesic equations, by functional minimization of the Riemannian length (or related cost) functional for curves with fixed endpoints, or (with some care) by inference from the Hamilton-Jacobi equation [28, 45, 50, 52].

Despite the appealing heuristics supporting the geometric paradigm, there are serious caveats we need to take into consideration:

1. In a geodesically complete space *any* pair of points is connected by at least one geodesic, raising the issue of 'false positives' (curves not corresponding to meaningful tracts).
2. A well-posed relation between geodesics as 'paths of least resistance' and meaningful neural tracts is not self-evident. Tissue microstructure, which remains unresolved at scanner resolution, induces mesoscopic diffusivity patterns involving more complex factors than plain presence or absence of nerve axons [41, 42, 44]. As a consequence, there may be many a priori equally viable microstructural explanations for any given DTI image,[1] so that a one-to-one mapping between DTI and Riemannian metric is unlikely to work for tractography.

We will address both concerns and outline our strategy towards an improved framework.

Ad caveat 1: Geodesic completeness , tantamount to (huge) redundancy, may be used to our advantage, provided two conditions are met:

(i) Meaningful tracts correspond, to acceptable approximation and at least piecewise, to geodesics.
(ii) One can identify true (or reject false) positives, based on some deterministic or probabilistic criterion.

The first condition is our main hypothesis, the second one is a constitutional part of the geodesic tractography problem. The problem then boils down to finding 'the right' metric together with effective connectivity criteria for pruning its geodesics [2, 50, 56]. Streamline tractography, in its simplest form a singular limit of geodesic

---

[1] This fact also implies that 'ground truth' simulations must be interpreted with great care in order not to penalise experimental results that deviate from the stipulated ground truth used to create a phantom if such results are equally compatible with data evidence.

tractography based on a degenerate metric, lacks completeness, a direct consequence of its first order nature, which prohibits generic endpoint constraints. As a result, odds are that two fiducial endpoints (picked by an authoritative expert, say) fail to be connected, leaving us in a quandary how to repair for 'false negatives'. Clearly, geodesic completeness requires (at least) second order schemes.

Ad caveat 2: To allow for prior knowledge or retrospective corrections we do not determine a unique metric in terms of DTI data evidence a priori. Instead we aim for optimal control parameters for a *family* of metrics and induced geodesics in a joint (semi-)automated procedure. The premise is that, due to unknown microstructural factors, apparent diffusivities reflect fiber orientations at best qualitatively. By investigating the parameter unfolding of the family of metrics along with geodesic pruning we may investigate whether a stable result compatible with ground truth can be achieved (bearing in mind footnote 1). To this end we have conducted a feasibility study on the Fibercup [13, 49] simulator to clarify all conjectured features of our approach in a simplified context. There are no fundamental obstructions for application to more sophisticated phantoms or to real data, but this elaboration is left for future work.

For computational reasons one could employ any coordinate basis on the tangent bundle TM induced by an arbitrary coordinate map. Such a basis is commonly denoted by $\{\partial_i \doteq \partial/\partial x^i\}_{i=1,2,3}$. (A natural choice would be to employ the same coordinates as those given by a Cartesian coordinate frame of the associated Euclidean space.) However, given a suitable metric g_{ij}, we may instead opt for a special, g-orthogonal (non-coordinate, or anholonomic) basis $\{e_a\}_{a=1,2,3}$. This is the triad alluded to in Sect. 1. If we write the new basis vectors in terms of linear combinations of the coordinate vectors,[2]

$$e_a = e_a^i \partial_i \,, \tag{1}$$

then the coefficients e_a^i define the transformation matrix relating the general coordinate basis to the triad. We may define a new metric holor with entries h_{ab} relative to the triad by the standard change-of-basis formula:

$$g_{ij} e_a^i e_b^j = h_{ab} \,. \tag{2}$$

The triad can be chosen so as to put h_{ab} into a convenient form. In order to illustrate this 'gauge fixing', let us assume that the dual metric g^{inv} is identified with the DTI matrix D, as originally proposed. Take $h = \mathrm{diag}(1/\lambda^1, 1/\lambda^2, 1/\lambda^3)$, in which the λ^a are the eigenvalues of D. In the parlance of classical matrix theory [27], e_a is then the eigenvector of D associated with λ^a, which, in turn, defines a rank-one matrix $Z_a = e_a \otimes e_a$ ('Frobenius covariant'). Equation (2) is the geometrical counterpart of the classical Lagrange-Sylvester matrix decomposition, $D = \sum_{a=1}^{3} \lambda^a Z_a$, and the eigenvalues are our controls. We may vary the λ^a and observe the effect on the

---

[2]The Einstein summation convention for identical upper/lower index pairs applies throughout in tensorial equations.

metric, induced geodesics, and ultimately fiber tracts. This parameter freedom unifies originally proposed models for the connection between DTI and Riemannian metric [21, 24, 25, 31, 43], since all instances can be obtained by slick (local or global) choice of λ^a. This includes streamline tractography via singular perturbation theory.

To find a geodesic we fix seed and target points, A and B say, and minimize the length functional for curves connecting these points. This can be extended to any pair of regions provided one employs an efficient algorithm. Solving the Hamilton-Jacobi equation (with the help of fast marching methods) seems attractive in this respect, but, by design, provides only global minimizers, and is therefore not likely to solve the shortcut problem. Instead we opt for direct, coarse-to-fine minimization of the parametrization invariant functional

$$L_g(\gamma) = \int_{t_A}^{t_B} \sqrt{g_{ij}(x(t))\dot{x}^i(t)\dot{x}^j(t)}\, dt \,, \tag{3}$$

in which

$$\gamma : [t_A, t_B] \to \mathbb{R}^3 : t \mapsto x(t) \tag{4}$$

is an arbitrarily parametrized curve connecting $A = x(t_A)$ and $B = x(t_B)$ in $\mathbb{R}^3$ and $\dot{x}$ denotes its derivative. We embed the metric in a multiresolution family, with scale parameter $\sigma \in \mathbb{R}^+$, according to the multiplicative scheme proposed by Florack et al. for positive symmetric matrices [16, 18], viz. if g is the Gram matrix with entries g_{ij}, then

$$g(x, \sigma) = \exp\left((\phi_\sigma * \ln g)(x)\right) \,, \tag{5}$$

in which exp and ln are matrix-exp and matrix-log, ϕ_σ is the L^1-normalised isotropic Gaussian kernel of scale σ [17, 19, 29, 34], and $*$ denotes entry-wise convolution. (This definition ensures that the dual of a blurred metric equals the blurred dual metric.) Since the metric can be shown to become Euclidean in the $\sigma \to \infty$ limit under suitable, weak conditions, the asymptotic minimizer is the straight line connecting A and B. As one gradually decreases σ, this minimizer (geodesic at resolution $1/\sigma$) is expected to deform likewise gradually, while adapting to the refined metric (at this level of rigor we ignore the problem of scale space bifurcations, cf. Damon [10, 11]). A higher resolution geodesic can then be found by adding a few control points roughly equidistantly along the curve in proportion to resolution, and minimizing the multivariate function of the control points representing our discretization of Eq. (3). Now suppose we have a (local) minimizer at some scale σ, then along with a scale refinement $\sigma \to \sigma - d\sigma$ (we take $d\sigma \propto \sigma$) we increment the number of control points along the curve, and seek a new minimizer in the vicinity of the previous one. In this way one can arbitrarily refine the geodesic curve (until grid scale, if needed), cf. Fig. 1.

The multiresolution scheme for geodesics sketched above applies to each fixed member of the $(\lambda^1, \lambda^2, \lambda^3)$-family of metrics, and should not be confused with optimization with respect to the unfolding of this family as a function of $(\lambda^1, \lambda^2, \lambda^3)$. The details of this are given in the next section, notably Eq. (6).

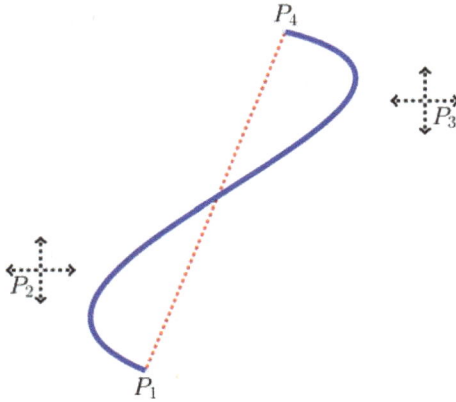

Fig. 1 Coarse-to-fine minimization of the length functional in Eq. (3) is implemented as an iterative multivariate function minimization in which the number of variables gradually increases in proportion to resolution. The control points represent a Bezier curve (blue) which by default we initialize by the unique Euclidean geodesic (dotted line), although this initialising curve may be manually overruled. In this example there are two control points, P_2 and P_3 (recall that the endpoints, P_1 and P_4 in this case, remain fixed). Note that the control points in general do not lie on the curve itself. These points are part of the 'behind-the-scenes' machinery in the minimization procedure and in themselves do not have any anatomical meaning. However, given k points in space, we may construct a Bezier curve through these points (although not uniquely). This may be useful for manual initialisation or correction by an expert, based on anatomical landmarks.

3 Experiments

In the following experiments we use an operational procedure to automatically select 'the right' (anisotropic) scaling of the control parameters λ^a for the triad, relative to the original eigenvalues of the diffusion tensor. To begin with, we have opted for a single, global parameter $\varepsilon \in (0, 1]$, introduced so as to rescale the diagonal metric h^{inv} (the dual of h in Eq. (2)) by[3] $\mathrm{diag}(1, \varepsilon, \varepsilon)$. The global character of ε allows us to leave λ^1 unscaled, since any scaling of λ^1 can be absorbed into ε. Effectively this yields an ε-parametrized family of Riemannian metrics g_ε, replacing the unscaled metric $g = g_{\varepsilon=1}$ in Eqs. (2), (3) and (5). However, we only use this scaling whenever the fractional anisotropy (FA) [6] of the diffusion tensor is large enough (here we choose FA $= 0.15$ as the ad hoc threshold) to ensure a well-defined main eigendirection.[4] Below this threshold the diffusion tensor is kept unscaled. As a result we obtain an ε-parametrized family of geodesics γ_ε, cf. Eq. (4). These geodesics are computed one

---

[3]In principle, the control parameter triple $(\lambda^1, \lambda^2, \lambda^3)$ permits local adaptation leading to a 3-parameter family in every voxel. The singular limit $\varepsilon = 0$ corresponds to streamline tractography and is excluded for its geodesic incompleteness. Values $\varepsilon > 1$ may affect the ordering of the eigenvalues of h^{inv}.

[4]This ad hoc anisotropy threshold calls for a more rigorously motivated alternative, but it serves our purpose in this feasibility study, viz. to ensure well-posedness.

by one by the coarse-to-fine scheme outlined at the end of Sect. 2 for each fiducial pair of endpoints and each setting of ε.

Subsequently each geodesic thus obtained is quantitatively evaluated in terms of the tract length unbiased, nonlinear connectivity functional [2, 50, 56]

$$C(\gamma_\varepsilon) = \frac{L_\eta(\gamma_\varepsilon)}{L_g(\gamma_\varepsilon)}, \qquad (6)$$

in which η denotes the standard Euclidean metric (so that $L_\eta(\gamma)$ is the usual length of γ), recall Eq. (3). Note that in the denominator we use the unscaled diffusion metric $g = g_{\varepsilon=1}$, for which the curve γ_ε is, in general, not a geodesic (unless $\varepsilon = 1$). Using the scaled length g_ε in the denominator would entail a bias towards $\varepsilon = 1$. Each geodesic γ_ε will almost surely not correspond to a streamline (unless both endpoints are on the streamline), which implies that there exists a $t \in (0, 1)$ such that the tangent vector $\dot{\gamma}_\varepsilon(t)$ has a component in the direction of a non-principal eigenvector of the diffusion tensor. Since this component scales with ε^{-1}, the connectivity would vanish for $\varepsilon \to 0$ even if the geodesic remains (approximately) the same curve. Instead, the ε control parameter merely serves to single out an optimal, metric-compatible geodesic through a suitable anisotropic scaling of the metric tensor. The connectivity criterion for the latter is based on data evidence, viz. average apparent diffusivity, which should not involve ε. Connectivity $C(\gamma_\varepsilon)$ is thus some average measure of diffusivity along γ_ε. Parameter values

$$\varepsilon^* \in \underset{\varepsilon \in [\delta, 1]}{\operatorname{argmax}}\ C(\gamma_\varepsilon), \qquad (7)$$

for some fixed $0 < \delta \ll 1$, at which *locally* [5] optimal connectivity is attained, are tract-specific sharpening parameters, affected in a still unknown manner, by underlying microstructure. The corresponding tracts γ_{ε^*} are called *optimal tracts*.

As illustrated in Fig. 2 for the Fiberfox-reconstructed Fibercup phantom with added Riccian noise [40] (cf. Fillard et al. for a detailed description [13] of the original Fibercup) only optimal tracts are retained. Since it is known that the inverse metric has certain shortcomings (e.g. see [24]), we also employed our automated procedure in combination with one of the other metric proposals in the literature, namely the adjugate metric [21, 22]. We chose this one, because it is derived from first principles of the underlying diffusion process and in addition it's very easy to implement. In Fig. 2, the optimal tracts found for the inverse (bottom-left) and adjugate DTI metric (bottom-right) emerge for different optimal parameter values, but are, remarkably, virtually indistinguishable. As the adjugate DTI metric is essentially a *local* isotropic scaling of the inverse DTI metric, it is a noteworthy feature that we are able to obtain essentially the same tracts in both cases with a family of metrics characterised by a single, *global* parameter. After all, the governing geodesic equations for the inverse and adjugate are not the same. In practice, a broad range of parameter values (the high plateaux in the graphs) can be used to single out essentially the same tracts, making this procedure robust with respect to the precise choice of ε. The experiment shows that, in all cases and by virtue of proper placement of endpoints, correct tracts

---

[5] We define $\operatorname{argmax}_x f(x) \doteq \{x \mid f(y) \le f(x)$ for all y in a neighbourhood of $x\}$.

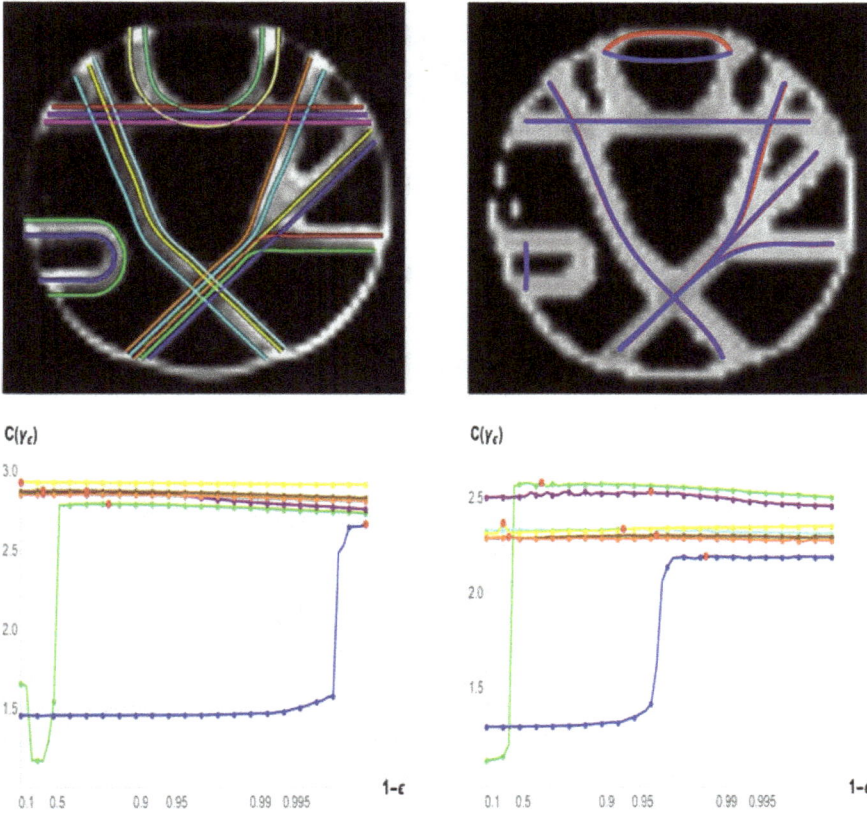

Fig. 2 (Top-left) Ground truth tracts (for various seed points); illustration adapted from Fillard et al. [13]. (top-right) Tracts obtained by using the unscaled inverse (red), respectively adjugate (blue) DTI tensor as the default metric for geodesic tractography. Both metrics produce shortcuts (false positives) instead of following the U-shape fibers, leaving false negatives. (middle-left) Per-tract connectivity for the ε-parameter unfolding of the inverse DTI metric, $g = D^{\mathrm{inv}}$. (middle-right) Idem for the adjugate DTI metric, $g = \det D D^{\mathrm{inv}}$. Colors correspond to those used for the tracts in the bottom two figures. The red dots indicate the tract-specific parameter values $\varepsilon^* \in (0, 1]$ for the metrics used for each optimal tract. This optimal parameter is not affected by a suitable (global) normalization of g, meaning that the absolute value of the connectivity is inherently meaningless. It is, however, the relative value that determines the optimal parameter. Note the logarithmic scale for ε. (bottom-left) Optimal results obtained by automatic optimization for inverse DTI metric ansatz. (bottom-right) Optimal results obtained by automatic optimization for adjugate DTI metric ansatz. The latter two results are virtually indistinguishable, but optimality is obtained at different control parameter settings.

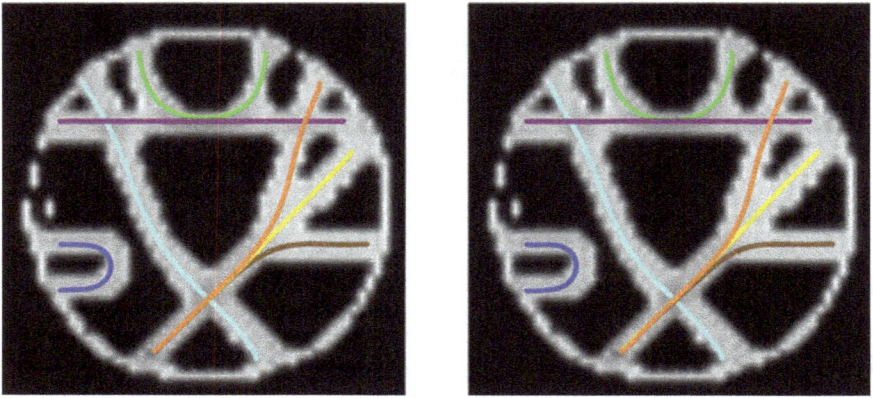

Fig. 2 (continued)

are found automatically, and, in particular, that shortcuts do not occur. To assess the robustness to placement of endpoints, in Fig. 3 the optimal tracts are computed for pairs of perturbed endpoints, demonstrated for the adjugate DTI metric.

Low connectivity typically corresponds to shortcuts. Figure 4 shows optimal tracts for each pair of corresponding (ground truth) seed and target points (top), as well as connectivities for optimal tracts between a selection of seed point and all remaining target points (bottom). For each pair of possible seed/target points, the optimal tract was determined by Eq. (7). Figure 4 shows two representative cases: one in which the stipulated ground truth yields the highest connectivity and a second one in which it is among the top but is still dominated by another tract.

Multiple local maxima of $C(\gamma_\varepsilon)$ generally exist, occurring at $\varepsilon \in \{\varepsilon_1^*, \ldots, \varepsilon_k^*\}$ say, corresponding to k plausible tracts (recall Eq. (7) and footnote 5) that invariably include the stipulated 'ground truth', although the latter does not necessarily correspond to the *global* maximum, recall Fig. 4b. Figure 5 illustrates this case by another simulation (based on Fiberfox [40]), emblematic of an ambiguous configuration of kissing or crossing fiber tracts . Both interpretations are indeed confirmed, in the sense of corresponding to local optima of $C(\gamma_\varepsilon)$. Our current coarse-to-fine implementation, however, cannot handle scale space bifurcations of geodesics automatically, and, due to our default initialization by the Euclidean geodesic, automatically zooms in on the kissing tract. The crossing tract requires a (simple) manual edit to overrule the default initialization and encourage convergence to another solution. This edit and its effect is shown in Fig. 5, and further illustrated in Fig. 6. Here the default initialization by the Euclidean geodesic has been replaced by a quadratic Bezier curve forced to pass through a manually selected inclusive ROI called an AND gate (green circle), while connectivity maximization for this control point has been constrained to this neighbourhood. Note that, for the inverse as well as the adjugate diffusion tensor ansatz, Fig. 6, boundary extrema γ_ε are found near $\varepsilon = 1$, illustrating the same inclination to produce shortcuts as γ_1 for the unscaled metric. Closer to the optimal $\varepsilon^* \in \mathrm{argmax}_{\varepsilon \in [\delta, 1]} C(\gamma_\varepsilon)$ we observe correct convergence, though.

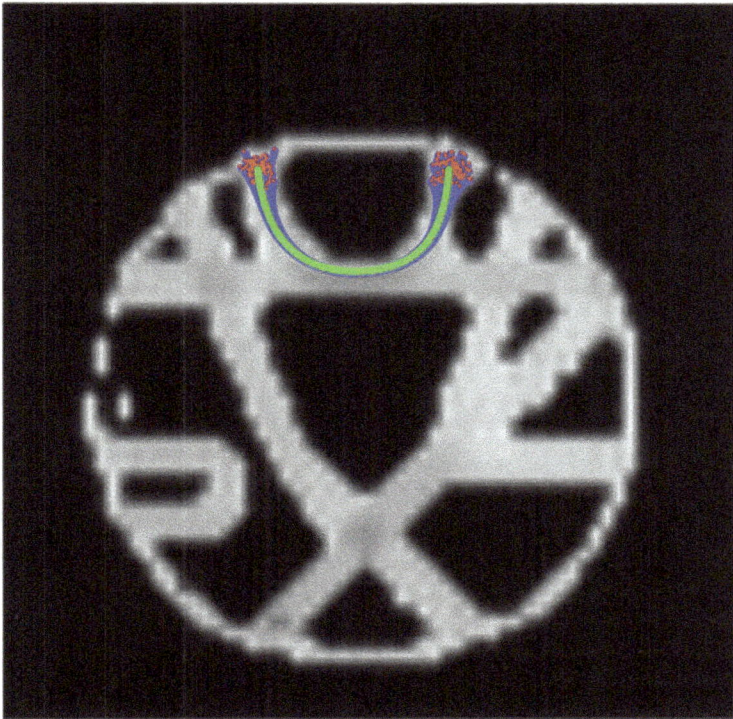

Fig. 3 For each endpoint of the optimal green curve (cf. Fig. 2) the red points are generated from a multivariate normal distribution centered around the endpoints with covariance matrix $\Sigma =$ diag$(0.5, 0.5)$. The blue tracts are the optimal tracts for pairs of such perturbed endpoints by using the adjugate DTI metric ansatz. The green curve has a connectivity value of 2.749 and the mean connectivity of all perturbed tracts is 2.708 with a standard deviation of 0.037. Visually, the green curve is a good representative of this bundle of tracts as a whole and its connectivity is approximately one standard deviation away from the mean.

In Fig. 7, the left Cortical Spinal Tract of the ISMRM 2015 Tractography challenge data [35] is used to illustrate the operational procedure on a 3D dataset. Quantitative measures used in the challenge are the overlap (OL), overreach (OR) and F_1 scores, see Côté et al. [9]. OL is defined as the number of voxels through which both the reconstruction and the ground truth pass, normalized by the number of ground truth voxels. OR is the number of voxels through which the reconstruction passes but the ground truth does not, normalized by the number of reconstructed voxels. These two quantities can be interpreted as the true positive rate and the false discovery rate. The F_1 score is the harmonic mean of OL and $1 - $ OR. Our tractography results in OR $= 0.324$, OL $= 0.375$, leading to $F_1 = 0.427$, ranking it in the top 40% of the submitted challenge results based solely on F_1 scores. It is particularly noteworthy that this DTI-based procedure has a performance comparable to HARDI type methods, outperforming other DTI-based methods applied to the CST [36].

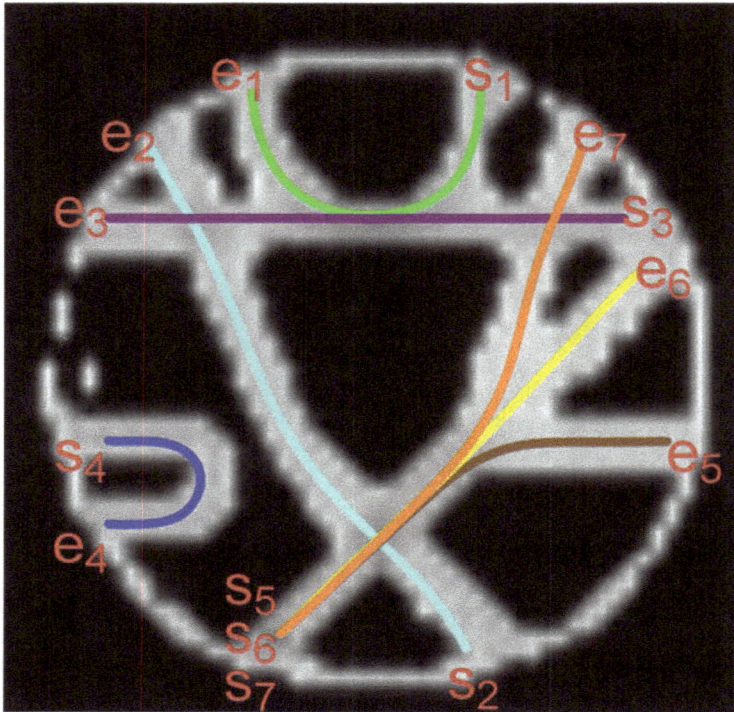

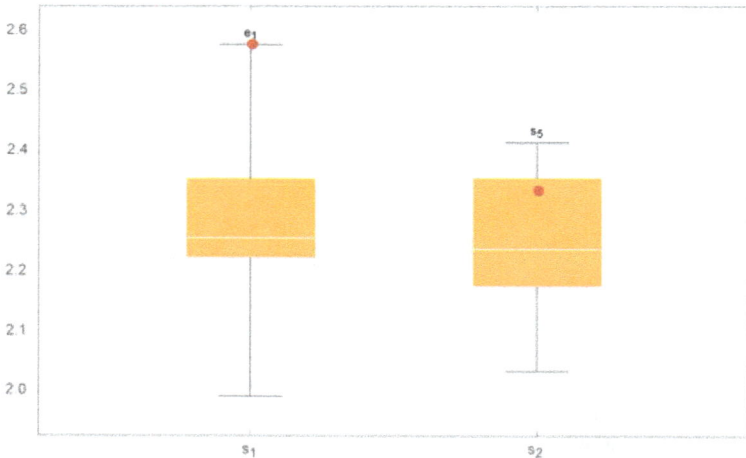

Fig. 4 (Top) Seed and target point labeling: s_i = start of tract, e_i = end of tract; the attached number $i \in \{1, \ldots, 7\}$, refers to the tract label. (bottom) Box-and-whisker plot of connectivity in the adjugate DTI metric versus seed point. For the seed points s_1 and s_2 the boxplot above it shows the optimal connectivity of that seed point to every other target point s_j and e_j. Endpoints that yield the highest connectivity are indicated on top on the boxplot. The red dots indicate the connectivity value for the stipulated ground truth, i.e. the tract running from s_i to e_i, for each tract label.

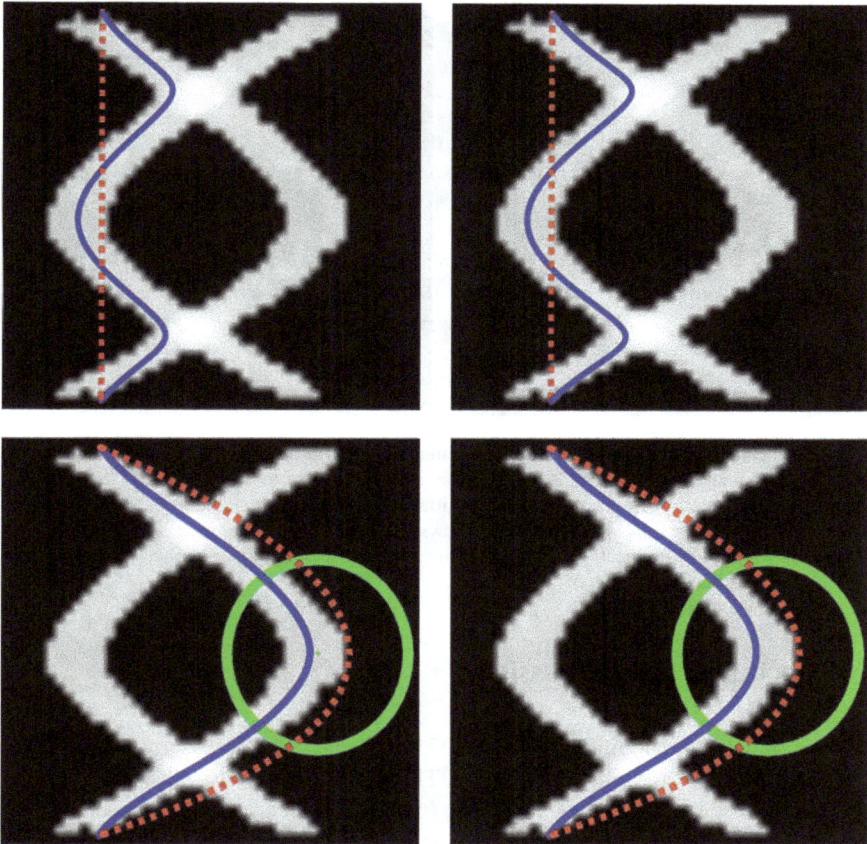

Fig. 5 Simulation with kissing/crossing fibers. Seed and target points are invariably taken to lie in upper and lower left corner. Note that there are two plausible solutions. The left two figures pertain to the (connectivity pruned) ε-family induced by the inverse diffusion metric, the right two figures to that of the adjugate diffusion metric. In either case, our automatic procedure only finds one solution (blue curves). (top) The equally viable crossing scenario is due to the coarse-to-fine scheme with default initialization (red dotted Euclidean geodesic in the upper two figures). (bottom) By employing a restricted neighbourhood search region (AND gate, green circle) with suitable initialization (red dotted quadratic Bezier curve), the algorithm converges to the blue crossing geodesics in the bottom two figures, resolving the false negative result.

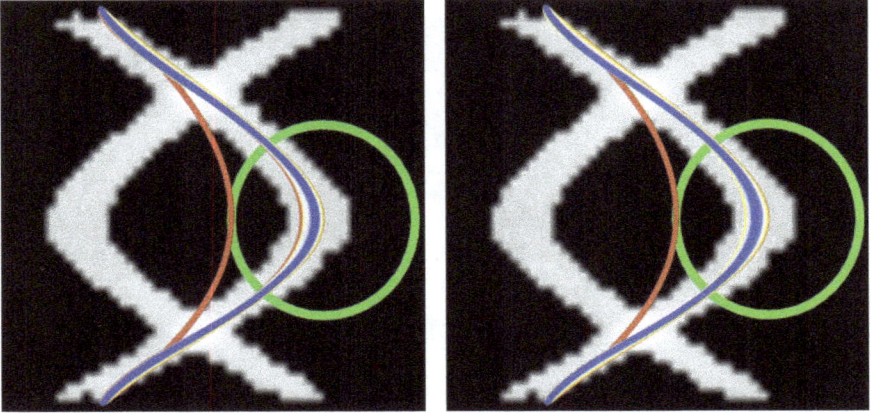

Fig. 6 Geodesics found by unfolding of the Riemannian metric family g_ε, Eq. (7) in combination with connectivity optimization, Eq. (6) for inverse (left) and adjugate (right) diffusion tensor ansatz $g = g_1$, the AND gate of Fig. 5 (green circle). Initialization is given by the quadratic Bezier curve in Fig. 5. Red-to-blue rainbow color coding shows the connectivity evolution as a function of ε, with red indicating low connectivity tracts and blue indicating the (unique) connectivity optimized result for $\varepsilon^* = \mathrm{argmax}_{\varepsilon \in [\delta, 1]} C(\gamma_\varepsilon)$.

4   Conclusion and Discussion

We have proposed a modified Riemann-DTI geodesic tractography framework. The modification entails the embedding of a DTI induced metric into a family of metrics designed for the purpose of tractography, furnished with an operational scheme for tract-specific optimization of metric and geodesic(s) for any given pair of endpoints based on a nonlinear connectivity functional.

In a simple experiment based on the Fibercup simulator we conclude that all local connectivity maxima correspond to plausible tracts that consistently follow high diffusivity pathways and avoid, as much as logically possible, inconsistent regions. For instance, despite a rather simplistic connectivity measure, all ground truth tracts are reconstructed provided seed and target points are (roughly) correctly placed, i.e. in such a way that a plausible fiber does indeed exist, recall Fig. 2. The robustness of our procedure with respect to the placement of endpoints is demonstrated in Fig. 3, where tracts reconstructed using slightly perturbed seed and target locations, remained in close proximity to the original one.

It is essential, in this respect, to appreciate the ill-posed nature of tractography. Although the generation of diffusivity patterns from a given tract configuration is robust (albeit model-dependent), this process is not invertible. The equivalence class of tract configurations consistent with a given diffusivity pattern may contain many 'metamers', so to speak. The actual power of the proposed method is precisely its ability to generate multiple representatives from this equivalence class, so that no a priori bias is introduced. In clinical practice this could help an expert to reject

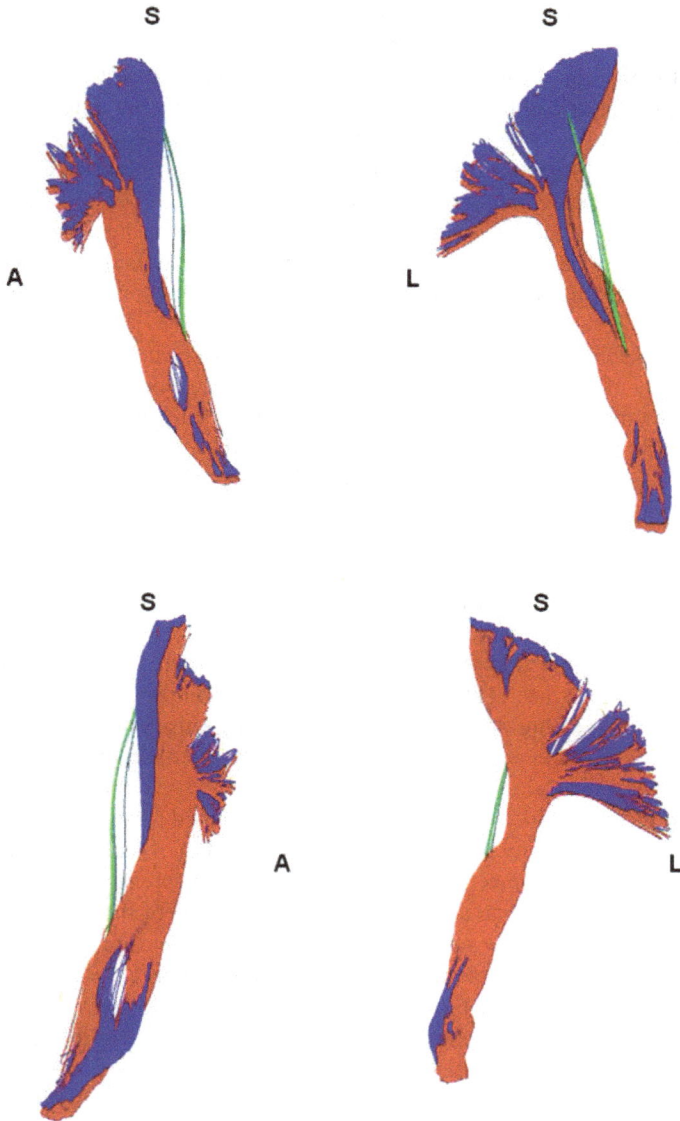

Fig. 7 Comparison of the Left Cortical Spinal Tract (CST) between the ground truth (red) and connectivity-optimal geodesics according to Eqs. (6) and (7) in which the base metric is the adjugate DTI-metric. The geodesics are color-coded according to their connectivity value, which increases from green to blue. The depicted views are labeled with S(uperior), A(nterior) and L(eft) to determine the plane and orientation of the cross-section. On visual inspection of the multiple views of the CST, the optimal geodesics lie close to the ground truth. Using more refined connectivity measures or a criterion to remove single tracts, the results can be improved. This can relieve us from those spurious tracts that are seemingly leaving the bundle and clearly have a lower connectivity than the rest of the tracts. A selection of 3000 ground truth tracts were extracted from the CST to determine seed and target points for the geodesics.

false positives (if not eliminated automatically), based on data-extrinsic anatomical insight, in favour of selecting one of the remaining plausible candidate tracts, i.e. without being confronted with false negatives. Ideally such candidates present themselves, viz. as one of the local optima actually found by our automatic scheme.

False negatives may nevertheless occur if local maxima are missed due to inadequate numerical optimization. We have indicated how this problem may be mitigated through a simple manual edit by an expert. The delineation of an AND gate through which one believes tentative tracts of interest should pass, or the sketch of a coarse initialization curve between the endpoints, overruling the default straight line segment, both make it more likely to drive numerical optimization to the correct local maximum.

To prevent scaling of the diffusion tensors in a non-sensical manner, their eigenvalues are only affected whenever an FA threshold was exceeded in order to ensure a well-defined main eigendirection of the diffusion tensor. This treshold was based on experiments with the Fiberfox-reconstructed Fibercup phantom, but needs to be adjusted for tractography on real data. This ad hoc approach should be considered a first stepping stone towards a more refined technique combining the scaling of eigenvalues with the anisotropy of the tensor and the noise level of the data. Highly anisotropic tensors have a relatively large first eigenvalue and may tolerate a more severe scaling, since they provide more evidence for an articulated orientation. Since many low FA regions contain complex fiber structures, not scaling the DTI tensors there may influence the shape of the geodesics. In these areas we do not expect to get accurate results, since DTI is not able to resolve complex fiber structures adequately. However, scaling of the tensors enhances any noise present in the data and may increase the number of false positives, especially when the signal-to-noise-ratio is very low.

We stress that 'most likely' fiber candidates are obtained for *any* pair of endpoints in terms of local optimality of connectivity. Recalling Fig. 4, note that even though tract s_2-e_2 has a high connectivity value, maximal connectivity is attained by the tract s_2-s_5. This effect can be ascribed to the construction of the family of metrics g_ε as well as the conservative nature of the connectivity measure in Eq. (6). At the intersection of two bundles the diffusion tensors are (nearly) isotropic, leading to similar diffusivities in all directions. This isotropy (FA ≤ 0.15) prevents anisotropic rescaling of the metric, as explained. Moreover, as the connectivity measure only accounts for average diffusion along a tract and does not penalize curvature, a bend in the tract does not decrease its connectivity, as the latter is based on diffusivity, not geometry. These effects combined result in a connectivity-optimal tract which differs from the ground truth one. This needs to be remedied by using more sophisticated connectivity measures, which will be application dependent and may require prior knowledge about the bundles of interest. Our generic connectivity measure has been motivated mainly by our desire to preserve true positives, not as a criterion to remove false positives, allowing a modular approach. Additional criteria for pruning false positives connecting a fiducial seed point to a volumetric target region (possibly the whole brain) will be studied in future work, including appropriate extensions of the proposed coarse-to-fine tractography algorithm.

In the ISMRM Tractography challenge reconstructed tracts are interpreted as centerlines of a fiber with a thickness/resolution of 1 mm isotropic voxel. The OL, OR and F_1 scores were calculated based on such 1 mm voxels [36]. The tractography, however, was performed on the diffusion weighted images consisting of 2 mm isotropic voxels. This way we can only resolve detail up to the grid scale of 2 mm and quantification measures should take this resolution difference into account when using the affine transformation to convert between the two different spaces. By downscaling this minimal resolution of 1 mm to 2 mm we increase the F_1 score to 0.526, ranking in the top 20% (as the only DTI based algorithm).

In this experiment details could be resolved up to grid scale, but if one would consider the effects of noise (e.g. in the acquisition process), this may lead to a different effective minimal resolution, determined by grid scale and noise level. The influence of different kinds of noise (e.g. acquisition noise, seeding errors) needs to be addressed in future studies in which the control-triad framework is combined with uncertainty quantification.

These experiments confirm our main conjecture that the potential of DTI has not yet been fully exploited in tractography.

Acknowledgments This work is part of the research programme 'Diffusion MRI Tractography with Uncertainty Propagation for the Neurosurgical Workflow' with project number 16338, (partly) financed by the Netherlands Organisation for Scientific Research (NWO).

The work of Andrea Fuster is part of the research programme of the Foundation for Fundamental Research on Matter (FOM), which is financially supported by the Netherlands Organisation for Scientific Research (NWO).

References

1. Arsigny, V., Fillard, P., Pennec, X., Ayache, N.: Log-Euclidean metrics for fast and simple calculus on diffusion tensors. Magn. Reson. Med. **56**(2), 411–421 (2006)
2. Astola, L., Florack, L., ter Haar Romeny, B.: Measures for pathway analysis in brain white matter using diffusion tensor images. In: Karssemeijer, N., Lelieveldt, B. (eds.) Proceedings of the Twentieth International Conference on Information Processing in Medical Imaging– IPMI 2007 (Kerkrade, The Netherlands), Lecture Notes in Computer Science, vol. 4584, pp. 642–649. Springer, Berlin (2007)
3. Astola, L., Fuster, A., Florack, L.: A Riemannian scalar measure for diffusion tensor images. Pattern Recognit. **44**(9), 1885–1891 (2011)
4. Aumentado-Armstrong, T., Kadivar, A., Savadjiev, P., Zucker, S.W., Siddiqi, K.: Conduction in the heart wall: helicoidal fibers minimize diffusion bias. Sci. Rep. **8**(1), 7165 (2018). https://doi.org/10.1038/s41598-018-25334-7
5. Basser, P.J., Pajevic, S., Pierpaoli, C., Duda, J., Aldroubi, A.: In vivo fiber tractography using DT-MRI data. Magn. Reson. Med. **44**(4), 625–632 (2000)
6. Basser, P.J., Pierpaoli, C.: Microstructural and physiological features of tissues elucidated by quantitative-diffusion-tensor MRI. J. Magn. Reson. Ser. B **111**(3), 209–219 (1996)
7. Batchelor, P.G., Moakher, M., Atkinson, D., Calamante, F., Connelly, A.: A rigorous framework for diffusion tensor calculus. Magn. Reson. Med. **53**, 221–225 (2005)
8. Beaulieu, C.: The basis of anisotropic water diffusion in the nervous system–a technical review. NMR Biomed. **15**(7–8), 435–455 (2002)
9. Côté, M.A., Girard, G., Boré, A., Garyfallidis, E., Houde, J.C., Descoteaux, M.: Tractometer: towards validation of tractography pipelines. Med. Image Anal. **17**, 844–857 (2013)

10. Damon, J.: Local Morse theory for solutions to the heat equation and Gaussian blurring. J. Differ. Equ. **115**(2), 368–401 (1995)
11. Damon, J.: Generic properties of solutions to partial differential equations. Arch. Ration. Mech. Anal. 353–403 (1997)
12. De Santis, S., Drakesmith, M., Bells, S., Assaf, Y., Jones, D.K.: Why diffusion tensor MRI does well only some of the time: variance and covariance of white matter tissue microstructure attributes in the living human brain. Neuroimage **89**, 35–44 (2014)
13. Fillard, P., Descoteaux, M., Goh, A., Gouttard, S., Jeurissen, B., Malcolm, J., Ramirez-Manzanares, A., Reisert, M., Sakaie, K., Tensaouti, F., Yo, T., Mangin, J.F., Poupon, C.: Quantitative evaluation of 10 tractography algorithms on a realistic diffusion MR phantom. Neuroimage **56**(1), 220–234 (2011)
14. Filler, A.: MR neurography and diffusion tensor imaging: Origins, history, and clinical impact of the first 50 000 cases with an assessment of efficacy and utility in a prospective 5000-patient study group. Neurosurgery **65**(4), 29–43 (2009)
15. Fletcher, P.T., Joshi, S.: Riemannian geometry for the statistical analysis of diffusion tensor data. Signal Process. **87**(2), 250–262 (2007)
16. Florack, L., van Assen, H.: Multiplicative calculus in biomedical image analysis. J. Math. Imaging Vis. **42**(1), 64–75 (2012)
17. Florack, L.M.J.: Image Structure, Computational Imaging and Vision Series, vol. 10. Kluwer Academic Publishers, Dordrecht, The Netherlands (1997)
18. Florack, L.M.J., Astola, L.J.: A multi-resolution framework for diffusion tensor images. In: Aja Fernández, S., de Luis Garcia, R. (eds.) CVPR Workshop on Tensors in Image Processing and Computer Vision, Anchorage, Alaska, USA, IEEE (2008). Digital proceedings
19. Florack, L.M.J., Haar Romeny, B.M.t., Koenderink, J.J., Viergever, M.A.: Linear scale-space. J. Math. Imaging Vis. **4**(4), 325–351 (1994)
20. Fuster, A., Astola, L.J., Florack, L.M.J.: A Riemannian scalar measure for diffusion tensor images. In: Jiang, X., Petkov, N. (eds.) Proceedings of the 13th International Conference on Computer Analysis of Images and Patterns, CAIP 2009 (September 2–4 2009, Münster, Germany), Lecture Notes in Computer Science, vol. 5702, pp. 419–426. Springer, Berlin (2009)
21. Fuster, A., Dela Haije, T., Tristán-Vega, A., Plantinga, B., Westin, C.F., Florack, L.: Adjugate diffusion tensors for geodesic tractography in white matter. J. Math. Imaging Vis. **54**(1), 1–14 (2016)
22. Fuster, A., Tristan-Vega, A., Dela Haije, T., Westin, C.F., Florack, L.: A novel Riemannian metric for geodesic tractography in DTI. In: O'Donnell, L., Schultz, T., Nedjati-Gilani, G., Panagiotaki, E. (eds.) MICCAI Workshop on Computational Diffusion MRI, pp. 47–54. Japan, Nagoya (2013)
23. Hagmann, P., Jonasson, L., Maeder, P., Thiran, J.P., Wedeen, V.J., Meuli, R.: Understanding diffusion MR imaging techniques: From scalar diffusion-weighted imaging to diffusion tensor imaging and beyond. RadioGraphics **26**, S205–S223 (2006)
24. Hao, X., Whitaker, R.T., Fletcher, P.T.: Adaptive Riemannian metrics for improved geodesic tracking of white matter. In: Székely, G., Hahn, H.K. (eds.) Proceedings of the Twenty-Second International Conference on Information Processing in Medical Imaging–IPMI 2011 (Kloster Irsee, Germany), Lecture Notes in Computer Science, vol. 6801, pp. 13–24. Springer, Berlin (2011)
25. Hao, X., Zygmunt, K., Whitaker, R.T., Fletcher, P.T.: Improved segmentation of white matter tracts with adaptive Riemannian metrics. Med. Image Anal. **18**, 161–175 (2014)
26. Hauberg, S., Schober, M., Liptrot, M., Hennig, P., Feragen, A.: A random Riemannian metric for probabilistic shortest-path tractography. In: Navab, N., Hornegger, J., Wells, W.M., Frangi, A. (eds.) Proceedings of the 18th International Conference on Medical Image Computing and Computer Assisted Intervention—MICCAI 2015 (Munich, Germany, October 5–9 2015), Lecture Notes in Computer Science, vol. 9349, pp. 597–604. Springer, Berlin (2015)
27. Higham, N.J.: Functions of Matrices: Theory and Computation. SIAM (2008)
28. Jackowski, M., Kao, C.Y., Qiu, M., Costable, R.T., Staib, L.H.: White matter tractography by anisotropic wavefront evolution and diffusion tensor imaging. Med. Image Anal. **9**, 427–440 (2005)

29. Koenderink, J.J.: The structure of images. Biol. Cybern. **50**, 363–370 (1984)
30. Le Bihan, D., Mangin, J.F., Poupon, C., Clark, C.A., Pappata, S., Molko, N., Chabriat, H.: Diffusion tensor imaging: concepts and applications. J. Magn. Reson. Imaging **13**, 534–546 (2001)
31. Lenglet, C., Deriche, R., Faugeras, O.: Inferring white matter geometry from diffusion tensor MRI: Application to connectivity mapping. In: Pajdla, T., Matas, J. (eds.) Proceedings of the Eighth European Conference on Computer Vision (Prague, Czech Republic, May 2004), Lecture Notes in Computer Science, vol. 3021–3024, pp. 127–140. Springer, Berlin (2004)
32. Lenglet, C., Prados, E., Pons, J.P.: Brain connectivity mapping using Riemannian geometry, control theory and PDEs. SIAM J. Imaging Sci. **2**(2), 285–322 (2009)
33. Lenglet, C., Rousson, M., Deriche, R., Faugeras, O.: Statistics on the manifold of multivariate normal distributions: Theory and application to diffusion tensor MRI processing. J. Math. Imaging Vis. **25**(3), 423–444 (2006)
34. Lindeberg, T.: Scale-Space Theory in Computer Vision. The Kluwer International Series in Engineering and Computer Science. Kluwer Academic Publishers, Dordrecht, The Netherlands (1994)
35. Maier-Hein, K., Neher, P., Houde, J.C., Caruyer, E., Daducci, A., Dyrby, T., Stieltjes, B., Descoteaux, M.: Tractography Challenge ISMRM 2015 High-resolution Data (2017). https://doi.org/10.5281/zenodo.579933,
36. Maier-Hein, K.H., Neher, P., Houde, J., Côté, M., Garyfallidis, E., Zhong, J., Chamberland, M., et al.: Tractography-based connectomes are dominated by false-positive connections. bioRxiv (2016). https://doi.org/10.1101/084137, https://www.biorxiv.org/content/early/2016/11/21/084137
37. Maier-Hein, K.H., Neher, P., Houde, J., Côté, M., Garyfallidis, E., Zhong, J., Chamberland, M., et al.: The challenge of mapping the human connectome based on diffusion tractography. Nat. Commun. **8**(1), 1349 (2017)
38. Mori, S.: Introduction to Diffusion Tensor Imaging. Elsevier, Amsterdam (2007)
39. Neher, P.F., Descoteaux, M., Houde, J.C., Stieltjes, B., Maier-Hein, K.H.: Strengths and weaknesses of state of the art fiber tractography pipelines - a comprehensive in-vivo and phantom evaluation study using Tractometer. Med. Image Anal. **26**(1), 287–305 (2015)
40. Neher, P.F., Laun, F.B., Stieltjes, B., Maier-Hein, K.H.: Fiberfox: Facilitating the creation of realistic white matter software phantoms. Magn. Reson. Med. **72**, 1460–1470 (2014)
41. Novikov, D.S., Fieremans, E., Jensen, J.H., Helpern, J.A.: Random walks with barriers. Nature **7**, 508–514 (2011)
42. Novikov, D.S., Kiselev, V.: Effective medium theory of a diffusion-weighted signal. NMR Biomed. **23**, 682–697 (2010)
43. O'Donnell, L., Haker, S., Westin, C.F.: New approaches to estimation of white matter connectivity in diffusion tensor MRI: Elliptic PDEs and geodesics in a tensor-warped space. In: Proceedings of Medical Imaging, Computing and Computer Assisted Intervention, Lecture Notes in Computer Science, vol. 2488, pp. 459–466. Springer, Berlin (2002)
44. Özarslan, E., Koay, C.G., Shepherd, T.M., Komlosh, M.E., İrfanoğlu, M.O., Pierpaoli, C., Basser, P.J.: Mean apparent propagator (MAP) MRI: a novel diffusion imaging method for mapping tissue microstructure. Neuroimage **78**, 16–32 (2013)
45. Parker, G.J.M., Wheeler-Kingshott, C.A.M., Barker, G.J.: Estimating distributed anatomical connectivity using fast marching methods and diffusion tensor imaging. IEEE Trans. Med. Imaging **21**(5), 505–512 (2002)
46. Pennec, X., Fillard, P., Ayache, N.: A Riemannian framework for tensor computing. Int. J. Comput. Vis. **66**(1), 41–66 (2006)
47. Pierpaoli, C., Jezzard, P., Basser, P.J., Barnett, A., Di Chiro, G.: Diffusion tensor MR imaging of the human brain. Radiology **201**(3), 637–648 (1996)
48. Piuze, E., Sporring, J., Siddiqi, K.: Maurer-Cartan forms for fields on surfaces: Application to heart fiber geometry. IEEE Trans. Pattern Anal. Mach. Intell. **37**(12), 2492–2504 (2015)
49. Poupon, C., Rieul, B., Kezele, I., Perrin, M., Poupon, F., Mangin, J.F.: New diffusion phantoms dedicated to the study and validation of HARDI models. Magn. Reson. Med. **60**, 1276–1283 (2008)

50. Prados, E., Soatto, S., Lenglet, C., Pons, J.P., Wotawa, N., Deriche, R., Faugeras, O.: Control theory and fast marching techniques for brain connectivity mapping. In: Proceedings of the IEEE Computer Society Conference on Computer Vision and Pattern Recognition (New York, USA, June 2006), vol. 1, pp. 1076–1083. IEEE Computer Society Press (2006)
51. Pujol, S., et al.: The DTI Challenge: towards standardized evaluation of diffusion tensor imaging tractography for neurosurgery. J. Neuroimaging **25**(6), 875–882 (2015)
52. Rund, H.: The Hamilton-Jacobi Theory in the Calculus of Variations. Robert E. Krieger Publishing Company, Huntington, N.Y. (1973)
53. Rutten, G.J.M., Kristo, G., Pigmans, W., Peluso, J., Verheul, H.B.: Het gebruik van MR-tractografie in de dagelijkse neurochirurgische praktijk. Tijdschrift voor Neurologie Neurochirurgie **115**(4), 204–211 (2014). With English abstract
54. Savadjiev, P., Strijkers, G.J., Bakermans, A.J., Piuze, E., Zucker, S.W., Siddiqi, K.: Heart wall myofibers are arranged in minimal surfaces to optimize organ function. Proc. Natl. Acad. Sci. **109**(24), 9248–9253 (2012)
55. Schober, M., Kasenburg, N., Feragen, A., Hennig, P., Hauberg, S.: Probabilistic shortest path tractography in DTI using Gaussian process ODE solvers. In: Golland, P., Hata, N., Barillot, C., Hornegger, J., Howe, R. (eds.) Proceedings of the 17th International Conference on Medical Image Computing and Computer Assisted Intervention—MICCAI 2014 (Boston, US, September 14–18 2014), Lecture Notes in Computer Science, vol. 8675, pp. 265–272. Springer, Berlin (2014)
56. Sebastiani, G., De Pasquale, F., Barone, P.: Quantifying human brain connectivity from diffusion tensor MRI. J. Math. Imaging Vis. **25**(2), 227–244 (2006)
57. Sen, P.N., Basser, P.J.: Modeling diffusion in white matter in the brain: a composite porous medium. J. Magn. Reson. Imaging **23**(2), 215–220 (2005)

11

Conceptual Parallels between Stochastic Geometry and Diffusion-Weighted MRI

Tom Dela Haije and Aasa Feragen

Abstract Diffusion-weighted magnetic resonance imaging (MRI) is sensitive to ensemble-averaged molecular displacements, which provide valuable information on e.g. structural anisotropy in brain tissue. However, a concrete interpretation of diffusion-weighted MRI data in terms of physiological or structural parameters turns out to be extremely challenging. One of the main reasons for this is the multi-scale nature of the diffusion-weighted signal, as it is sensitive to the microscopic motion of particles averaged over macroscopic volumes. In order to analyze the geometrical patterns that occur in (diffusion-weighted measurements of) biological tissue and many other structures, we may invoke tools from the field of stochastic geometry. Stochastic geometry describes statistical methods and models that apply to random geometrical patterns of which we may only know the distribution. Despite its many uses in geology, astronomy, telecommunications, etc., its application in diffusion-weighted MRI has so far remained limited. In this work we review some fundamental results in the field of diffusion-weighted MRI from a stochastic geometrical perspective, and discuss briefly for which other questions stochastic geometry may prove useful. The observations presented in this paper are partly inspired by the Workshop on Diffusion MRI and Stochastic Geometry held at Sandbjerg Estate (Denmark) in 2019, which aimed to foster communication and collaboration between the two fields of research.

T. Dela Haije (✉)
Department of Computer Science, University of Copenhagen, Copenhagen, Denmark
e-mail: tom@di.ku.dk

A. Feragen
Department of Applied Mathematics and Computer Science, Technical University of Denmark, Lyngby, Denmark
e-mail: afhar@dtu.dk

1 Introduction

Diffusion-weighted magnetic resonance imaging (MRI) [22] is one of the few imaging modalities that is capable of mapping the immensely complex *micro-structural* architecture of the human brain in a non-destructive manner. This is achieved by applying a specific sequence of diffusion-sensitizing magnetic field gradients during the MRI acquisition, producing a signal with a decay rate that is dependent on the relative mobility of water molecules in the tissue [30]. As the overall mobility of the molecules is decreased by the presence of any material barrier, the acquired signal effectively provides an indirect probe of the ambient structure. At the micrometer length scales accessible in current scanners, the dominant barriers to the diffusing molecules in the brain are the fiber-like *neurites* that transmit information between different regions of the brain [3], and whose properties are (naturally) of profound importance in neurology. Aside from the anticipated and present clinical value of this modality, diffusion-weighted MRI also stands to provide unique information about the evolution, morphogenesis, and function of the brain, already being pursued through the tracking of macroscopic fibers in *tractography* [2, 4, 20]. The challenge faced by the diffusion-weighted MRI community is to identify the relevant structural parameters determining the signal decay, and—to the extent that this is possible—to invert the relation between them. This is where we believe *stochastic geometry* could play a role.

Stochastic geometry [29] is an area of statistics that provides modeling and inference techniques for complex spatial objects whose structure can be described effectively as random patterns. One of the first cases studied in what is now called stochastic geometry—the problem of *stereology* [1]—included the inference of geometric properties of 3D objects from their intersections with a small number of 2D planes. This topic gained traction in the 1960s, as its solutions alleviated the challenging and computationally demanding task of actually computing 3D reconstructions. Stereology found a wealth of applications ranging from geology to for example microscopy of neuroanatomy. Since then stochastic geometry has evolved into a mature field of mathematics that offers a rich toolbox of rigorously developed techniques, including models with potential relevance for diffusion-weighted MRI such as random tessellations and *fiber processes*. Skimming the stochastic geometry literature one quickly comes across a number of concepts that have obvious analogues independently developed in the diffusion-weighted MRI literature, and yet the two research domains have had very limited contact so far.

This chapter is written from the perspective of diffusion-weighted MRI as a field working on a set of challenging modeling problems, focusing on the potential utility of stochastic geometry in addressing them. Our aim is not to provide an exhaustive overview of relevant theory in either field, but simply to highlight some well-known results where the conceptual links between the two become apparent. A comprehensive introduction to stochastic geometry can be found in the book by Stoyan et al. [29], while the books by Jones [16] and by Johansen-Berg and Behrens [15] provide a good entryway to diffusion-weighted MRI.

To initiate an exchange of ideas between these two fields, the first *Workshop on Diffusion MRI and Stochastic Geometry* was co-organized by the authors and Eva B. Vedel Jensen at Sandbjerg Estate (Denmark, January 20–24, 2019). The observations presented in this chapter partly inspired this workshop, but the chapter likewise builds on the valuable discussions held at the lively and interactive workshop. We are not aware of any previous works pointing out the parallels described here, although some of them will undoubtedly have been noticed before.

Overview

The molecular dynamics relevant for diffusion-weighted MRI are conveniently described by a displacement probability density function—the *diffusion propagator* $P(r, t)$. The propagator $P(r, t)$ represents the probability of a displacement r at a *diffusion time* t, and the diffusion-weighted signal is related to the characteristic function S of the propagator given by

$$S(q) = \int e^{-iqr} P(r, t)\, dr, \tag{1}$$

where qr denotes the inner product between q and r. The Fourier parameter q can be considered an experimental parameter, determined in practice by the diffusion-sensitizing gradients of the acquisition.

In the following sections we give examples of connections to stochastic geometry for three different 'limiting regimes' of the diffusion-weighted signal, where expressions for the relevant molecular dynamics can be simplified significantly. Sections 2 and 3, which feature the short and long diffusion time limits respectively, can be well-understood in terms of the time-dependent *diffusion coefficient* $D(t)$, which represents the mean squared molecular displacement at a time t. This diffusion coefficient appears as the first non-trivial coefficient in the Taylor series of $\log S$, the cumulant expansion.[1]

$$\log S(q) \simeq -D(t)\, q^2 + \dots \tag{2}$$

From this expansion it can already be gathered that results based on the diffusion coefficient are mostly applicable when q is relatively small—i.e., when the gradients are weak enough for the D-dependent term in this expansion to dominate—and in Sect. 4 we consider instead the limit where the gradient strength parameter $\sim q$ becomes large. In this regime the diffusion coefficient no longer provides an adequate vehicle for the description and analysis of the diffusion process, and we have to rely on other descriptors. Where necessary, additional details on these concepts will be given in the text, although technicalities will be skipped in favor of accessibility. To simplify the exposition further, we will make implicit use of the narrow

---

[1] We restrict ourselves in Eq. (2) to the one-dimensional case, to keep technicalities to a minimum.

pulse approximation [6, 27] throughout. The paper concludes with an outlook on the possible future uses of stochastic geometry in Sect. 5.

2   Specific Volumes and the Short-Time Limit

...the time-dependent diffusion coefficient $D(t)$ of mobile molecules confined in pores or cells carries information about the confining geometry. At early times, [a perturbative expansion of $D(t)$] gives, irrespective of details, the pore surface to volume ratio ..., which is a measure of microscopic length.

Sen (2004)

The first regime we consider is the *short-time limit*, described in the seminal works by Mitra, Sen, and others [18, 19, 26], and with important insights dating back to Kac [17]. The situation analyzed in these works is the diffusion of molecules in the neighborhood of obstructive geometrical features, Fig. 1. At the boundaries between the diffusive medium and the geometry, the diffusion is prescribed by boundary conditions that can depend on e.g. the *permeability*, leaving the spatially averaged, time-dependent diffusion coefficient $D(t)$ to be solved. While this is a very challenging problem for any non-zero, finite time t, the limiting behavior for $t \to 0$ can be expressed in terms of a relatively small set of practically useful structural parameters. As $D(t)$ can be measured in the scanner, we can use diffusion-weighted measure-

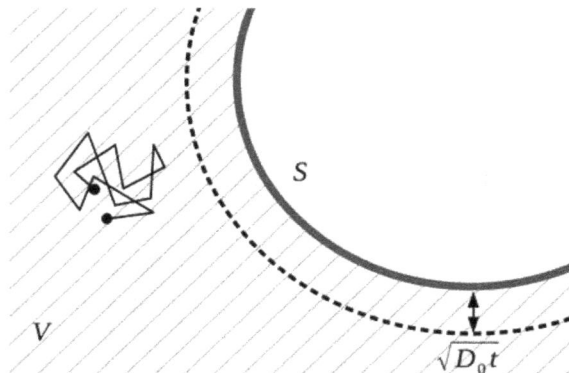

Fig. 1 A schematic showing a particle (black solid line) diffusing in a volume V, marked by the blue hatch pattern. The gray boundary S acts as a barrier to the diffusion, resulting in a decreased time-dependent diffusion coefficient $D(t)$ compared to the free diffusion coefficient D_0. At short times only particles in a small neighborhood of the boundary (within a distance $\sim \sqrt{D_0 t}$) are significantly impacted by this effect, resulting in a characteristic decrease of $D(t)$ proportional to the ratio between the surface area S and the volume V. Note that the time-dependent diffusion coefficient measured in diffusion-weighted MRI is acquired by averaging over volumes much larger than the typical diffusion length $\sqrt{D_0 t}$

ments to obtain estimates of these parameters. The emergence of these structural parameters in the diffusion-weighted signal can be understood as follows.

In the limit $t \to 0$ the diffusing molecules do not have enough time to interact with the geometry, and the diffusion coefficient naturally approaches the medium's free diffusion coefficient $D(0) = D_0$—the diffusion coefficient of the medium in the absence of any geometry. At times close to the limit, the particles move a typical distance in the order of $\sqrt{D_0 t}$ (by virtue of the definition of D in terms of the mean squared displacement), and so roughly speaking only the fraction of particles within some distance $\sim \sqrt{D_0 t}$ can 'see' the geometry. As the geometry essentially acts as a barrier to the diffusion, the time-dependent diffusion coefficient decreases from its free diffusion limit, and this decrease is more significant if a larger fraction of the total number of particles can interact with the boundaries. For diffusion times approaching 0, this fraction becomes exactly proportional to the surface area of the geometry. Formalizing this notion, Mitra et al. [19] then showed that the first order correction describing the approach of the limit becomes proportional to the *surface-to-volume ratio* S/V of a smooth geometry, cf. Fig. 1, according to

$$D(t) \sim D_0 - \frac{4}{3 d \sqrt{\pi}} \frac{S}{V} \sqrt{D_0 t} \, D_0 \quad (t \to 0), \tag{3}$$

where d is the spatial dimension. From this relation, the surface-to-volume ratio can be estimated reliably from the diffusion-weighted measurements [13]. The order t terms in this expansion depend to varying degrees on the boundary condition parameters such as the permeability, as well as on the average curvature of the geometry, while higher order terms depend on even more intricate details of the environment.

Although the complex interactions between particles undergoing random displacements and their surrounding structures have not been considered in stochastic geometry as such—this problem is closer to mathematical physics [5]—the surface-to-volume ratio uncovered in Eq. (3) is a commonly estimated quantity in for example stereology. The surface-to-volume ratio is also referred to as the *surface density* or as the *specific surface area* in stereology and stochastic geometry. An example of a question in this scenario for which stochastic geometry could be helpful is: "what is the smallest observation window (voxel) that has an acceptable error when estimating a given characteristic?" [29, Sect. 6.4.6].

3   Stationarity and the Long-Time Limit

That is, we assume here that the voxel is *statistically homogeneous*. This macroscopic uniformity allows us to go from averaging over the contributions from all parts of the system ... to ensemble averaging over all disorder realizations, leading to the description of the signal in terms of the statistical properties embodied in the correlators ...

Novikov and Kiselev (2010)

In the short-time limit of the previous section, particles could not explore their surroundings outside a vanishingly small window, leading to the simple relation between the surface-to-volume ratio and the time-dependent diffusion coefficient. In the *long-time limit*, on the other hand, we can assume that particles in fact see *all* the surrounding structures. Under fairly general conditions, the diffusion process in this limit can still be described in terms of a single diffusion coefficient—the limiting coefficient $D_\infty = \lim_{t \to \infty} D(t)$. While determining D_∞ for a given sample is very difficult, Novikov et al. [21, 23, 24] showed that once again there are basic geometrical properties of the structure that completely determine the limiting behavior of $D(t)$. In this case, the diffusion coefficient approaches its limit as

$$D(t) \sim D_\infty + c\,t^{-\vartheta} \quad (t \to \infty), \tag{4}$$

where c is a constant, and where the exponent $\vartheta = (p+d)/2$ is determined by the spatial dimension d and a *structural exponent* p. The structural exponent p is the exponent of the limiting power law behavior of the structure correlation function, which roughly speaking quantifies the presence of long-range correlations in the diffusion barriers. While the constant c varies significantly between different configurations of barriers, we know of only a few possible values for the exponent p, with many vastly different configurations of barriers producing the same critical exponent in experiments. The structural exponent p effectively distinguishes between different *structural universality classes* [23].

A central concept in the cited works by Novikov et al. is that *because* the diffusing particles can explore their entire surroundings, the exact configuration of the barriers becomes irrelevant. Consequently, one can theoretically replace the complicated original structure with a simpler-to-analyze *effective medium*, permitting the derivation of e.g. Eq. (4). As explained in the quote above, the practical application of this simplifying methodology requires that we assume that the subsets of a sample explored by different particles can be viewed as random samples ('disorder realizations') generated by some basic statistical properties of the global structure. In the context of stochastic geometry, this same assumption is more commonly called (spatial) *stationarity*, and it is a key assumption in many classical proofs in the field. In particular, the assumption of stationarity enabled the first proofs of the *fundamental formulas of stereology*. In a number of stochastic geometry results it is now known that a weaker 'first order stationarity' assumption is sufficient [1], and it might be interesting to see if the same is possible for the long-time limit results discussed here.

4   Directional Measures and the Strong-Gradient Limit

The concept of disordered media and statistical averaging can be particularly valuable to deal with the geometric complexity of biological tissues. We believe that further progress in the field can be achieved by merging microscopic geometric models, statistical [descriptions] of disordered media, and high-gradient features of the signal formation.

Grebenkov (2016)

The gradient strength parameter set during a diffusion-weighted MRI experiment determines—roughly speaking—the scanner's sensitivity to diffusive motion. The mobility of the water molecules in the sample affects the decay rate of the diffusion-weighted signal, cf. Sect. 1, and stronger gradients make this effect correspondingly stronger. Although stronger gradients thus produce weaker, harder to detect signals, these signals contain information about interesting features of the diffusion that cannot be observed at lower gradient strengths. The results for the short-time and long-time limits described in the previous subsections are mainly used at low to moderate gradient strength acquisitions.

When we increase the gradient strength in a diffusion-weighted MRI experiment we first notice that the *anisotropy*, i.e., the orientation-dependence, becomes more significant. While we omitted this before, anisotropy is already a factor at the lower gradient strength experiments used for the concepts discussed in Sects. 2 and 3. In the general case, the long-time limit D_∞ of $D(t)$ considered in Eq. (3) is, for example, in fact a *tensorial quantity* related to the mean squared molecular displacements along different orientations. This anisotropy in the diffusion reflects the anisotropy in the sample's micro-structure, so for example in the brain it is predominantly determined by the orientation distribution of neurites, cf. Sect. 1. At higher gradient strengths, the orientation-dependence of the diffusion can no longer be represented by a simple tensor, and a more complete set of orientational features becomes accessible.

The neurite orientations are generally characterized by a so-called *fiber orientation distribution function* (ODF), which specifies the likelihood that a neurite is locally tangent to a given orientation, and the estimation of this object from strong-gradient experiments is a common problem considered in diffusion-weighted MRI. A large number of techniques have been proposed to deal with this question [14, 31, 32], but we will not discuss them in detail here. Instead we will point out that the ODF also has a stochastic geometrical analogue: the *rose of directions* [29]. The rose of directions is defined for *fiber processes*—a well-defined (*locally finite*) collection of randomly placed fibers. The rose of directions is the distribution of the direction tangent to a typical point on a fiber, where the meaning of a 'typical point' can be made explicit using *Palm distribution theory*. A stationary fiber process is an example of a stochastic geometrical model in the spirit of the work by Novikov et al. [21] that could be useful to model biological tissues. Furthermore, the rose of directions has a dual—the *rose of intersections*. These two objects are related through the *Funk–Radon* integral transform [9], which also appears naturally in for example the diffusion-weighted signal expression for narrow cylinders (the 'fiber ball' ODF) [14].

At even stronger gradients we move to another limiting regime—the strong-gradient limit, or *localization regime* [11]. Here we can still make use of the intuition developed in Sect. 2: any structure in the sample will act as a barrier for particles in its vicinity, and thus slow down the average motion in that region. Slower diffusion in turn implies less signal decay, and so the main contributions to the signal now come from particles *localized* near the barriers [28]. The localization effect carries non-trivial consequences for the diffusion time-dependence of the signal, but for this chapter we are more interested in the orientation dependence in this regime.

A simple way to understand the impact of the geometry's anisotropy on the signal in the strong-gradient limit, is to look at a diffusion propagator with a *compact support* Ω. At strong gradients the integrand in Eq. (1) oscillates rapidly, and as a result the integral indeed tends to 0. However, the rate at which it decays is in fact governed by the support Ω, which is of course determined by the geometry. We can consider as an illustration the 'indicator' propagator which is 1 everywhere inside a sufficiently smooth and convex Ω, and 0 elsewhere. The signal decay is then exponential with an exponent $\sup_{r \in \Omega} qr$ [12], where the exponent, viewed as a function in q, determines Ω completely. We could for example use this observation to recover the shape of a convex pore enveloping a diffusing substance, provided that the pore is much smaller than the typical diffusion length $\sqrt{D_0 t}$ [6], but this approach is not limited to closed pores.

The function $q \mapsto \sup_{r \in \Omega} qr$ is called the *support function* of Ω in stochastic geometry, where it can be associated with the rose of directions of a fiber process [29]. A similar interpretation can be given to the support function as it occurs here, leading to the definition of the *barrier ODF* [7, 8]. It must be mentioned that as these developments are very recent, their practical utility is currently still being investigated.

5   Perspectives

We have given three examples of concepts in diffusion-weighted MRI that also occur in stochastic geometry: the surface-to-volume ratio, which is related to the specific surface area commonly estimated in stereology; statistical homogeneity, which is related to stationarity; and the fiber orientation distribution function, whose stochastic geometry analogue is the rose of directions. In the context of neuroimaging, the common thread between them appears to be the concept of stationary fiber and surface processes from stochastic geometry, which are completely characterized by the specific length/area and the rose of directions.

We believe further investigations in this direction could lead to new and useful methods for the analysis of diffusion-weighted MRI data, as well as novel research problems in stochastic geometry. In particular, we hope that the link to the mature statistical theory found in stochastic geometry may offer practical tools for the analysis of uncertainty and variation in applications of diffusion weighted MRI: How much faith can we put in estimated anatomical quantities? Which quantities can we expect to be able to derive from given data? What are the conditions we should put on our data acquisition in order to be able to draw sound conclusions for the hypotheses that we ultimately hope to investigate?

Acknowledgments The authors thank the Villum Foundation for funding this work through a block stipendium and a Villum Experiment grant, as well as for funding the Centre for Stochastic Geometry and Advanced Bioimaging (CSGB) which fostered 10 years of interactions between the authors and experts in stochastic geometry, including the 2019 Workshop on Diffusion MRI and

Stochastic Geometry. In particular, the authors wish to thank Eva B. Vedel Jensen and Markus Kiderlen, for their engagement, input, and encouragement in our ideas for collaboration, as well as their kind patience with us when beginning to engage with stochastic geometry. We are also indebted to Andrea Fuster and Luc Florack for their contributions to the PhD project of Tom Dela Haije, which included the introduction of the barrier ODF. The authors would like to acknowledge Schloss Dagstuhl and the organizers and participants of Dagstuhl seminars 16142 and 18442 for facilitating discussions that further supported developing the ideas discussed in this work. Finally, the authors wish to thank all the participants in the 2019 Workshop on Diffusion MRI and Stochastic Geometry.

References

1. Baddeley, A., Jensen, E.B.V.: Stereology for statisticians. CRC Press (2004)
2. Basser, P.J., Pajevic, S., Pierpaoli, C., Duda, J., Aldroubi, A.: In vivo fiber tractography using DT-MRI data. Magn. Reson. Med. **44**(4), 625–632 (2000)
3. Beaulieu, C.: The basis of anisotropic water diffusion in the nervous system–a technical review. NMR Biomedi. **15**(7–8), 435–455 (2002). https://doi.org/10.1002/nbm.782
4. Behrens, T., Berg, H., Jbabdi, S., Rushworth, M., Woolrich, M.: Probabilistic diffusion tractography with multiple fibre orientations: what can we gain? NeuroImage **34**(1), 144–155 (2007)
5. Bogachev, L.: Random walks in random environments. In: Encyclopedia of Mathematical Physics, pp. 353–371. Elsevier (2006). https://doi.org/10.1016/B0-12-512666-2/00063-8
6. Callaghan, P.T., MacGowan, D., Packer, K.J., Zelaya, F.O.: High-resolution q-space imaging in porous structures. J. Magn. Reson. (1969) **90**(1), 177–182 (1990). https://doi.org/10.1016/0022-2364(90)90376-k
7. Dela Haije, T.C.J.: Finsler geometry and diffusion MRI. Ph.D. thesis, Eindhoven University of Technology, Eindhoven (2017)
8. Dela Haije, T.C.J., Fuster, A., Florack, L.M.J.: A new fiber orientation distribution function. In: Proceedings of the 25th Annual Meeting of the ISMRM, p. 3368. Honolulu, HI (2017)
9. Funk, P.: Über Flächen wen geodätischen Linien. Mathematische Annalen **74**(2), 278–300 (1913). https://doi.org/10.1007/BF01456044
10. Grebenkov, D.S.: From the microstructure to diffusion NMR, and Back. In: Valiullin, R. (ed.) New Developments in NMR, pp. 52–110. Royal Society of Chemistry, Cambridge (2016). https://doi.org/10.1039/9781782623779-00052
11. Grebenkov, D.S.: Diffusion MRI/NMR at high gradients: challenges and perspectives. Microporous Mesoporous Mater. **269**, 79–82 (2018). https://doi.org/10.1016/j.micromeso.2017.02.002
12. Herz, C.S.: Fourier transforms related to convex sets. Ann. Math. **75**(1), 81 (1962). https://doi.org/10.2307/1970421
13. Hurlimann, M., Helmer, K., Latour, L., Sotak, C.: Restricted diffusion in sedimentary rocks. determination of surface-area-to-volume ratio and surface relaxivity. J. Magn. Reson. Ser. A **111**(2), 169–178 (1994). https://doi.org/10.1006/jmra.1994.1243
14. Jensen, J.H., Russell Glenn, G., Helpern, J.A.: Fiber ball imaging. NeuroImage **124**, 824–833 (2016). https://doi.org/10.1016/j.neuroimage.2015.09.049
15. Johansen-Berg, H., Behrens, T.E.J.: Diffusion MRI: From quantitative measurement to in-vivo neuroanatomy. Elsevier Science, Amsterdam (2014)
16. Jones, D.K.: Diffusion MRI: Theory, Methods, and Application. Oxford University Press, Oxford, New York (2010)
17. Kac, M.: Can one hear the shape of a drum? Am. Math. Mon. **73**(4), 1 (1966). https://doi.org/10.2307/2313748
18. Mitra, P.P., Sen, P.N., Schwartz, L.M.: Short-time behavior of the diffusion coefficient as a geometrical probe of porous media. Phys. Rev. B **47**(14), 8565–8574 (1993). https://doi.org/10.1103/PhysRevB.47.8565

19. Mitra, P.P., Sen, P.N., Schwartz, L.M., Le Doussal, P.: Diffusion propagator as a probe of the structure of porous media. Phys. Rev. Lett. **68**(24), 3555–3558 (1992). https://doi.org/10.1103/PhysRevLett.68.3555
20. Mori, S., van Zijl, P.C.M.: Fiber tracking: principles and strategies–a technical review. NMR Biomed. **15**(7–8), 468–480 (2002). https://doi.org/10.1002/nbm.781
21. Novikov, D.S., Fieremans, E., Jensen, J.H., Helpern, J.A.: Random walks with barriers. Nat. Phys. **7**(6), 508–514 (2011). https://doi.org/10.1038/nphys1936
22. Novikov, D.S., Fieremans, E., Jespersen, S.N., Kiselev, V.G.: Quantifying brain microstructure with diffusion MRI: theory and parameter estimation. NMR Biomed. **32**(4), e3998 (2019). https://doi.org/10.1002/nbm.3998
23. Novikov, D.S., Jensen, J.H., Helpern, J.A., Fieremans, E.: Revealing mesoscopic structural universality with diffusion. Proc. Natl. Acad. Sci. **111**(14), 5088–5093 (2014). https://doi.org/10.1073/pnas.1316944111
24. Novikov, D.S., Kiselev, V.G.: Effective medium theory of a diffusion-weighted signal. NMR Biomed. **23**(7), 682–697 (2010). https://doi.org/10.1002/nbm.1584
25. Sen, P.N.: Time-dependent diffusion coefficient as a probe of geometry. Concepts Magn. Reson. **23A**(1), 1–21 (2004). https://doi.org/10.1002/cmr.a.20017
26. Sen, P.N., Schwartz, L.M., Mitra, P.P., Halperin, B.I.: Surface relaxation and the long-time diffusion coefficient in porous media: Periodic geometries. Phys. Rev. B **49**(1), 215–225 (1994). https://doi.org/10.1103/PhysRevB.49.215
27. Stejskal, E.O.: Use of spin echoes in a pulsed magnetic-field gradient to study anisotropic, restricted diffusion and flow. J. Chem. Phys. **43**(10), 3597–3603 (1965). https://doi.org/10.1063/1.1696526
28. Stoller, S.D., Happer, W., Dyson, F.J.: Transverse spin relaxation in inhomogeneous magnetic fields. Phys. Rev. A **44**(11), 7459–7477 (1991). https://doi.org/10.1103/PhysRevA.44.7459
29. Stoyan, D., Kendall, W.S., Mecke, J.: Stochastic Geometry and Its Applications, 2nd edn. Wiley series in probability and statistics. Wiley, Chichester, New York (1995)
30. Torrey, H.C.: Bloch equations with diffusion terms. Phys. Rev. **104**(3), 563 (1956)
31. Tournier, J.D., Calamante, F., Connelly, A.: Robust determination of the fibre orientation distribution in diffusion MRI: non-negativity constrained super-resolved spherical deconvolution. NeuroImage **35**(4), 1459–1472 (2007). https://doi.org/10.1016/j.neuroimage.2007.02.016
32. Tuch, D.S.: Q-ball imaging. Magn. Reson. Med. **52**(6), 1358–1372 (2004). https://doi.org/10.1002/mrm.20279

Measuring Anisotropy

12

Anisotropy in the Human Placenta in Pregnancies Complicated by Fetal Growth Restriction

Paddy J. Slator, Alison Ho, Spyros Bakalis, Laurence Jackson, Lucy C. Chappell, Daniel C. Alexander, Joseph V. Hajnal, Mary Rutherford, and Jana Hutter

Abstract The placenta has a unique structure, which enables the transfer of oxygen and nutrients from the mother to the developing fetus. Abnormalities in placental structure are associated with major complications of pregnancy; for instance, changes in the complex branching structures of fetal villous trees are associated with fetal growth restriction. Diffusion MRI has the potential to measure such fine placental microstructural details. Here, we present in-vivo placental diffusion MRI scans from controls and pregnancies complicated by fetal growth restriction. We find that after 30 weeks' gestation fractional anisotropy is significantly higher in placentas associated with growth restricted pregnancies. This shows the potential of diffusion MRI derived measures of anisotropy for assessing placental function during pregnancy.

1 Introduction

Fetal growth restriction (FGR) is a condition where the developing fetus does not reach its full growth potential in-utero [13]. It constitutes a major pregnancy complication and is associated with a high degree of fetal mortality, morbidity and life-long complications [19]. Early onset FGR (defined as that diagnosed before 32 weeks' gestation) affects 0.5–1% of pregnancies and late onset FGR (diagnosed after 32

P. J. Slator (✉) · D. C. Alexander
Centre for Medical Image Computing, Department of Computer Science, University College London, London, UK
e-mail: p.slator@ucl.ac.uk

A. Ho · S. Bakalis · L. C. Chappell
Women's Health Department, King's College London, London, UK

L. Jackson · J. V. Hajnal · M. Rutherford · J. Hutter
Centre for the Developing Brain, School of Biomedical Engineering and Imaging Sciences, King's College London, London, UK
e-mail: jana.hutter@kcl.ac.uk

L. Jackson · J. V. Hajnal · J. Hutter
Biomedical Engineering Department, School of Biomedical Engineering and Imaging Sciences, King's College London, London, UK

weeks') affects 5–10% of pregnancies [6, 9]. Early diagnosis and close monitoring are essential to optimize the outcome for pregnancies affected by FGR. Recently, pharmacological treatments have shown promise for severe early onset FGR cases [31]. However, while routine antenatal monitoring utilising symphysis-fundal height and ultrasound measurements can identify a significant proportion of FGR cases [10], detection remains a major problem—only 31% of cases were diagnosed during a previous study [11]. Detection is crucial as FGR is a leading cause of late unexpected stillbirth [11].

Post-delivery histopathological analysis shows a significant degree of characteristic pathologies in FGR placentas [15, 33], emphasising its importance and involvement in the cascade of events leading ultimately to sub-optimal growth. However, while post-delivery detection or confirmation plays an important role in increasing knowledge and possible causes of FGR, it comes too late to influence clinical care for these pregnancies. Therefore, recent novel developments, mainly using Magnetic Resonance Imaging (MRI) focus on studying the placenta during pregnancy to complement available antenatal screening.

1.1 Placental Microstructure

The placenta constitutes the key connection between mother and baby in utero and acts as the life support system for the growing fetus. Among its many functions are the exchange of oxygen and nutrients from the maternal blood circulation to the fetal blood, and the removal of waste products. The placenta comprises 10–40 individual lobules each constituting one key exchange unit. The transfer relies on a delicate and dynamically evolving microstructure within these units, focused around the fetal villi and depicted in Fig. 1A. These tree-like structures originate from the umbilical cord and contain fetal arteries and veins. They are bathed in maternal blood that enters the intervillous spaces from the spiral arteries at the level of the basal plate. A thin membrane called the syncytiotrophoblast separates maternal and fetal circulation, and allows the transport of oxygen and nutrients through it. Several histopathological features are associated with FGR, including elongated villous trees without the appropriate branching patterns [21] as illustrated schematically in Fig. 1B.

1.2 Placental MRI

Specific challenges of in-utero placental MRI include motion, such as maternal breathing and fetal movements, various air-tissue interfaces such as amniotic fluid, abdominal fat and bowel gas, and the suboptimal position of the imaging coil with respect to the organ of interest, especially for placentas located on the posterior wall of the uterus [5]. Another important limitation specifically for diffusion MRI arises

A   B

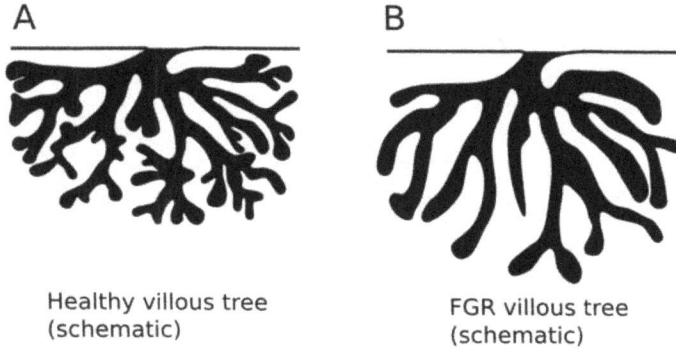

Healthy villous tree   FGR villous tree
(schematic)   (schematic)

Fig. 1 Schematic representation of single fetal villous trees in the placenta. Fetal blood flows through the convoluted branching structures, allowing nutrient exchange with the surrounding maternal blood. Panel **A**: complex branching structure of a healthy placenta. Panel **B**: pathological branching structure, associated with prengancy complications such as fetal growth restriction

from the safety requirement to reduce the acoustic sound level of the sequence to protect the fetal hearing. Therefore, typically the gradient slew rate is reduced, in consequence increasing the read-out length and thus the echo time resulting in lower Signal-to-Noise Ratios (SNR) [5, 16].

Diffusion MRI can reveal fine tissue microstructure details through sensitivity to the diffusion of water. By varying the strength and direction of diffusion gradients, the MRI sequence can be specifically tailored to microstructural features of interest. For example, sequences with a high number of distinct gradient directions can inform on tissue anisotropy and directionality. Such diffusion MRI sequences are commonly used to image white matter fibre tracts in the brain, however applications elsewhere in body exist, including in the placenta [3, 20, 27].

2 Methods

Pregnant women, recruited as part of a larger cohort for the Placental Imaging Project (PIP), underwent an MRI scan between 20 and 40 weeks of gestation. Informed consent was obtained (REC 14/LO/1169) and the scan was performed on a clinical 3T Philips Achieva (Best, The Netherlands) scanner using the 32-channel cardiac coil. All women were scanned in supine position under frequent monitoring of heart rate, saturation and blood pressure throughout the scan. Dedicated padding was provided to increase maternal comfort and verbal interaction was maintained.

After initial structural T2-weighted sequences of the entire uterus and the fetal brain in multiple orientations, a B0 map was acquired. An in-house developed tool for image based shimming was employed [12] to focus the shim on the placental parenchyma, avoiding air-tissue interfaces with bowel gas as much as possible. We next acquired a diffusion-prepared spin echo sequence. For some participants we used a modified sequence which acquires additional gradient-echos after the initial Spin Echo (e.g. [28]), although in this study we only utilised the first spin echo measurement. Conventional Stejskal-Tanner diffusion preparation was performed, with a total of 65 b-value/b-vector combinations, optimized specifically for the placenta as previously described [17, 26], with three diffusion gradient directions at $b = [5, 10, 25, 50, 100, 200, 400, 600, 1200, 1600]$ s/mm^2, eight directions at $b = 18$ s/mm^2, seven at $b = 36$ s/mm^2, fifteen at $b = 800$ s/mm^2, and six $b = 0$ volumes. Further parameters include FOV $= [300–340] \times 320 \times 84$ mm, TR $= 7$ s, SENSE $= 2.5$, halfscan $= 0.6$, resolution $= 3$ mm^3 for scans with additional gradient-echos and 2 mm^3 otherwise. The total acquisition time was 8 min 30 s. Fat was suppressed with SPIR saturation pulses. The acquisition plane was coronal to the mother, chosen to assure that the in-plane direction coincides with the longest placental dimensions in mostly anterior and posterior placentas.

To correct for motion, the diffusion weighted volumes were registered non-rigidly to a common template using the ANTs multivariate template construction tool with the cross-correlation similarity metric [1]. Subsequently, a region of interest (ROI) comprising the whole placenta and adjacent basal placenta was manually segmented in all slices of the first $b = 0$ volume. We estimated diffusion tensor, mean diffusivity (MD), and fractional anisotropy (FA) maps—using all diffusion weightings—for the motion corrected scans using MRTrix [32].

2.1 Recruitment

We include a total of twenty-nine participants in this study, who were categorised as follows. Sixteen women were normal uncomplicated control pregnancies; their outcomes were obtained and checked to ensure that no new diagnosis of pre-eclampsia, gestational hypertension, fetal growth restriction or gestational diabetes had occurred, and their birthweight was greater than the 5th centile (by INTERGROWTH-21st). Seven women were diagnosed with fetal growth restriction, detected from antenatal ultrasound assessments. We also include six women recruited as uncomplicated control pregnancies who gave birth to a baby under the 5th centile, but did not have a formal antenatal FGR diagnosis. Although these could simply be constitutionally small babies, they could also be undiagnosed FGR cases, so we hence analysed them as a separate cohort. Five of the six below the 5th centile had co-morbidities, such as chronic hypertension in pregnancy (CHTN), or pre-eclampsia (PE). The patient population characteristics are given in Table 1.

Table 1 Scan details for all participants. FGR denotes participants diagnosed with fetal growth restriction according to guidelines [13]. Under 5% denotes participants with no FGR diagnosis who delivered a baby weighting under the 5th percentile

Participant ID	Cohort	Gestational age at scan
1	Control	23.86
2	Control	26.14
3	Control	26.72
4	Control	26.72
5	Control	27.14
6	Control	27.14
7	Control	28.29
8	Control	28.86
9	Control	28.86
10	Control	29.67
11	Control	29.86
12	Control	31.29
13	Control	33.43
14	Control	35.57
15	Control	36.29
16	Control	36.43
21	FGR	22.0
22	FGR	23.42
23	FGR	28.57
24	FGR	29.57
25	FGR	30.85
26	FGR	32.85
28 (Scan 1)	FGR + CHTN	30.71
28 (Scan 2)	FGR + PE	34.14
29	Under 5%	21.29
30	Under 5%	25.72
31	Under 5% + CHTN	38.0
32	Under 5% + CHTN	19.86
33	Under 5% + PE	28.71
34 (Scan 1)	Under 5% + PE	31.42
34 (Scan 2)	Under 5% + PE	33.42

3   Results

Figures 2 and 3 display FA maps, ordered by gestational age (GA) at scanning time, for control and growth restricted participants respectively. We next examine the evolution of FA values over gestation (Fig. 4). Finally, we compare FA values

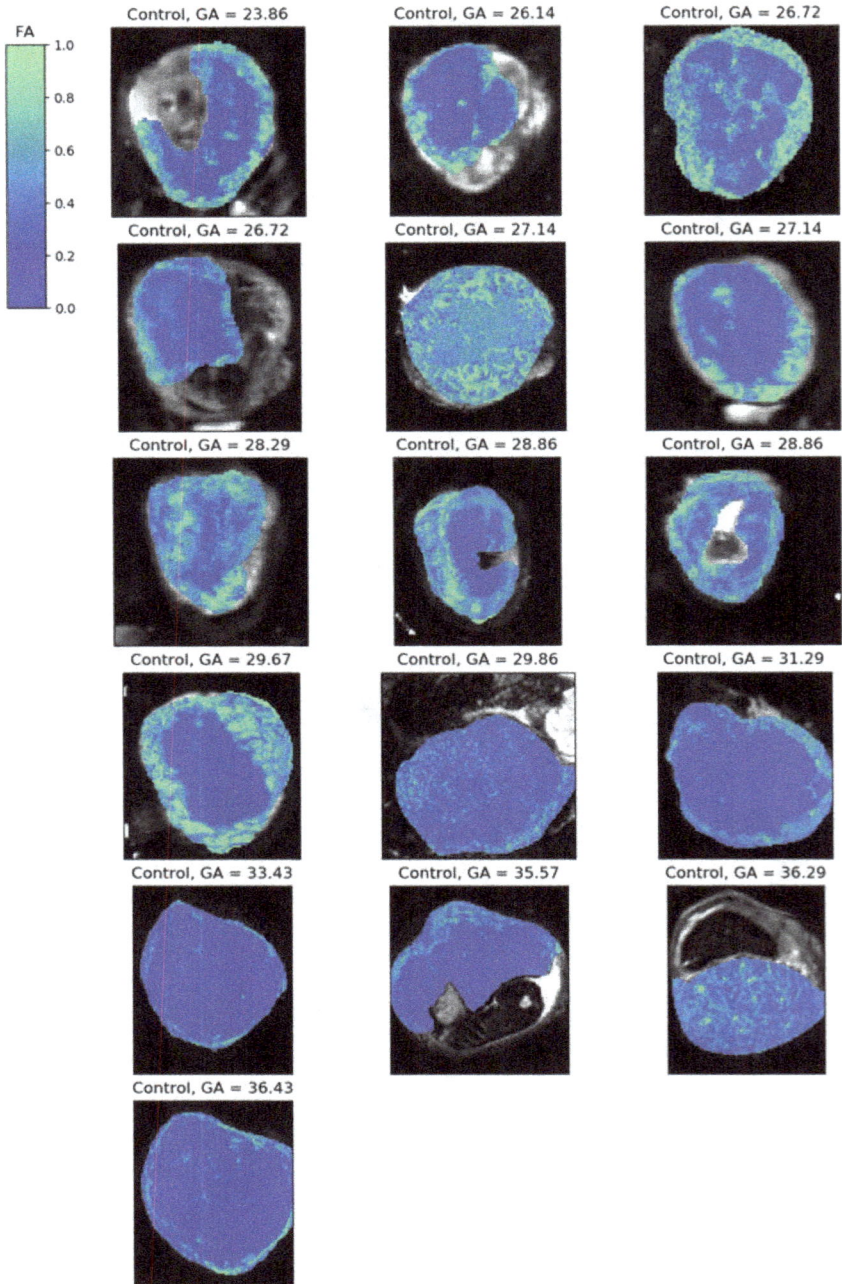

Fig. 2 Fractional anisotropy maps for all control participants. The boundary between the uterine wall (higher FA) and placenta (lower FA) is clear in most placentas with gestational age less than 30 weeks

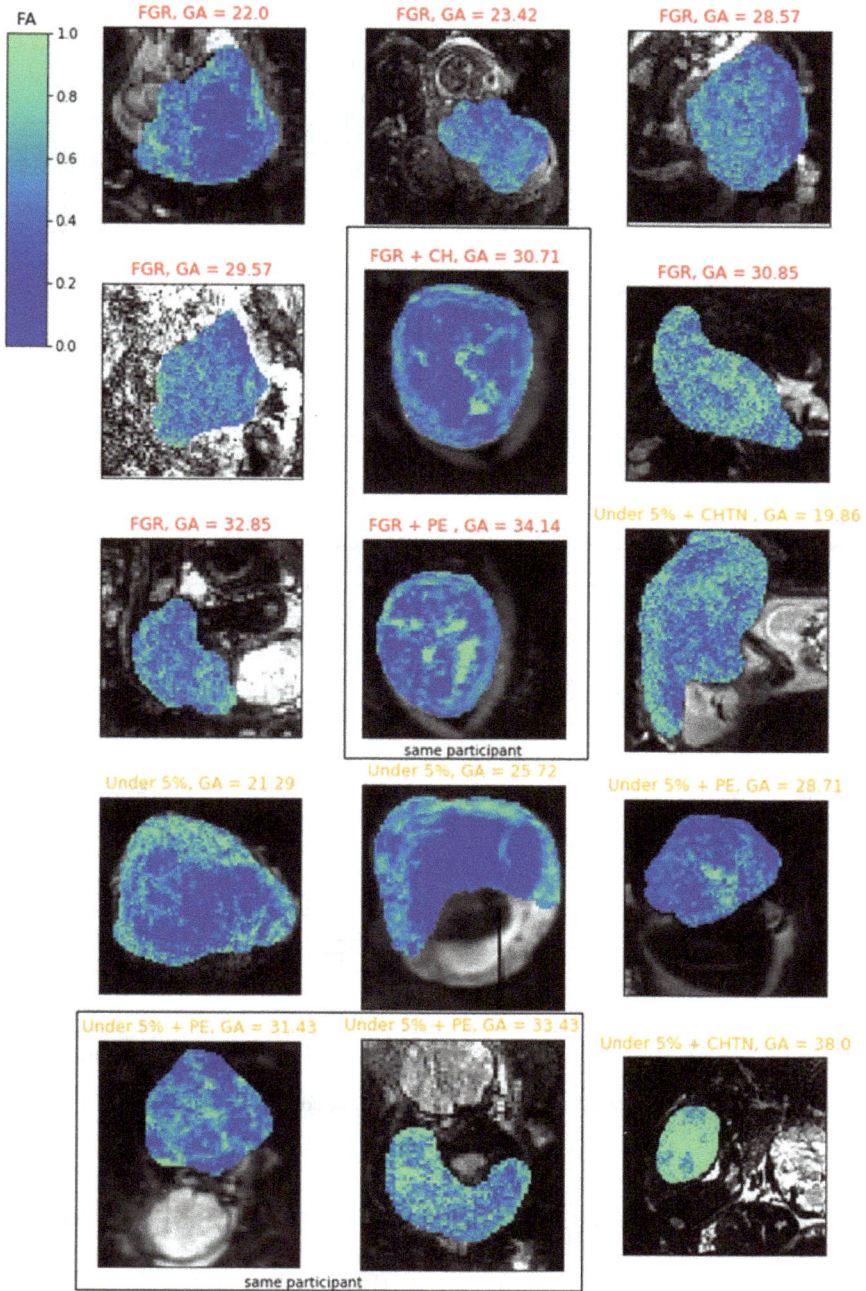

Fig. 3 Fractional anisotropy maps for FGR and low birthweight cohorts

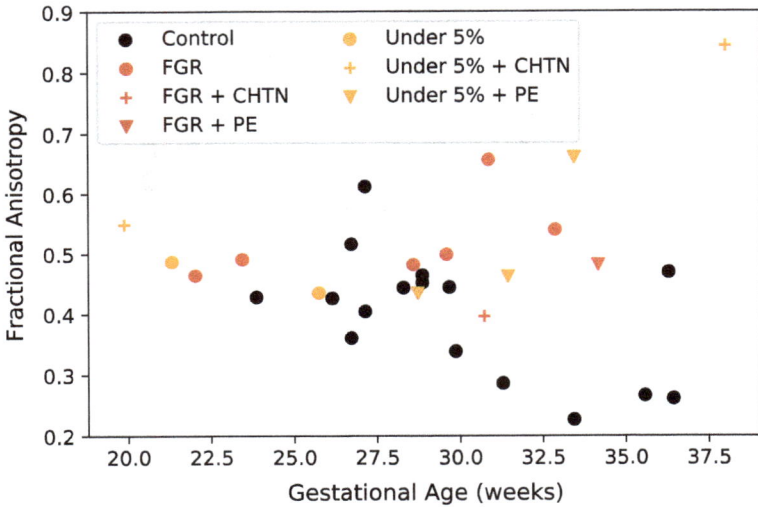

Fig. 4 Mean FA across placenta and uterine wall as a function of gestational age

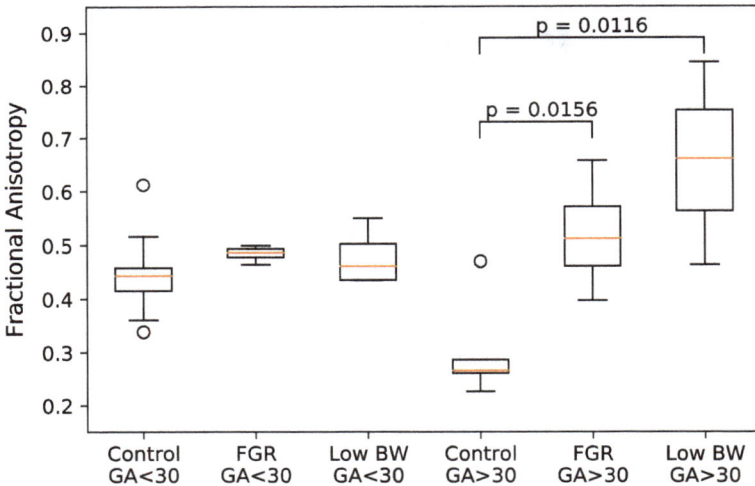

Fig. 5 Mean FA across placenta and uterine wall for the three cohorts, split at 30 weeks gestational age

between control and growth-restricted pregnancies (Fig. 5), assessing the potential of FA to inform on FGR-associated placental abnormalities.

The FA maps for all control participants (Fig. 2) show distinctive patterns in agreement with the literature [27]—the FA is lower in the placenta, and higher in the adjacent uterine wall. These observations likely represent isotropic nature of maternal blood pools and fetal villous tree branching within the placenta, and the prevalence of anisotropic fibrous muscle tissue in the uterine wall. Another factor in the high FA in the uterine wall may be the coherent orientation of vasculature in these areas. On the other hand, the distinction between the uterine wall and placenta is less clear in the growth restriction placentas (Fig. 3), and many of these maps show areas of high FA within the placenta.

Figure 4 plots the mean FA value within the ROI comprising the placenta and uterine wall. Control scans (Fig. 4, black dots) show an apparent decrease in mean FA over gestation, potentially reflecting microstructural changes during normal placenta maturation. On the other hand, growth restriction scans (Fig. 4, red and yellow dots), do not show a clear trend over gestation. At early gestational ages, control and compromised placentas have comparable mean FA values. Due to the downward FA trend in control placentas, from around 30 weeks gestation growth restriction placentas appear to have considerably lower FA values than control placentas. This difference was statisically significant, both when comparing controls with the combined FGR and low birth weight cohorts ($p = 0.005$, independent samples t-test), and when comparing controls to these two cohorts separately (Fig. 5). We tested if trends over GA differ between controls and growth restriction (i.e. FGR and under 5%) cohorts by calculating a linear regression to predict mean FA based on GA, cohort, and the interaction between GA and cohort. The coefficients and p-values are given in Table 2, and Fig. 6 visualises the fits. The coefficient of the interaction term between GA and low birth weight is statistically significant ($p = 0.002$), suggesting a different trend over gestation.

Table 2 Coefficients and corresponding statistics from linear regression to predict mean FA based on GA, cohort, and GA × cohort (interaction term)

Coefficient	Value	Standard error	t	$P > |t|$
Intercept	0.8457	0.193	4.384	0.000
Cohort (FGR)	−0.4541	0.313	−1.449	0.160
Cohort (Under5)	−0.7159	0.259	−2.765	0.011
GA	−0.0150	0.006	−2.328	0.028
GA × Cohort (FGR)	0.0188	0.011	1.769	0.089
GA × Cohort (Under5)	0.0299	0.009	3.412	0.002

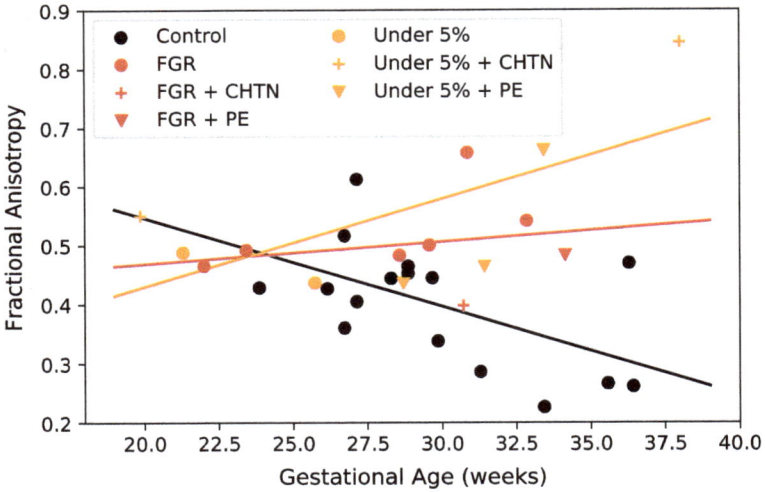

Fig. 6 Linear regression to predict mean FA based on GA, cohort, and GA × cohort

4 Discussion and Conclusion

This study visualizes and quantifies fractional anisotropy in placentas of both control pregnancies and those affected by growth restriction. Encouragingly we find that the FA is significantly different across gestation in growth restricted pregnancies compared to controls. The fractional anisotropy is a coarse measurement that averages over a number of tissue properties. There are hence a number of microstructural or functional changes that could explain this observation. Our initial speculations of plausible factors behind the changes in FA are as follows.

We observed a clear pattern of decreasing FA with gestational age for control placentas, suggesting that normal placental development causes a reduction in coherently orientated tissue. A consistent hypothesis is that this reduction in FA reflects the normal onset of terminal villi formation along the surfaces of intermediate villi, which occurs predominately during the third trimester [2]. On the other hand, we observed higher FA for placentas associated with growth restricted pregnancies after 30 gestational weeks. This is consistent with histological findings that FGR placentas show a lack of side-branching terminal villi [2].

In this study, we utilised an MR protocol with multiple b-values and gradient directions. However, we only fit a relatively simple model to the data, the diffusion tensor. This data can support more complex models, such as anisotropic IVIM-type models, which separately consider the perfusion (i.e. low b-value) and diffusion (high b-value) signal components (e.g. [26]), and may help disentangle the tissue microstructure changes underlying the observed difference in FA. Another potential approach is to use orientation distribution functions, which might reveal differences in complexity of the orientation of villi (e.g. [23]).

There are a number of MRI-derived biomarkers that show promise for detecting placental dysfunction. These include: T2* relaxometry [18, 24, 25], which relates to oxygenation levels; structural measures [7]; apparent diffusion coefficient (ADC) [4, 14, 30], relating to global tissue structure; and intravoxel incoherent motion (IVIM) MRI perfusion fraction [8, 22, 29], which relates to blood flow. Thus far, dMRI-derived potential biomarkers (ADC and perfusion fraction) have considered tissue to be isotropic. Our results strongly suggest that the FA is also sensitive to placental dysfunction, and hence that quantifying tissue anisotropy is an additional important avenue for assessing placental health. It may be the case that multiple biomarkers need to be combined in order to best assess the health of an individual placenta. The fact that we see higher FA values for placentas with birth weight under the 5th percentile, as well as those diagnosed with FGR is interesting and merits further investigation. It is likely that a significant proportion of cases under the 5th percentile are undiagnosed FGR. Our results suggest that quantifying tissue anisotropy in the placenta could have a role to play in the detection of FGR, We will investigate this by combining further scanning with post-delivery placental histology to test the ability to distinguish FGR cases from small but otherwise healthy babies.

Acknowledgments We thank all mothers, midwives, obstetricians, and radiographers who played a key role in obtaining the datasets. This work was supported by the NIH Human Placenta Project grant 1U01HD087202-01 (Placenta Imaging Project [PiP]); Wellcome Trust (201374/Z/16/Z); EPSRC (N018702, M020533, EP/N018702/1); NIHR (RP-2014-05-019); the Wellcome EPSRC Centre for Medical Engineering at Kings College London (WT 203148/Z/16/Z) and by the National Institute for Health Research (NIHR) Biomedical Research Centre based at Guy's and St Thomas' NHS Foundation Trust and Kings' College London. The views expressed are those of the authors and not necessarily those of the NHS, the NIHR or the Department of Health.

References

1. Avants, B.B., Tustison, N.J., Song, G., Cook, P.A., Klein, A., Gee, J.C.: A reproducible evaluation of ANTs similarity metric performance in brain image registration. NeuroImage **54**(3), 2033–2044 (2011). https://doi.org/10.1016/j.neuroimage.2010.09.025
2. Benirschke, K., Burton, G.J., Baergen, R.N.: Pathology of the Human Placenta. Springer, Heidelberg (2012). https://doi.org/10.1007/978-3-642-23941-0
3. Bonel, H.M., Stolz, B., Diedrichsen, L., Frei, K., Saar, B., Tutschek, B., Raio, L., Surbek, D., Srivastav, S., Nelle, M., Slotboom, J., Wiest, R.: Diffusion-weighted MR imaging of the placenta in fetuses with placental insufficiency. Radiology **257**(3), 810–9 (2010). https://doi.org/10.1148/radiol.10092283
4. Bonel, H.M., Stolz, B., Diedrichsen, L., Frei, K., Saar, B., Tutschek, B., Raio, L., Surbek, D., Srivastav, S., Nelle, M., Slotboom, J., Wiest, R.: Diffusion-weighted MR Imaging of the Placenta in Fetuses with Placental Insufficiency. Radiology **257**(3), 810–819 (2010). https://doi.org/10.1148/radiol.10092283
5. Christiaens, D., Slator, P.J., Cordero-Grande, L., Price, A.N., Deprez, M., Alexander, D.C., Rutherford, M., Hajnal, J.V., Hutter, J., : In Utero Diffusion MRI. Top. Magn. Reson. Imaging **28**(5), 255–264 (2019). https://doi.org/10.1097/RMR.0000000000000211
6. Crovetto, F., Triunfo, S., Crispi, F., Rodriguez-Sureda, V., Roma, E., Dominguez, C., Gratacos, E., Figueras, F.: First-trimester screening with specific algorithms for early- and late-onset fetal growth restriction. Ultrasound Obstet. Gynecol. **48**(3), 340–348 (2016). https://doi.org/10.1002/uog.15879

7. Damodaram, M., Story, L., Eixarch, E., Patel, A., McGuinness, A., Allsop, J., Wyatt-Ashmead, J., Kumar, S., Rutherford, M.: Placental MRI in intrauterine fetal growth restriction. Placenta **31**(6), 491–8 (2010). https://doi.org/10.1016/j.placenta.2010.03.001

8. Derwig, I., Lythgoe, D., Barker, G., Poon, L., Gowland, P., Yeung, R., Zelaya, F., Nicolaides, K.: Association of placental perfusion, as assessed by magnetic resonance imaging and uterine artery Doppler ultrasound, and its relationship to pregnancy outcome. Placenta **34**(10), 885–891 (2013). https://doi.org/10.1016/j.placenta.2013.07.006

9. Figueras, F., Caradeux, J., Crispi, F., Eixarch, E., Peguero, A., Gratacos, E.: Diagnosis and surveillance of late-onset fetal growth restriction. Am. J. Obstet. Gynecol. **218**(2), S790–S802.e1 (2018). https://doi.org/10.1016/j.ajog.2017.12.003

10. Gardosi, J., Kady, S.M., McGeown, P., Francis, A., Tonks, A.: Classification of stillbirth by relevant condition at death (ReCoDe): population based cohort study. BMJ (Clinical research ed.) **331**(7525), 1113–7 (2005). https://doi.org/10.1136/bmj.38629.587639.7C

11. Gardosi, J., Madurasinghe, V., Williams, M., Malik, A., Francis, A.: Maternal and Fetal Risk Factors for Stillbirth: population based study. BMJ: Br. Med. J. **346**, f108 (2013). https://doi.org/10.1136/bmj.f108

12. Gaspar, A.S., Nunes, R.G., Ferrazzi, G., Hughes, E.J., Hutter, J., Malik, S.J., McCabe, L., Baruteau, K.P., Rutherford, M.A., Hajnal, J.V., Price, A.N.: Optimizing maternal fat suppression with constrained image-based shimming in fetal MR. Magn. Reson. Med. (2018). https://doi.org/10.1002/mrm.27375

13. Gordijn, S.J., Beune, I.M., Thilaganathan, B., Papageorghiou, A., Baschat, A.A., Baker, P.N., Silver, R.M., Wynia, K., Ganzevoort, W.: Consensus definition of fetal growth restriction: a Delphi procedure. Ultrasound Obstet. Gynecol. **48**(3), 333–339 (2016). https://doi.org/10.1002/uog.15884

14. Gorkem, S.B., Coskun, A., Eslik, M., Kutuk, M.S., Ozturk, A.: Diffusion-weighted imaging of placenta in intrauterine growth restriction with worsening Doppler US findings. Diagn. Interv. Radiol. **25**(4), 280–284 (2019). https://doi.org/10.5152/dir.2019.18358

15. Haeussner, E., Schmitz, C., Frank, H.G., Edler von Koch, F.: Novel 3D light microscopic analysis of IUGR placentas points to a morphological correlate of compensated ischemic placental disease in humans. Sci. Rep. **6**(April), 24004 (2016). https://doi.org/10.1038/srep24004

16. Hutter, J., Harteveld, A.A., Jackson, L.H., Franklin, S., Bos, C., van Osch, M.J.P., O'Muircheartaigh, J., Ho, A., Chappell, L., Hajnal, J.V., Rutherford, M., De Vita, E.: Perfusion and apparent oxygenation in the human placenta (PERFOX). Magn. Reson. Med. **83**(2), 549–560 (2019). https://doi.org/10.1002/mrm.27950

17. Hutter, J., Slator, P.J., Jackson, L., Gomes, A.D.S., Ho, A., Story, L., O'Muircheartaigh, J., Teixeira, R.P., Chappell, L.C., Alexander, D.C., Rutherford, M.A., Hajnal, J.V.: Multi-modal functional MRI to explore placental function over gestation. Magn. Reson. Med. **81**(2), 1191–1204 (2019). https://doi.org/10.1002/mrm.27447

18. Ingram, E., Morris, D., Naish, J., Myers, J., Johnstone, E.: MR Imaging Measurements of Altered Placental Oxygenation in Pregnancies Complicated by Fetal Growth Restriction. Radiology **285**(3), 953–960 (2017). https://doi.org/10.1148/radiol.2017162385

19. Malhotra, A., Allison, B.J., Castillo-Melendez, M., Jenkin, G., Polglase, G.R., Miller, S.L.: Neonatal Morbidities of Fetal Growth Restriction: Pathophysiology and Impact. Front. Endocrinol. **10**, (2019). https://doi.org/10.3389/fendo.2019.00055

20. Melbourne, A., Aughwane, R., Sokolska, M., Owen, D., Kendall, G., Flouri, D., Bainbridge, A., Atkinson, D., Deprest, J., Vercauteren, T., David, A., Ourselin, S.: Separating fetal and maternal placenta circulations using multiparametric MRI. Magn. Reson. Med. **81**(1), 350–361 (2018). https://doi.org/10.1002/mrm.27406

21. Mifsud, W., Sebire, N.J.: Placental pathology in early-onset and late-onset fetal growth restriction. Fetal Diagn. Therapy **36**(2), 117–28 (2014). https://doi.org/10.1159/000359969

22. Moore, R., Strachan, B., Tyler, D., Duncan, K., Baker, P., Worthington, B., Johnson, I., Gowland, P.: In utero Perfusing Fraction Maps in Normal and Growth Restricted Pregnancy Measured Using IVIM Echo-Planar MRI. Placenta **21**(7), 726–732 (2000). https://doi.org/10.1053/plac.2000.0567

23. Seunarine, K.K., Alexander, D.C.: Multiple fibers. Beyond the diffusion tensor. In: Diffusion MRI: From Quantitative Measurement to In vivo Neuroanatomy, 2nd edn., pp. 105–123. Elsevier (2013). https://doi.org/10.1016/B978-0-12-396460-1.00006-8

24. Sinding, M., Peters, D.A., Frøkjaer, J.B., Christiansen, O.B., Petersen, A., Uldbjerg, N., Sørensen, A.: Placental magnetic resonance imaging T2* measurements in normal pregnancies and in those complicated by fetal growth restriction. Ultrasound Obstet. Gynecol. **47**(6), 748–754 (2016). https://doi.org/10.1002/uog.14917

25. Sinding, M., Peters, D.A., Poulsen, S.S., Frøkjær, J.B., Christiansen, O.B., Petersen, A., Uldbjerg, N., Sørensen, A.: Placental baseline conditions modulate the hyperoxic BOLD-MRI response. Placenta **61**, 17–23 (2018). https://doi.org/10.1016/J.PLACENTA.2017.11.002

26. Slator, P.J., Hutter, J., Ianuş, A., Panagiotaki, E., Rutherford, M., Hajnal, J.V., Alexander, D.C.: A framework for calculating time-efficient diffusion MRI protocols for anisotropic IVIM and an application in the placenta. In: Bonet-Carne, E., Grussu, F., Ning, L., Sepehrband, F., Tax, C. (eds.) Computational Diffusion MRI. Springer, Berlin (2019)

27. Slator, P.J., Hutter, J., McCabe, L., Gomes, A.D.S., Price, A.N., Panagiotaki, E., Rutherford, M.A., Hajnal, J.V., Alexander, D.C.: Placenta microstructure and microcirculation imaging with diffusion MRI. Magn. Reson. Med. **80**(2), 756–766 (2018). https://doi.org/10.1002/mrm.27036

28. Slator, P.J., Hutter, J., Palombo, M., Jackson, L.H., Ho, A., Panagiotaki, E., Chappell, L.C., Rutherford, M.A., Hajnal, J.V., Alexander, D.C.: Combined diffusion-relaxometry MRI to identify dysfunction in the human placenta. Magn. Reson. Med. **82**(1), 95–106 (2019). https://doi.org/10.1002/mrm.27733

29. Sohlberg, S., Mulic-Lutvica, A., Olovsson, M., Weis, J., Axelsson, O., Wikström, J., Wikström, A.K.: Magnetic resonance imaging-estimated placental perfusion in fetal growth assessment. Ultrasound Obstet. Gynecol. **46**(6), 700–705 (2015). https://doi.org/10.1002/uog.14786

30. Song, F., Wu, W., Qian, Z., Zhang, G., Cheng, Y.: Assessment of the Placenta in Intrauterine Growth Restriction by Diffusion-Weighted Imaging and Proton Magnetic Resonance Spectroscopy. Reprod. Sci. **24**(4), 575–581 (2017). https://doi.org/10.1177/1933719116667219

31. Spencer, R., Ambler, G., Brodszki, J., Diemert, A., Figueras, F., Gratacós, E., Hansson, S.R., Hecher, K., Huertas-Ceballos, A., Marlow, N., Marsál, K., Morsing, E., Peebles, D., Rossi, C., Sebire, N.J., Timms, J.F., David, A.L.: EVERREST Consortium: EVERREST prospective study: a 6-year prospective study to define the clinical and biological characteristics of pregnancies affected by severe early onset fetal growth restriction. BMC Pregnancy Childbirth **17**(1), 43 (2017). https://doi.org/10.1186/s12884-017-1226-7

32. Tournier, J.D., Smith, R., Raffelt, D., Tabbara, R., Dhollander, T., Pietsch, M., Christiaens, D., Jeurissen, B., Yeh, C.H., Connelly, A.: MRtrix3: a fast, flexible and open software framework for medical image processing and visualisation. NeuroImage **202**, 116137 (2019). https://doi.org/10.1016/j.neuroimage.2019.116137

33. Vedmedovska, N., Rezeberga, D., Teibe, U., Melderis, I., Donders, G.G.G.: Placental pathology in fetal growth restriction. Eur. J. Obstet. Gynecol. Reprod. Biol. **155**(1), 36–40 (2011). https://doi.org/10.1016/j.ejogrb.2010.11.017

13

Magnetic Resonance Imaging of T_2- and Diffusion Anisotropy using a Tiltable Receive Coil

Chantal M. W. Tax, Elena Kleban, Muhamed Baraković, Maxime Chamberland, and Derek K. Jones

Abstract The anisotropic microstructure of white matter is reflected in various MRI contrasts. Transverse relaxation rates can be probed as a function of fibre-orientation with respect to the main magnetic field, while diffusion properties are probed as a function of fibre-orientation with respect to an encoding gradient. While the latter is easy to obtain by varying the orientation of the gradient, as the magnetic field is fixed, obtaining the former requires re-orienting the head. In this work we deployed a tiltable RF-coil to study T_2- and diffusional anisotropy of the brain white matter simultaneously in diffusion-T_2 correlation experiments.

C. M. W. Tan and E. Kleban share first authorship.

C. M. W. Tax (✉) · E. Kleban · M. Chamberland · D. K. Jones
Cardiff University Brain Research Imaging Centre (CUBRIC), Cardiff University, Cardiff, UK
e-mail: TaxC@cardiff.ac.uk

E. Kleban
e-mail: KlebanE@cardiff.ac.uk

M. Chamberland
e-mail: ChamberlandM@cardiff.ac.uk

D. K. Jones
e-mail: JonesD27@cardiff.ac.uk

M. Baraković
Cardiff University Brain Research Imaging Centre (CUBRIC), Cardiff University, Cardiff, UK
e-mail: muhamed.barakovic@epfl.ch

Signal Processing Laboratory 5, Ecole Polytechnique Federale de Lausanne, Lausanne, Switzerland

Translational Imaging in Neurology Basel, Department of Biomedical Engineering, University Hospital Basel, Basel, Switzerland

1 Introduction

1.1 Background

Magnetic Resonance Imaging (MRI) allows us to probe structural anisotropy of tissue *in vivo* by studying the magnetic properties and translational motion of for instance hydrogen protons. In the human body, hydrogen is naturally abundant in various compounds, with the highest MR signal amplitude detected from hydrogen in water molecules. Hydrogen protons possess spin angular momentum, which is an intrinsic quantum property that allows the occurrence of magnetic interactions and resonance. When placed in a magnetic field, there is a slight preference for spins to be aligned with the field, resulting in a net magnetisation **M** aligned with the main magnetic field $\mathbf{B}_0$ [1]. Upon the application of a radiofrequency field at the Larmor frequency $\mathbf{B}_1$, **M** can be tipped out of alignment with $\mathbf{B}_0$; most commonly into the perpendicular plane. The measured signal is the result of the ensemble of spins precessing coherently in the plane, and signal loss (decay) occurs when such coherence reduces.

This signal decay, which results from a progressive loss of coherence of precessional phase, i.e. *dephasing*, is also called spin-spin relaxation. The spin-spin relaxation rate is usually denoted by $R_2 = 1/T_2$, where T_2 is the time taken for the magnetization to decay to $1/e$ of its initial value [2, 3]. In addition to the irreversible spin-spin relaxation, dephasing can be caused by local variations in the magnetic field, a reversible process if the spins are static. Such local variations in the magnetic field arise from a difference in interaction of different substances with the magnetic field (*susceptibility effects*). If the spins experience Brownian molecular motion, they will experience various magnetic field strengths, and their dephasing due to local field changes will be effectively irreversible. Hence, the measured apparent T_2 of signal decay will no longer be purely induced by dephasing due to spin-spin interactions, but will depend on the amplitude and spatial characteristics of the local field variations, and the mean displacement of molecules per unit of time due to incoherent motion.

In diffusion-weighted MRI, additional (and typically much stronger) magnetic field variations are induced intentionally by applying *magnetic field gradients* which cause the strength of the main magnetic field $\mathbf{B}_0$ to vary linearly in space [4–6]. As such, the Brownian molecular motion can be encoded in the signal in a controlled way.

The interplay of susceptibility and diffusion effects leads to the anisotropy of tissue being reflected in different MRI contrasts and hence the combination of contrasts can give a more complete picture of tissue microstructure. In the following paragraphs we will describe anisotropy in the brain and these processes in more detail.

Structural anisotropy of human brain tissue. The dominant tissue exhibiting anisotropy in the brain is the *white matter* (WM). It is predominantly composed of the long extensions of neuronal cells—the axons, which are are grouped into *fibre bundles* and inter-connect different areas of the brain. The main function of axons is to transmit electric impulses between and within brain areas. Axons can be insulated by a *myelin sheath* which is formed of lipid chains and allows a faster signal transmission. The size, density, length of the axons and their myelination levels vary with age and may alter with pathology. Therefore, investigation of the white matter microstructure is of high importance in understanding the functionality of the healthy brain, but also in studying the mechanisms of normal/abnormal development and pathology.

Diffusion effects. Measurements of Brownian molecular motion in biological tissue can reflect not only the temperature and the viscosity of the medium it is occurring in, but most importantly will be sensitive to the underlying geometry and anisotropy of the tissue. For instance, in the brain white matter water molecules can propagate much easier along fibre bundles than perpendicular to them because of obstacles such as the cell membrane and myelin [7]. This property can be used to estimate the main orientation of fibre bundles and their virtual reconstruction by means of *fibre tractography* [8, 9]. Additionally, if water molecules are trapped inside the axon and cannot penetrate the boundaries (restricted diffusion), the mean displacement perpendicular to the axon will be similar to its diameter at long diffusion times. The diffusion of water molecules which reside outside axons is commonly thought of as not being fully restricted but hindered. An example of the differences between the movement of water molecules residing inside and outside of axons is visualised in Fig. 1a.

In MRI, Brownian motion of water molecules is most commonly encoded using a pair of pulsed magnetic field gradients, a dephasing and a rephasing gradient [6] (Fig. 2). If spins are stationary, these gradients would have no additional effect on the signal decay. However if spins change their positions during or between the application of the gradients, the rephasing will be incomplete which will result in signal loss. The signal loss due to diffusion can be enhanced by increasing the magnitude of the gradients, the time during which the gradients are on, and/or the time between the gradients. The strength of the diffusion weighting is described by the b-value; a parameter which combines the information on the diffusion gradient strength and timings.

Magnetic susceptibility effects. Any material placed inside a strong magnetic field interacts with it—it can become magnetised itself. The proportionality constant linking magnetic field strength and the magnetisation induced inside the material is called *magnetic susceptibility*. At the boundaries between materials with different magnetic properties the magnetic flux density is spatially inhomogeneous and its distribution will depend on the boundary orientation to the magnetic field and the difference in susceptibility between the materials. Additionally, the magnetisation induced in some materials may also depend on the orientation of the sample to the magnetic field, i.e. the magnetic susceptibility of those materials is anisotropic.

As mentioned above, the nerve fibres in human brain are insulated by myelin sheath. Myelin is more diamagnetic than water, i.e., it is repelled more strongly by a magnetic field. Additionally, several studies suggested that the magnetic suscepti-

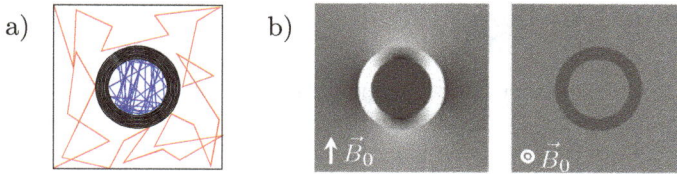

Fig. 1 A white matter nerve fibre can be modelled as a hollow cylinder composed of a myelin sheath. **a** Brownian motion prependicular to the nerve fibre is restricted inside the cylinder (the molecules are "trapped" inside the cylinder) and hindered outside of it (the mean displacement of the molecules per time-point is larger, but nevertheless lower than 'free' water). **b** The myelin sheath is more diamagnetic than the surrounding tissue and perturbs the local magnetic field. The field perturbation is inhomogeneous and depends on the nerve fibre orientation to the magnetic field $\mathbf{B_0}$

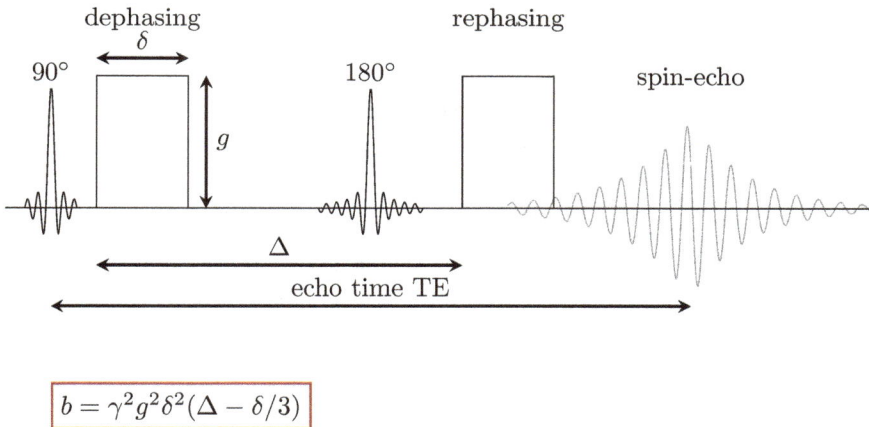

$$b = \gamma^2 g^2 \delta^2 (\Delta - \delta/3)$$

Fig. 2 Pulsed-gradient spin-echo sequence: the echo time TE is defined as the time between the radio-frequency pulse and the centre of the spin-echo. The diffusion-weighting strength, the b-value, can be calculated as $b = \gamma^2 g^2 \delta^2 (\Delta - \delta/3)$

bility of the myelin sheath has an anisotropic component [10–12]. It has been shown that the signal decay from white matter varies as a function of orientation to $\mathbf{B_0}$ [13–20] and can be well explained using a hollow cylinder fibre model (Fig. 1b) [17] and is often represented by a 3-pool model [18, 21, 22].

The strongest contribution to this signal anisotropy arises from the water trapped between the layers of myelin sheath, called the myelin water. However, the myelin water signal decays very quickly, on account of its short T_2 ($\sim$10–30 ms), and is usually negligible at the echo times used in a typical diffusion-weighted MRI experiment. Nevertheless, it has also been observed that signal decay rates may still be orientation-dependent even without this myelin-water component [13, 20, 23–25].

1.2   Scope of This Work

In myelinated white matter it has been reported that T_2^* (where $1/T_2^* = 1/T_2 + 1/T_2'$ and T_2' can be understood as capturing the reversible effects of field inhomogeneities) depends on the orientation of the fibre with respect to $\mathbf{B}_0$ due to microscopic susceptibility effects [13, 16, 18, 19, 26]. Orientation-dependence of T_2 was also reported recently [20, 23, 24, 27] and could potentially characterise microscopic-susceptibility more reliably and reflect effects related to axon diameter [24, 28]. Experiments designed to probe relaxation-anisotropy commonly involve reorienting the head inside the scanner, and are thus challenged by unintended signal-to-noise ratio (SNR) variations across orientations caused by differences in proximities to the receiver coil, and by increased susceptibility to motion and artefacts due to patient-discomfort. In this work, we re-purpose a tiltable RF coil (originally designed for patient comfort) to investigate T_2-orientational dependence within the context of a diffusion-T_2 correlation experiment. The coil can be tilted around the left-right axis by $0°$, $9°$ and $18°$ to $\mathbf{B}_0$, which: (1) minimises patient-discomfort and thus improves reliability; (2) offers a new degree of freedom as tilting around the left-right axis is otherwise difficult to achieve; (3) fixes the coil-to-brain distance across orientations and thus reduces SNR variations; and (4) increases the reproducibility of the experiment.

Finally, instead of studying global variations across the whole brain volume, we adopt an along-tract profiling *tractometry* approach (i.e., which is the mapping of measures along pathways reconstructed with tractography [29, 30]) to assess spatial variations in more detail.

2   Methods

2.1   Data Acquisition

The study was approved by the Cardiff University School of Psychology Ethics Committee and written informed consent was obtained. Two healthy volunteers (female, 30 y.) were scanned on a 3 T 300 mT/m Connectom scanner equipped with a modified 20-channel head/neck tiltable coil (Siemens Healthcare, Erlangen, Germany). Each subject was in supine, head first position and the direction of the magnetic field $\mathbf{B}_0$ was along the superior-inferior radiological axis. MRI data were acquired in the default ($0°$) and tilted ($18°$) orientations of the tiltable coil (Fig. 3a, b). One of the subjects underwent a second scan in the default head orientation to examine test-retest variability.

Diffusion-T_2 correlation data were acquired using a pulsed-gradient spin-echo echo-planar-imaging (PGSE-EPI) sequence [6] (Fig. 2), with different echo-times TE to probe T_2, and diffusion-weighting strengths or b-values to probe diffusion, (Fig. 3c). The timings of the diffusion encoding gradients were fixed for all echo

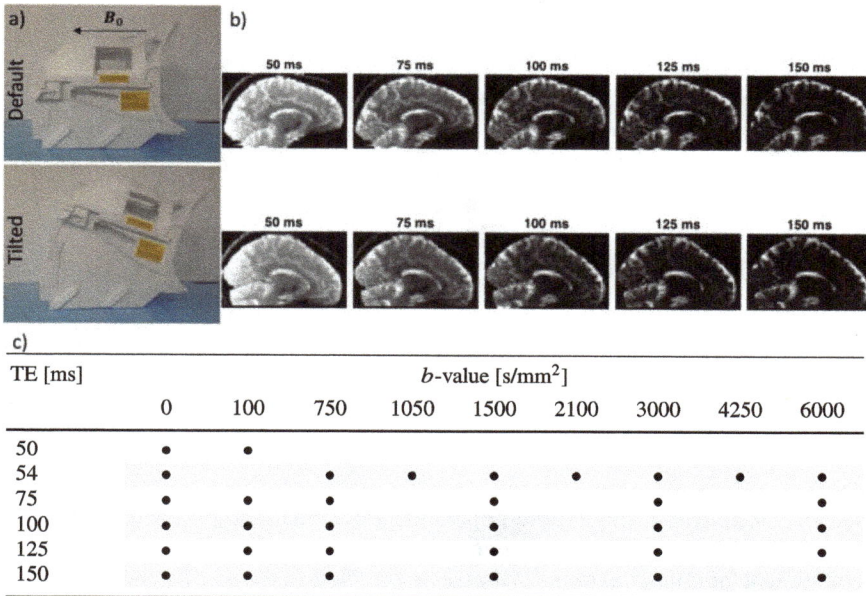

TE [ms]	b-value [s/mm^2]								
	0	100	750	1050	1500	2100	3000	4250	6000
50	•	•							
54	•		•	•	•	•	•	•	•
75	•	•	•		•		•		•
100	•	•	•		•		•		•
125	•	•	•		•		•		•
150	•	•	•		•		•		•

Fig. 3 **a** The coil in default (0°) and tilted (18°) position. **b** $b = 0\,\text{s/mm}^2$ images at different TE. **c** Acquisition parameters for the diffusion-correlation experiment

times. The signal in voxel **x** can then be denoted as $S_\mathbf{x}(b, \text{TE})$. Additional b_0-images were acquired in the halfway-tilted (9°) position (not shown). Remaining parameters were repetition time TR 3.5 s and voxel size $3 \times 3 \times 3\,\text{mm}^3$.

2.2 MRI Signal Processing

The diffusion-T_2 data were preprocessed to correct for subject motion, eddy current effects, Gibbs ringing and gradient non-linearities [31–33] for each subject and each head orientation. Spatial correspondence between the tilted and default head orientations was obtained in two ways: 1) by nonlinear registration [34] to the halfway-tilted (9°) space, and 2) by a tractometry approach in native space.

The tractometry [29] approach in native space of each head orientation relied on the quantitative mapping of measures along reconstructed brain pathways. First, fibre orientation distribution functions (fODFs) [35, 36] were estimated at each voxel using multi-tissue multi-shell constrained spherical deconvolution (MSMT-CSD) [37]. For each coil position, peaks were extracted from the resulting fODFs and used as input to perform streamline tractography on the TE = 54 ms data. Bundles were automatically segmented [38], a representative core-streamline was computed [39], and the bundles were subsequently subdivided [40, 41] into n = 20 segments (*s*)

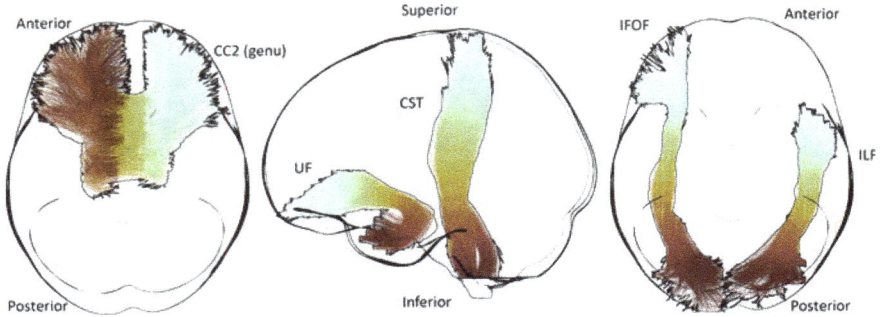

Fig. 4 Extracted brain pathways of interest from subject 2. Bundles were segmented into 20 sections (0: ligth-blue, 20: dark-brown). CC: corpus callosum, UF: uncinate fasciculus, CST: cortiospinal-tract, IFOF: inferior fronto-occipital fasciculus, ILF: inferior longitudinal fasciculus. Visualisation was performed using FiberNavigator [43]

(Fig. 4). Voxels with a single-fibre-population were identified [42] and assigned to s_i if their location was inside the segment and their orientation was within 30° of the orientation of the core-streamline in that segment. Note that only tract-segments with minimal fanning (assessed visually) were considered. T_2 values were then profiled for each bundle independently by taking the mean and standard error of the mean within each tract segment.

2.3 Estimation

SNR estimates were obtained from the background of the images acquired at TE $=$ 54 ms and $b = 0$ s/mm^2 for both the (0°) and (18°) coil-orientations [44]. The voxel-wise T_2 was estimated from the b_0-signals as $S(0, \text{TE}) = S(0, 0)e^{-\text{TE}/T_2}$, using a nonlinear least-squares trust-region-reflective algorithm in Matlab.

Fibre orientation θ to the main magnetic field $\mathbf{B}_0$ was estimated for the voxels with single fibre population. With prior knowledge of θ orientation-dependent (anisotropic) and -independent (isotropic) components of R_2 can be estimated as follows [20, 45]: $R_2(\theta) = R_{2,\text{isotropic}} + R_{2,\text{anisotropic}} \cdot \sin^4 \theta$.

3 Results

Signal-to-noise ratio. The estimated noise standard deviation was similar for the two coil-orientations. Figure 5a compares the signals in the default and the tilted position after registration to the common space; the signal (and thus SNR) between coil-orientations varied within a similar range as the range seen between test-retest scans in the same orientation. Globally, the images are aligned but some local misalignments can still be observed.

Voxel-wise T_2-estimates. Figure 5b highlights differences in estimated T_2 for different coil-orientations. Globally, the difference in T_2-values estimated from the data in default and tilted head orientations is larger than or equal to T_2-values estimated from test-retest scans in default head orientation for subject 1 (Fig. 5b, left half). Local differences in T_2 can be observed between the default and tilted position, as indicated by red arrows in the T_2-maps in Fig. 5b, for example. The inverse T_2-values in white matter mostly range between $9\,s^{-1}$ and $21\,s^{-1}$ for both subjects and head orientation.

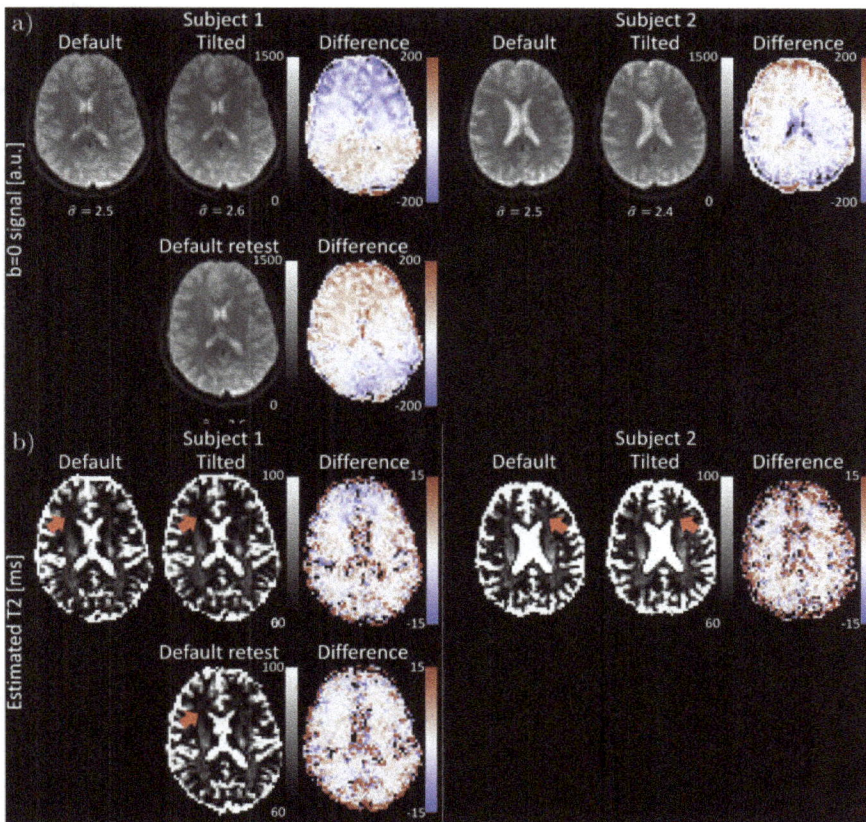

Fig. 5 a $b = 0\,s/mm^2$ image for the default (0°) and tilted (18°) orientation registered to the halfway-tilted (9°) space, and their difference (tilted—default). $\hat{\sigma}$ is the estimated standard deviation in the background of the image. **b** Estimated T_2 for the default (0°) and tilted (18°) orientation registered to the halfway-tilted (9°) space, and their difference. Red arrows indicate regions of visible difference

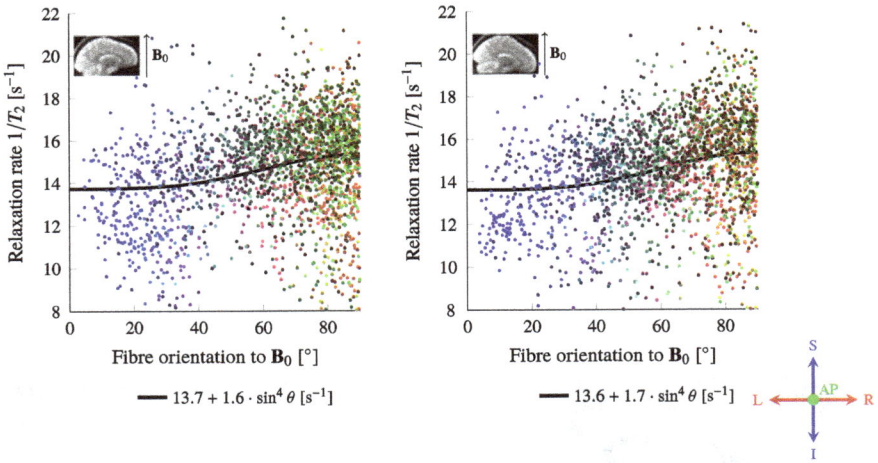

Fig. 6 Relaxation rate $1/T_2$ as a function of the fibre-orientation θ with the main magnetic field, in the default (left) and tilted (right) position. Each point represents a single-fibre-population voxel and is colour-coded according to its orientation (red, blue, and green correspond to the left-right, superior-inferior and anterior-posterior axis, respectively). The black line represents a model-fit of $1/T_2$ as a function of θ as described in previous literature and references therein [13, 19, 20]

Global WM isotropic and anisotropic T_2-stimates. The total isotropic (orientation-independent) component of relaxation rate of $R_2 = 1/T_2$ in WM was estimated at $13.7 \pm 0.2\,\mathrm{s}^{-1}$ and $13.6 \pm 0.2\,\mathrm{s}^{-1}$, in default and tilted head orientations, respectively, for one of the subjects (Fig. 6). For the same subject the total WM anisotropic components of the inverse T_2 were $1.6 \pm 0.25\,\mathrm{s}^{-1}$ and $1.7 \pm 0.25\,\mathrm{s}^{-1}$, for default and tilted head positions, respectively. The range of T_2-values in white matter and their total isotropic and anisotropic components are consistent with previously reported values in [20, 45].

Tractometry analysis. Figures 7, 8 and 9 show the along-tract profiles of the estimated T_2 (top plot) and angle w.r.t. $\mathbf{B_0}$ (bottom plot) for different tracts. Globally, the angular profiles show comparable characteristics between subjects in the default and tilted position. For Subject 1, the angular profiles remain similar in the default and default-retest acquisition, except for the inferior parts of the corticospinal tract (CST).

Profiling of T_2 in the CST, which runs along the z-axis (i.e., inferosuperior), reveals a significant increase in T_2, for example in segment 16 which experiences a change in orientation w.r.t. $\mathbf{B_0}$ from ~35° in the default position to ~55° in the tilted position (Fig. 7). This is in the regime where the derivative of the relaxation rate $1/T_2$ as a function of angle is the largest (Fig. 6). The uncinate fasciculus (UF) shows a noisier pattern, likely because the number of single-fibre voxels per segment is generally lower (~5−10 in the UF compared to ~15−20 in the CST).

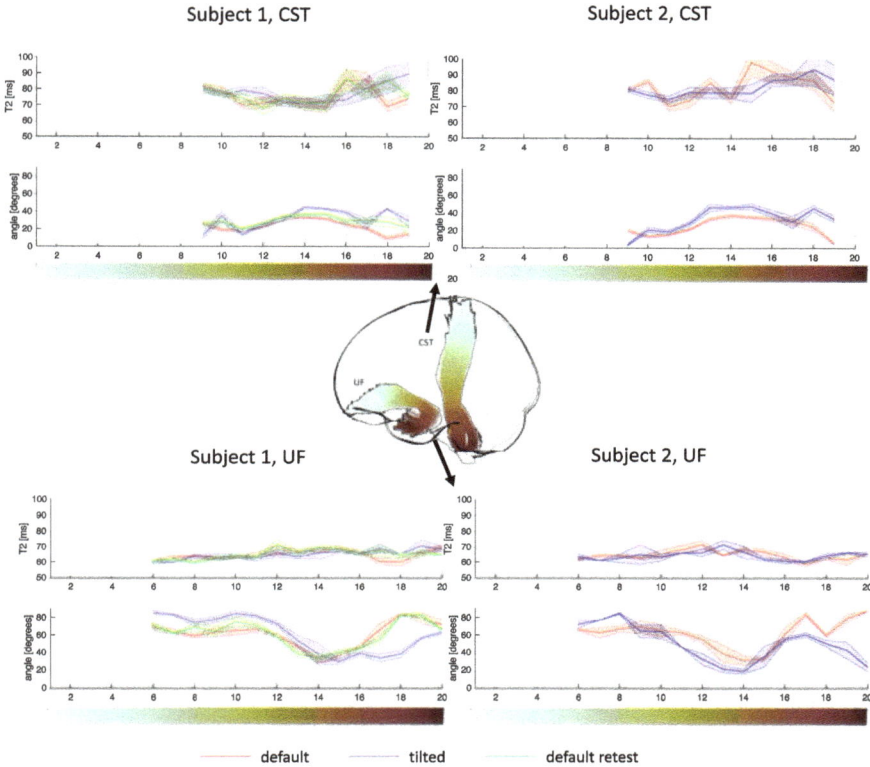

Fig. 7 Tractometry results (mean and standard error) for the corticospinal tract (CST, top plots) and the uncinate fasciculus (UF, bottom plots). T_2-values and fibre orientation to $\mathbf{B_0}$ are plotted against segment numbers for each fibre tract and subject. Colour bar indicates tract-segment location, corresponding to the colour-encoding on the visualised tracts. Note that only tract-segments with minimal fanning (assessed visually) were considered

In the inferior longitudinal fasciculus (ILF) (Fig. 8) a global decrease in angle w.r.t. $\mathbf{B_0}$ from default to tilted position leads to an overall, yet subtle, increase in T_2. In the inferior fronto-occipital fasciculus (IFOF) the pattern is less clear.

In the callosal midbody the angle w.r.t. $\mathbf{B_0}$ remains relatively unchanged and so does T_2.

4 Discussion

In this work we have incorporated a tiltable coil in T_2-diffusion-correlation experiment to modulate the white matter fibre orientation with respect to the main magnetic field $\mathbf{B_0}$.

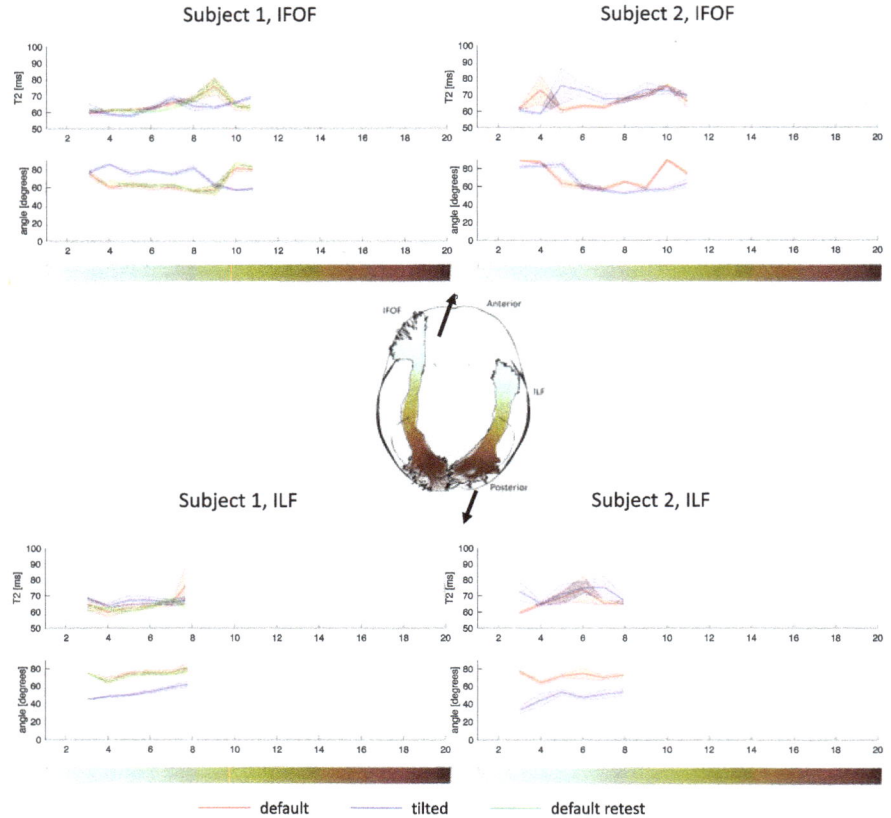

Fig. 8 Tractometry results (mean and standard error) for the inferior fronto-occipital fasciculus (IFOF, top plots) and the inferior longitudinal fasciculus (ILF, bottom plots). T_2-values and fibre orientation to $\mathbf{B}_0$ are plotted against segment numbers for each fibre tract and subject. Colour bar indicates tract-segment location, corresponding to the colour-encoding on the visualised tracts. Note that only tract-segments with minimal fanning (assessed visually) were considered

We observed changes in the T_2-tract-profile after the participants' heads were re-oriented in the scanner, with up to ~10 ms difference in T_2-values between the default and the tilted head orientation. The test-retest T_2-tract profiles in the default head position are more similar to each-other than to the tract profiles in the tilted head position, as for example evident from the results for the CST and IFOF. These initial results suggest variation of T_2 as a function of fibre orientation to $\mathbf{B}_0$ and that tilting the participant's head by 18° can reveal those variations.

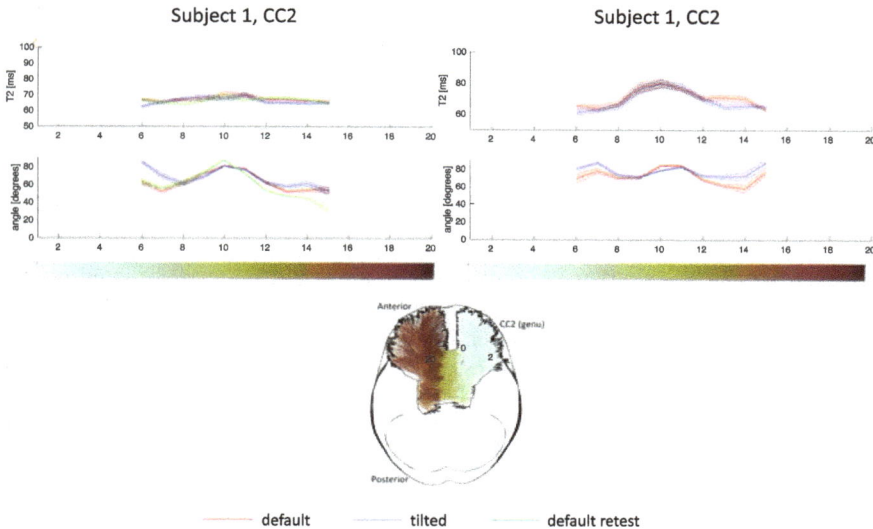

Fig. 9 Tractometry results for the genu of the corpus callosum (CC2). T_2-values and fibre orientation to $\mathbf{B}_0$ are plotted against segment numbers for each subject. Colour bar corresponds to tract-segment location, and this is colour-encoded on the tracts themselves. Note that only tract-segments with minimal fanning (assessed visually) were considered

4.1 Incorporating Tiltable Coil in T2-diffusion correlation experiments

Robustness of experimental setup when using a tiltable coil. The tiltable coil allows us to control the pitch orientation of the participant's head. We have demonstrated an overall test-retest similarity of estimated fibre-tract orientations in the default position, whereas there were clear differences with the tilted head position. However, some differences in fibre orientation of the CST between the test and retest could be observed, particularly in the inferior segments, while differences in fibre orientation in the IFOF tract were less obvious. Given the apparent stability of the IFOF tract profiling and insensitivity of fibre orientation to the head rotation around $\mathbf{B}_0$ in the default head position (yaw), the test-retest differences in the CST could be a result of an additional roll head orientation with respect to the coil between the two scans. Such additional differences in orientation could be mitigated by further restricting head position within the coil, e.g. additional padding or performing the experiments with differences in tilt immediately after each other without a break. Whereas this introduces a confound in assessing test-retest variability, the estimation of fibre-orientations is done in each coil orientation independently and as such we do not expect this to be detrimental in the overall assessment of T_2-anisotropy.

Range of orientations. By re-purposing a coil that was designed to maximise patient comfort in clinical situations means that the range of coil orientations was limited. A larger range of orientations would allow the θ versus $1/T_2$ relationship to be elucidate more robustly. Nevertheless, these initial experiments demonstrate the utility of this hardware design for uncovering orientational-anisotropy effects in vivo.

4.2   Origin of T_2-Contrast and -Anisotropy in WM

The water signal from each WM voxel is a superposition of the signal from different water pools (e.g. intra- and extra-axonal, and at short TE also myelin water), therefore the macroscopic T_2-values measured in this study could be approximated as the weighted average of the T_2-values from each of the two compartments. This means that the macroscopic T_2 will depend on individual T_2-values of the compartments, but also on the relative contribution to the signal. As such, tracts with the same apparent T_2 but differences in signal fractions and T_2-values between the intra- and the extra-axonal compartments could exhibit very different apparent orientational anisotropy.

A clear separation of macroscopic T_2-values in some tracts and segments between the default and the tilted head orientations suggests that macroscopic T_2 is also a function of orientation to $\mathbf{B_0}$. Fibre tract re-orientation in the magnetic field will cause local changes in the local magnetic field due to differences between the magnetic susceptibility of the myelin sheath and the surrounding tissue. Following the hollow-cylinder model [26], the magnitude of these microscopic $\mathbf{B_0}$-field perturbations is expected to be larger in the extra-axonal compartment and, in combination with molecular Brownian motion, will cause an additional faster relaxation in the extra- relative to the intra-axonal compartment [46]. Future work will explore the effects of fibre orientation on individual compartments [47, 48].

4.3   Considerations in Data Processing

Tractometry versus image registration In this work we adopted tract profiling for comparing the T_2-diffusion-correlation data in default and tilted head positions, instead of a voxel-based analysis; image registration was used for visualisation purposes only. Voxel-based analysis is known to suffer from confounds related to mis-registration and data interpolation, an effect we visually observed in our study likely amplified by the large imaging voxels (3 mm isotropic) and imperfect correction for geometric distortions due to gradient nonlinearities. With the tractometry approach, spatial correspondence between coil orientations was established while keeping the data in native space, and each segment consitutes information from multiple voxels improving robustness to noise. We were able to reproduce T_2-values from test-retest along the tracts.

Correction for subject motion The individual images have been corrected for subject motion, which could also involve head re-orientation with respect to the magnetic field. Subjects who participated in this study were experienced MR scanner participants, therefore the maximum rotation around any axis rarely exceeded 1 5°

However, when considering the application of this method on less compliant subjects, subject motion could become a confounding factor.

5 Conclusion

Including a tiltable coil into the experimental set-up for diffusion-T_2 correlation measurements paves the way for a more reliable assessment of orientational T_2 dependence. Microstructural origins of the differences in T_2 could include differences in pathway properties (e.g. axon diameter) or susceptibility effects. T_2 orientational-dependence would furthermore impact analyses frameworks that assume constant T_2 along pathways [49]. Voxel-wise comparison of T_2 from different head-orientations remains challenging due to complications in experiment setup, imperfect correction for geometric distortions, and intrinsic scan-variability. Using a tractometry framework, we found indications of regional changes in T_2 upon tilting of the head. Studying this effect in a larger population is necessary to increase statistical power. In future work, the diffusion-T_2 correlation experiments can be used to study compartmental T_2 orientation-dependence.

Acknowledgments CMWT is supported by a Rubicon grant (680-50-1527) from the Netherlands Organisation for Scientific Research (NWO) and a Sir Henry Wellcome Fellowship (215944/Z/19/Z). DKJ, CMWT and MC were all supported by a Wellcome Trust Investigator Award (096646/Z/11/Z) and DKJ was supported by a Wellcome Strategic Award (104943/Z/14/Z).

The data were acquired at the UK National Facility for In Vivo MR Imaging of Human Tissue Microstructure funded by the EPSRC (grant EP/M029778/1), and The Wolfson Foundation.

We are grateful to Fabrizio Fasano, Peter Gall, and Matschl Volker from Siemens Healthcare GmbH for their support.

References

1. Hanson, L.G.: Concepts Magn. Reson. Part A **32A**(5), 329 (2008). https://doi.org/10.1002/cmr.a.20123
2. Bloch, F.: Phys. Rev. (1946). https://doi.org/10.1103/PhysRev.70.460
3. Hahn, E.L.: Phys. Rev. **80**(4), 580 (1950). https://doi.org/10.1103/PhysRev.80.580
4. Carr, H.Y., Purcell, E.M.: Phys. Rev. (1954). https://doi.org/10.1103/PhysRev.94.630
5. Torrey, H.C.: Phys. Rev. (1956). https://doi.org/10.1103/PhysRev.104.563
6. Stejskal, E.O., Tanner, J.E.: J. Chem. Phys. **42**(1), 288 (1965). https://doi.org/10.1063/1.1695690
7. Beaulieu, C.: NMR Biomed. Int. J. Devoted Develop. Appl. Magn. Reson. Vivo **15**(7–8), 435 (2002)
8. Conturo, T.E., Lori, N.F., Cull, T.S., Akbudak, E., Snyder, A.Z., Shimony, J.S., McKinstry, R.C., Burton, H., Raichle, M.E.: Proc. Natl. Acad. Sci. **96**(18), 10422 (1999)
9. Jeurissen, B., Descoteaux, M., Mori, S., Leemans, A.: NMR Biomed. **32**(4), e3785 (2019)
10. Liu, C.: Magn. Reson. Med. **63**(6), 1471 (2010). https://doi.org/10.1002/mrm.22482, https://doi.org/10.1002/mrm.22482doi.wiley.com/10.1002/mrm.22482
11. Lee, J., Shmueli, K., Fukunaga, M., van Gelderen, P., Merkle, H., Silva, A.C., Duyn, J.H.: Proc. Natl. Acad. Sci. **107**(11), 5130 (2010). https://doi.org/10.1073/pnas.0910222107
12. van Gelderen, P., Mandelkow, H., de Zwart, J.A., Duyn, J.H.: Magn. Reson. Med. **74**(5), 1388 (2015). https://doi.org/10.1002/mrm.25524, https://onlinelibrary.wiley.com/doi/abs/10.1002/mrm.25524

13. Bender, B., Klose, U.: NMR Biomed. **23**(9), 1071 (2010). https://doi.org/10.1002/nbm.1534
14. Denk, C., Torres, E.H., MacKay, A., Rauscher, A.: NMR Biomed. **24**(3), 246 (2011). https://doi.org/10.1002/nbm.1581
15. Lee, J., van Gelderen, P., Kuo, L.W., Merkle, H., Silva, A.C., Duyn, J.H.: NeuroImage **57**(1), 225 (2011). https://doi.org/10.1016/j.neuroimage.2011.04.026
16. Van Gelderen, P., De Zwart, J.A., Lee, J., Sati, P., Reich, D.S., Duyn, J.H.: Magn. Reson. Med. **67**(1), 110 (2012). https://doi.org/10.1002/mrm.22990
17. Wharton, S., Bowtell, R.: NeuroImage **83**, 1011 (2013). https://doi.org/10.1016/j.neuroimage.2013.07.054
18. Sati, P., van Gelderen, P., Silva, A.C., Reich, D.S., Merkle, H., De Zwart, J.A., Duyn, J.H.: NeuroImage **77**, 268 (2013). https://doi.org/10.1016/j.neuroimage.2013.03.005
19. Rudko, D.A., Klassen, L.M., De Chickera, S.N., Gati, J.S., Dekaban, G.A., Menon, R.S.: Proc. Natl. Acad. Sci. U. S. Am. **111**, 1 (2014). https://doi.org/10.1073/pnas.1306516111
20. Gil, R., Khabipova, D., Zwiers, M., Hilbert, T., Kober, T., Marques, J.P.: NMR Biomed. **29**(12), 1780 (2016). https://doi.org/10.1002/nbm.3616
21. Thapaliya, K., Vegh, V., Bollmann, S., Barth, M.: NeuroImage **182**, 407 (2018). https://doi.org/10.1016/j.neuroimage.2017.11.029, http://www.sciencedirect.com/science/article/pii/S1053811917309436
22. Tendler, B.C., Bowtell, R.: Magn. Reson. Med. **81**(5), 3017 (2019). https://doi.org/10.1002/mrm.27626
23. Knight, M.J., Wood, B., Couthard, E., Kauppinen, R.: Biomed. Spectro. Imaging **4**(3), 299 (2015). https://doi.org/10.3233/BSI-150114, http://www.medra.org/servlet/aliasResolver?alias=iospress&doi=10.3233/BSI-150114
24. McKinnon, E.T., Jensen, J.H.: Magn. Reson. Med. **81**(5), 2985 (2019). https://doi.org/10.1002/mrm.27617, http://doi.wiley.com/10.1002/mrm.27617
25. Kleban, E., Tax, C.M., Rudrapatna, U.S., Jones, D.K., Bowtell, R.: NeuroImage **116793**, (2020). https://doi.org/10.1016/j.neuroimage.2020.116793, http://www.sciencedirect.com/science/article/pii/S1053811920302809
26. Wharton, S., Bowtell, R.: Proc. Natl. Acad. Sci. U. S. A. **109**(45), 18559 (2012). https://doi.org/10.1073/pnas.1211075109
27. Birkl, C., Doucette, J., Fan, M., Hernandez-Torres, E., Rauscher, A.: bioRxiv (2020). https://doi.org/10.1101/2020.03.11.987925, https://www.biorxiv.org/content/early/2020/03/12/2020.03.11.987925
28. Kaden, E., Alexander, D.C.: Proceedings of the International Conference on Information Processing in Medical Imaging, vol. 23, p. 607 (2013). https://doi.org/10.1007/978-3-642-38868-2_51
29. Bells, S., Cercignani, M., Deoni, S., Assaf, Y., Pasternak, O., Evans, C., Leemans, A., Jones, D.K.: Proceedings of the International Society for Magnetic Resonance in Medicine, p. 678 (2011)
30. Jones, D.K., Travis, A.R., Eden, G., Pierpaoli, C., Basser, P.J.: Magn. Reson. Med. Official J. Int. Soc. Magn. Reson. Med. **53**(6), 1462 (2005)
31. Andersson, J.L.: Diffusion MRI, pp. 63–85. Elsevier Inc., (2014). https://doi.org/10.1016/B978-0-12-396460-1.00004-4
32. Kellner, E., Dhital, B., Kiselev, V.G., Reisert, M.: Magn. Reson. Med. **76**(5), 1574 (2016). https://doi.org/10.1002/mrm.26054
33. Rudrapatna, S.U., Parker, G.D., Roberts, J., Jones, D.K.: ISMRM, p. 1206 (2018)
34. Avants, B.B., Tustison, N.J., Stauffer, M., Song, G., Wu, B., Gee, J.C.: Front. Neuroinform. **8**(APR) (2014). https://doi.org/10.3389/fninf.2014.00044
35. Tournier, J.D., Calamante, F., Connelly, A.: NeuroImage **35**(4), 1459 (2007). https://doi.org/10.1016/j.neuroimage.2007.02.016, https://linkinghub.elsevier.com/retrieve/pii/S1053811907001243
36. Descoteaux, M., Deriche, R., Knosche, T.R., Anwander, A.: IEEE Trans. Med. Imaging **28**(2), 269 (2008)
37. Jeurissen, B., Tournier, J.D., Dhollander, T., Connelly, A., Sijbers, J.: NeuroImage **103**, 411 (2014). https://doi.org/10.1016/j.neuroimage.2014.07.061

38. Wasserthal, J., Neher, P., Maier-Hein, K.H.: NeuroImage **183**, 239 (2018). https://doi.org/10. 1016/j.neuroimage.2018.07.070

39. Chamberland, M., St-Jean, S., Tax, C.M., Jones, D.K.: International Conference on Medical Image Computing and Computer-Assisted Intervention, pp. 359–366. Springer, Berlin (2018)

40. Cousineau, M., Jodoin, P.M., Morency, F.C., Rozanski, V., Grand'Maison, M., Bedell, B.J., Descoteaux, M.: NeuroImage: Clin. **16**, 222 (2017). https://doi.org/10.1016/j.nicl.2017.07.020

41. Chamberland, M., Raven, E.P., Genc, S., Duffy, K., Descoteaux, M., Parker, G.D., Tax, C.M., Jones, D.K.: NeuroImage **200**, 89 (2019)

42. Tax, C., Jeurissen, B., Vos, S., Viergever, M., Leemans, A.: NeuroImage **86**, (2014). https:// doi.org/10.1016/j.neuroimage.2013.07.067

43. Chamberland, M., Whittingstall, K., Fortin, D., Mathieu, D., Descoteaux, M.: Front. Neuroinform. **8**, 59 (2014)

44. Koay, C.G., Özarslan, E., Pierpaoli, C.: J. Magn. Reson. **199**(1), 94 (2009). https://doi.org/10.1016/J.JMR.2009.03.005, https://www.sciencedirect.com/science/article/ pii/S1090780709000767?via%3Dihub

45. Knight, M.J., Dillon, S., Jarutyte, L., Kauppinen, R.A.: Biophys. J. **112**(7), 1517 (2017). https:// doi.org/10.1016/j.bpj.2017.02.026

46. Cowan, B.: Contemp. Phys. **56**, 1 (2015). https://doi.org/10.1080/00107514.2015.1005683

47. Tax, C., Rudrapatna, U., Witzel, T., Jones, D.: ISMRM, p. 0838 (2017)

48. Veraart, J., Novikov, D.S., Fieremans, E.: NeuroImage **182**, 360 (2018). https://doi. org/10.1016/J.NEUROIMAGE.2017.09.030, https://www.sciencedirect.com/science/article/ pii/S1053811917307784?via%3Dihub

49. Daducci, A., Canales-Rodríguez, E.J., Zhang, H., Dyrby, T.B., Alexander, D.C., Thiran, J.P.: NeuroImage **105**, 32 (2015). https://doi.org/10.1016/J.NEUROIMAGE.2014.10.026, https:// www.sciencedirect.com/science/article/pii/S1053811914008519?via%3Dihub

Permissions

All chapters in this book were first published by Springer; hereby published with permission under the Creative Commons Attribution License or equivalent. Every chapter published in this book has been scrutinized by our experts. Their significance has been extensively debated. The topics covered herein carry significant findings which will fuel the growth of the discipline. They may even be implemented as practical applications or may be referred to as a beginning point for another development.

The contributors of this book come from diverse backgrounds, making this book a truly international effort. This book will bring forth new frontiers with its revolutionizing research information and detailed analysis of the nascent developments around the world.

We would like to thank all the contributing authors for lending their expertise to make the book truly unique. They have played a crucial role in the development of this book. Without their invaluable contributions this book wouldn't have been possible. They have made vital efforts to compile up to date information on the varied aspects of this subject to make this book a valuable addition to the collection of many professionals and students.

This book was conceptualized with the vision of imparting up-to-date information and advanced data in this field. To ensure the same, a matchless editorial board was set up. Every individual on the board went through rigorous rounds of assessment to prove their worth. After which they invested a large part of their time researching and compiling the most relevant data for our readers.

The editorial board has been involved in producing this book since its inception. They have spent rigorous hours researching and exploring the diverse topics which have resulted in the successful publishing of this book. They have passed on their knowledge of decades through this book. To expedite this challenging task, the publisher supported the team at every step. A small team of assistant editors was also appointed to further simplify the editing procedure and attain best results for the readers.

Apart from the editorial board, the designing team has also invested a significant amount of their time in understanding the subject and creating the most relevant covers. They scrutinized every image to scout for the most suitable representation of the subject and create an appropriate cover for the book.

The publishing team has been an ardent support to the editorial, designing and production team. Their endless efforts to recruit the best for this project, has resulted in the accomplishment of this book. They are a veteran in the field of academics and their pool of knowledge is as vast as their experience in printing. Their expertise and guidance has proved useful at every step. Their uncompromising quality standards have made this book an exceptional effort. Their encouragement from time to time has been an inspiration for everyone.

The publisher and the editorial board hope that this book will prove to be a valuable piece of knowledge for researchers, students, practitioners and scholars across the globe.

List of Contributors

Talha Bin Masood and Ingrid Hotz
Department of Science and Technology (ITN), Linköping University, Norrköping, Sweden

Evren Özarslan
Department of Biomedical Engineering, Linköping University, Linköping, Sweden
Center for Medical Image Science and Visualization, Linköping University, Linköping, Sweden

Carl-Fredrik Westin
Department of Radiology Brigham and Women's Hospital, Harvard Medical School, Boston, MA, USA

Yue Zhang, Hongyu Nie and Eugene Zhang
School of Electrical Engineering and Computer Science 3117 Kelley Engineering Center, Oregon State University, Corvallis, OR 97331, USA

Faizan Siddiqui
Delft University of Technology, Delft, The Netherlands

Thomas Höllt
Delft University of Technology, Delft, The Netherlands
Leiden University Medical Center, Leiden, The Netherlands

Anna Vilanova
Eindhoven University of Technology, Eindhoven, The Netherlands
Delft University of Technology, Delft, The Netherlands

Rafael Ballester-Ripoll
IE University, Madrid, Spain

Daniel Jörgens and Rodrigo Moreno
Department of Biomedical Engineering and Health Systems, KTH Royal Institute of Technology, Hälsovägen 11C, 14157 Huddinge, Stockholm, Sweden

Maxime Descoteaux
Université de Sherbrooke, Sherbrooke Connectivity Imaging Laboratory, 2500 Boulevard de l'Université, Sherbrooke, QC J1K 0A5, Canada

Ikram Jumakulyyev and Thomas Schultz
B-IT and Department of Computer Science II, University of Bonn, Friedrich-Hirzebruch-Allee 5, 53115 Bonn, Germany

Cem Yolcu
Department of Biomedical Engineering, Linköping University, Linköping, Sweden

Magnus Herberthson
Department of Mathematics, Linköping University, Linköping, Sweden

Renato Pajarola and Haiyan Yang
University of Zürich, Zürich, Switzerland

Susanne K. Suter
Zurich University of Applied Sciences, Zürich, Switzerland

Maëliss Jallais and Demian Wassermann
Université Paris-Saclay, Inria, CEA, 91120 Palaiseau, France

Luc Florack, Rick Sengers, Stephan Meesters, Lars Smolders and Andrea Fuster
Eindhoven University of Technology, Department of Mathematics & Computer Science, NL-5600 Eindhoven, MB, The Netherlands

Tom Dela Haije
Department of Computer Science, University of Copenhagen, Copenhagen, Denmark

Aasa Feragen
Department of Applied Mathematics and Computer Science, Technical University of Denmark, Lyngby, Denmark

Paddy J. Slator and Daniel C. Alexander
Centre for Medical Image Computing, Department of Computer Science, University College London, London, UK

Alison Ho, Spyros Bakalis and Lucy C. Chappell
Women's Health Department, King's College London, London, UK

Laurence Jackson, Joseph V. Hajnal and Jana Hutter
Centre for the Developing Brain, School of Biomedical Engineering and Imaging Sciences, King's College London, London, UK
Biomedical Engineering Department, School of Biomedical Engineering and Imaging Sciences, King's College London, London, UK

Mary Rutherford
Centre for the Developing Brain, School of Biomedical Engineering and Imaging Sciences, King's College London, London, UK

Chantal M. W. Tax, Elena Kleban, Maxime Chamberland and Derek K. Jones
Cardiff University Brain Research Imaging Centre (CUBRIC), Cardiff University, Cardiff, UK

Muhamed Baraković
Cardiff University Brain Research Imaging Centre (CUBRIC), Cardiff University, Cardiff, UK
Signal Processing Laboratory 5, Ecole Polytechnique Federale de Lausanne, Lausanne, Switzerland
Translational Imaging in Neurology Basel, Department of Biomedical Engineering, University Hospital Basel, Basel, Switzerland

Index